Intelligent Electronic Devices

Intelligent Electronic Devices

Special Issue Editors

Teen-Hang Meen
Wenbing Zhao
Cheng-Fu Yang

MDPI • Basel • Beijing • Wuhan • Barcelona • Belgrade • Manchester • Tokyo • Cluj • Tianjin

Special Issue Editors

Teen-Hang Meen
Department of Electronic
Engineering, National Formosa
University
Taiwan

Wenbing Zhao
Department of Electrical
Engineering and Computer
Science, Cleveland State
University
USA

Cheng-Fu Yang
Department of Chemical and
Materials Engineering, National
University of Kaohsiung
Taiwan

Editorial Office
MDPI
St. Alban-Anlage 66
4052 Basel, Switzerland

This is a reprint of articles from the Special Issue published online in the open access journal *Electronics* (ISSN 2079-9292) (available at: https://www.mdpi.com/journal/electronics/special_issues/Intelligent_Electronic_Devices).

For citation purposes, cite each article independently as indicated on the article page online and as indicated below:

LastName, A.A.; LastName, B.B.; LastName, C.C. Article Title. *Journal Name* **Year**, *Article Number*, Page Range.

ISBN 978-3-03928-973-8 (Pbk)
ISBN 978-3-03928-974-5 (PDF)

Contents

About the Special Issue Editors

Teen-Hang Meen was born in Tainan, Taiwan, in 1967. He received his B.S. degree from the Department of Electrical Engineering, National Cheng Kung University (NCKU), Tainan, Taiwan, in 1989; and M.S. and Ph.D. degrees from the Institute of Electrical Engineering, National Sun Yat-Sen University (NSYSU), Kaohsiung, Taiwan, in 1991 and 1994, respectively. He was the chairman of the Department of Electronic Engineering from 2005 to 2011 at the National Formosa University, Yunlin, Taiwan. He received excellent research awards from National Formosa University in 2008 and 2014. Currently, he is a distinguished professor with the Department of Electronic Engineering, National Formosa University, Yunlin, Taiwan. He is also the president of the International Institute of Knowledge Innovation and Invention (IIKII) and the chair of IEEE Tainan Section Sensors Council. He has published more than 100 SCI, SSCI, and EI papers in recent years.

Wenbing Zhao is a full Professor of Electrical Engineering and Computer Science (EECS) at Cleveland State University (CSU), Cleveland, Ohio, USA. He obtained his B.S. and M.S. degrees in Physics from Peking University, Beijing, China, in 1990 and 1993, respectively, and his M.S. and Ph.D. degrees in Electrical and Computer Engineering from University of California, Santa Barbara, in 1998 and 2002, respectively. Prior to joining Cleveland State University in 2004, Dr. Zhao worked as a post-doctoral researcher at the University of California, Santa Barbara, and as a senior research engineer/chief architect at Eternal Systems, Inc. (now dissolved), which he co-founded in 2000. Dr. Zhao has conducted research in several different areas, including fault tolerance computing, computer and network security, smart and connected healthcare, machine learning, Internet of Things, quantum optics, and superconducting physics. Currently, his research focuses on smart and connected healthcare. Dr. Zhao's recent research has been funded by the National Science Foundation, Ohio Bureau of Workers' Compensation, Ohio Department of Higher Education, Ohio Advancement Office (via the Ohio Third Frontier Program), the U.S. Department of Transportation (via CSU Transportation Center), Cleveland State University, and private companies.

Cheng-Fu Yang was born in Taiwan in 1964. Dr. Yang gained his B.S., M.S., and Ph.D. in 1988 and 1993 from the Department of Electrical Engineering of Cheng Kung University, Taiwan. Yang entered academic life in 1993, first in the Department of Electronic Engineering, Chinese Air Force Academy, and since February 2000 as a professor. In February 2004, he joined the faculty of the National University of Kaohsiung (NUK). Currently, he is a Professor of Chemical and Materials Engineering at NUK. He obtained the Outstanding Contribution Award of the Chinese Ceramic Society in 2009. In 2010, he was the first one (and the only one) to obtain the distinguished professor title at NUK. In 2014, he was a Fellow of ythe Taiwanese Institute of Knowledge Innovation (TIKI) and was a fellow of The Institution of Engineering and Technology (IET) in 2015. He was the Mingjiang Scholar and Chair Inviting Professor at Jimei University, Xiamen, Fujian, China.

 *electronics*

Editorial

Special Issue on Intelligent Electronic Devices

Teen-Hang Meen [1,*], Wenbing Zhao [2] and Cheng-Fu Yang [3,*]

[1] Department of Electronic Engineering, National Formosa University, Yunlin 632, Taiwan
[2] Department of Electrical Engineering and Computer Science, Cleveland State University, Cleveland, OH 44011, USA; w.zhao1@csuohio.edu
[3] Department of Chemical and Materials Engineering, National University of Kaohsiung, Kaohsiung 811, Taiwan
* Correspondence: thmeen@nfu.edu.tw (T.-H.M.); cfyang@nuk.edu.tw (C.-F.Y.)

Received: 14 April 2020; Accepted: 14 April 2020; Published: 15 April 2020

Abstract: The second IEEE International Conference on Knowledge Innovation and Invention 2019 (IEEE ICKII 2019) was held in Seoul, South Korea, 12–15 July 2019. This special issue "Intelligent Electronic Devices" selects 13 excellent papers form 260 papers presented in IEEE ICKII 2019 conference about the topics of Intelligent Electronic Devices. The main goals of this special issue are to encourage scientists to publish their experimental and theoretical results in as much detail as possible, and to discover new scientific knowledge relevant to the topics of electronics.

Keywords: electrical circuits and devices; computer science and engineering; communications and information processing

---

1. Introduction

In a modern technological society, electronic engineering and design innovations are both academic and practical engineering fields that involve systematic technological materialization through scientific principles and engineering designs. Engineers and designers must work together with a variety of other professionals in their quest to find systems solutions to complex problems. Fast advances in science and technology have broadened the horizons of engineering, whilst simultaneously creating a multitude of challenging problems in every aspect of modern life. Current research is interdisciplinary in nature, reflecting a combination of concepts and methods that often span several areas of mechanics, mathematics, electrical engineering, control engineering, and other scientific disciplines. In addition, the second IEEE Conference on Knowledge Innovation and Invention 2019 (IEEE ICKII 2019) was held in Seoul, South Korea, 12–15 July 2019, and it provided a unified communication platform for researchers in the topics of information technology, innovation design, communication science and engineering, industrial design, creative design, applied mathematics, computer science, electrical and electronic engineering, mechanical and automation engineering, green technology and architecture engineering, material science, and other related fields. This special issue on "Intelligent Electronic Devices" selected 13 excellent papers from 260 papers presented in IEEE ICKII 2019 on the topics of intelligent electronic devices. The fields include as follows: electrical circuits and devices, computer science and engineering, and communications and information processing. The main goals of this special issue are to encourage scientists to publish their experimental and theoretical results in as much detail as possible, and to discover new scientific knowledge relevant to the topics of electronics.

2. The Topics of Intelligent Electronic Device and Its Applications

This special issue on "Intelligent Electronic Devices" selected 13 excellent papers from 260 papers presented in IEEE ICKII 2019 on the topics of electronics. The published papers are introduced as follows:

Lei et al. reported "Learning Effective Skeletal Representations on RGB Video for Fine-Grained Human Action Quality Assessment" [1]. In this paper, the authors propose an integrated action classification and regression learning framework for the fine-grained human action quality assessment of RGB videos. On the basis of 2D skeleton data obtained per frame of RGB video sequences, the authors present an effective representation of joint trajectories to train action classifiers and a class-specific regression model for a fine-grained assessment of the quality of human actions. To manage the challenge of view changes due to camera motion, the authors develop a self-similarity feature descriptor extracted from joint trajectories and a joint displacement sequence to represent dynamic patterns of the movement and posture of the human body. To weigh the impact of joints for different action categories, a class-specific regression model is developed to obtain effective fine-grained assessment functions. The experimental results show that the proposed method achieved an improved performance, which is measured by the mean rank correlation coefficient between the predicted regression scores and the ground truths.

Lee et al. reported "A Hybrid Tabu Search and 2-opt Path Programming for Mission Route Planning of Multiple Robots under Range Limitations" [2]. In this study, the application of an unmanned vehicle system allows for accelerating the performance of various tasks. Due to limited capacities, such as battery power, it is almost impossible for a single unmanned vehicle to complete a large-scale mission area. An unmanned vehicle swarm has the potential to distribute tasks and coordinate the operations of many robots/drones with very little operator intervention. Therefore, multiple unmanned vehicles are required to execute a set of well-planned mission routes, in order to minimize time and energy consumption. A two-phase heuristic algorithm was used to pursue this goal. In the first phase, a tabu search and the 2-opt node exchange method were used to generate a single optimal path for all target nodes; the solution was then split into multiple clusters according to vehicle numbers as an initial solution for each. In the second phase, a tabu algorithm combined with a 2-opt path exchange was used to further improve the in-route and cross-route solutions for each route. This diversification strategy allowed for approaching the global optimal solution, rather than a regional one with less CPU time. After these algorithms were coded, a group of three robot cars was used to validate this hybrid path programming algorithm.

Chien et al. reported "Research on Anti-Radiation Noise Interference of High Definition Multimedia Interface Circuit Layout of a Laptop" [3]. In this paper, several aspects were studied, including the effect of an electromagnetic interference (EMI) noise interference strategy with High Definition Multimedia Interface (HDMI) 1.4, the analysis of a test on a printed circuit board (PCB) layout, and a comparison of the near field intensity radiation distribution between an EMI with a modified HDMI layout and an original layout. The near field detection instrument of APREL EM-ISight was employed to analyze the distribution of the strength of an electromagnetic noise field. After the practical validation, we found that the PCB layout complies with the standards after the modifications. Meanwhile, the PCB layout satisfies the requirements of most laptop HDMI-related products for EMI.

Du et al. reported "Stereo Vision-Based Object Recognition and Manipulation by Regions with Convolutional Neural Network" [4]. This paper develops a hybrid algorithm of adaptive network-based fuzzy inference system (ANFIS) and regions with convolutional neural network (R-CNN) for stereo vision-based object recognition and manipulation. The stereo camera at an eye-to-hand configuration firstly captures the image of the target object. Then, the shape, features, and centroid of the object are estimated. Similar pixels are segmented by the image segmentation method, and similar regions are merged through selective search. The eye-to-hand calibration is based on ANFIS to reduce computing burden. A six-degree-of-freedom (6-DOF) robot arm with a gripper will conduct experiments to demonstrate the effectiveness of the proposed system.

Perng et al. reported "An Electromagnetic Lock Actuated by a Mobile Phone Equipped with a Self-Made Laser Pointer" [5]. The main purpose of this study was to create an acousto-optic control lock device to convert electrical signals with a specific sound command using an acousto-optic conversion module, thereby improving the reliability and safety of opening or closing remote controlled door locks, such as car central locks or rolling doors. We used music playing through a smart phone speaker to create a special laser pointer to connect with the auxiliary input of the smart phone. The laser pointer (wavelength of 630 to 650 nm and maximum output of 5 mw) lights up when the music of the smart phone starts playing at a music frequency matching the light frequency. When the solar panel receives light, it converts the frequency of the light signal into an electrical frequency signal. The current is amplified using the power amplifier and then the amplified current flows to the sound recognition module. The sound recognition module performs audio comparison on the set sound signal, and once the comparison is correct, the output voltage activates the electromagnetic switch on the door to open or close it.

Lin et al. reported "Electrostatic-Discharge-Immunity Impacts in 300 V nLDMOS by Comprehensive Drift-Region Engineering" [6]. This paper focuses on comprehensive drift-region engineering for ultra-high-voltage (UHV) circular n-channel lateral diffusion metal-oxide-semiconductor transistor (nLDMOS) devices used to investigate impacts on ESD ability. Under the condition of fixed layout area, there are four kinds of modulation in the drift region. First, by floating a polysilicon stripe above the drift region, the breakdown voltage and secondary breakdown current of this modulation can be increased. Second, adjusting the width of the field-oxide layer in the drift region when the width of the field-oxide layer is 5.8 μm will result in the minimum breakdown voltage (105 V), but the best secondary breakdown current (6.84 A). Third, by adjusting the discrete unit cell and its spacing, the corresponding improved trigger voltage, holding voltage, and secondary breakdown current can be obtained. According to the experimental results, the holding voltage of all devices under test (DUTs) is greater than that of the reference group, so the discrete HV N-Well (HVNW) layer can effectively improve its latch-up immunity. Finally, by embedding different P-Well lengths, the findings suggest that when the embedded P-Well length is 9 μm, it will have the highest ESD ability and latch-up immunity.

Lam et al. reported "Generative Noise Reduction in Dental Cone-Beam CT by a Selective Anatomy Analytic Iteration Reconstruction Algorithm" [7]. In this paper, the authors propose a new algorithm called the selective anatomy analytic iteration reconstruction (SA2IR) algorithm for the sparse-projection set. The algorithm was simulated on a phantom structure analogous to a patient's head for geometric similarity. The proposed algorithm is projection-based. Interpolated set enrichment and trio-subset enhancement were used to reduce the generative noise and maintain the scan's clinical diagnostic ability. The results show that the proposed method was highly applicable in medico-dental imaging diagnostics fusion for the computer-aided treatment planning, because it had significant generative noise reduction and lowered computational cost when compared to the other common contemporary algorithms for sparse projection, which generate a low-dosed CBCT reconstruction.

Shen et al. reported "Dual-Input Isolated DC-DC Converter with Ultra-High Step-Up Ability Based on Sheppard Taylor Circuit" [8]. A dual-input high step-up isolated converter (DHSIC) is proposed in this paper, which incorporates Sheppard Taylor circuit into power stage design so as to step up voltage gain. In addition, the main circuit adopts boosting capacitors and switched capacitors, based on which the converter voltage gain can further be improved significantly. Since the proposed converter possesses an inherently ultra-high step-up feature, it is capable of processing low input voltages. The DHSIC also has the important features of leakage energy recycling, switch voltage clamping, and continuous input-current obtaining. These characteristics have an advantage on converter efficiency and benefit the DHSIC for high power applications. The structure of the proposed converter is concise. That is, it can lower cost and simplifies control approach. The operation principle and theoretical derivation of the proposed converter are discussed thoroughly in this paper. Simulations and hardware implementation are carried out to verify the correctness of theoretical analysis and to validate feasibility of the converter as well.

Tseng et al. reported "Secondary Freeform Lens Device Design with Stearic Acid for A Low-Glare Mosquito-Trapping System with Ultraviolet Light-Emitting Diodes" [9]. This study is dedicated to the development of a new mosquito-trapping system. Research has shown that specific wavelengths, colors, and temperatures are highly attractive to both Aedes aegypti and Aedes albopictus. The authors create equipment which effectively improves the trapping capabilities of mosquitoes in a wider field. The design of the special Secondary Freeform Lens Device (SFLD) is used to expand the range for trapping mosquitoes and create illumination uniformity; it also directs light downward for the protection of users' eyes. In addition, we use the correct amount of stearic acid as a mosquito attractant to allow a better control effect against mosquitoes during the day. In summary, when the UV LED mosquito trapping system is combined with a quadratic free-form lens, the experimental results show that the system can extend the capture range to $100\,\pi\,m^2$, in which the number of captured mosquitoes is increased by about 350%.

Wei et al. reported "Using Different Ions in the Hydrothermal Method to Enhance the Photoluminescence Properties of Synthesized ZnO-Based Nanowires" [10]. In this study, ZnO films with a thickness of ~200 nm were deposited on SiO_2/Si substrates as the seed layer. Then, $Zn(NO_3)_2$-$6H_2O$ and $C_6H_{12}N_4$ containing different concentrations of $Eu(NO_3)_2$-$6H_2O$ or $In(NO_3)_2$-$6H_2O$ were used as precursors, and a hydrothermal process was used to synthesize pure ZnO as well as Eu-doped and In-doped ZnO nanowires at different synthesis temperatures. The effect of different concentrations of Eu^{3+} and In^{3+} ions on the physical and optical properties of ZnO-based nanowires was well investigated. FESEM observations found that the undoped ZnO nanowires could be grown at 100 ∘C. The temperatures required to grow the Eu-doped and In-doped ZnO nanowires decreased with increasing concentrations of Eu^{3+} and In^{3+} ions. XRD patterns showed that with the addition of Eu^{3+} (In^{3+}), the diffraction intensity of the (002) peak slightly increased with the concentration of Eu^{3+} (In^{3+}) ions and reached a maximum at 3 (0.4) at%. It is revealed that the concentrations of Eu^{3+} and In^{3+} ions have considerable effects on the synthesis temperatures and photoluminescence properties of Eu^{3+}-doped and In^{3+}-doped ZnO nanowires.

Ulansky et al. reported "Electronic Circuit with Controllable Negative Differential Resistance and its Applications" [11]. In this paper, a new NDR circuit that comprises a combination of a field effect transistor (FET) and a simple bipolar junction transistor (BJT) current mirror (CM) with multiple outputs is proposed. A distinctive feature of the proposed circuit is the ability to change the magnitude of the NDR by increasing the number of outputs in the CM. Mathematical expressions are derived to calculate the threshold currents and voltages of the N-type current-voltage characteristics for various types of FET. The calculated current and voltage thresholds are compared with the simulation results. The possible applications of the proposed NDR circuit for designing single-frequency oscillators and voltage-controlled oscillators (VCO) are considered. The designed NDR VCO has a very low level of phase noise and has one of the best values of a standard figure of merit (FOM) among recently published VCOs. The effectiveness of the proposed oscillators is confirmed by the simulation results and the implemented prototype.

Zhang et al. reported "Low Cost Test Pattern Generation in Scan-Based BIST Schemes" [12]. This paper proposes a low-cost test pattern generator for scan-based built-in self-test (BIST) schemes. Our method generates broadcast-based multiple single input change (BMSIC) vectors to fill more scan chains. The proposed algorithm, BMSIC-TPG, is based on our previous work multiple single-input change (MSIC)-TPG. The broadcast circuit expends MSIC vectors, so that the hardware overhead of the test pattern generation circuit is reduced. Simulation results with ISCAS'89 benchmarks and a comparison with the MSIC-TPG circuit show that the proposed BMSIC-TPG reduces the circuit hardware overhead at about 50%, ensuring low power consumption and high fault coverage.

Pamungkas et al. reported "Overview: Types of Lower Limb Exoskeletons" [13]. In order to provide information about which actuator location is more suitable; a review study on the design of actuator locations is presented in this paper. The location of actuators is an important factor because it is related to the analysis of the design and the control system. This factor affects the entire lower limb exoskeleton's performance and functionality. In addition, the disadvantages of several types of lower limb exoskeletons in terms of actuator locations and the challenges of the lower limb exoskeleton in the future are also presented in this paper.

Author Contributions: Writing and reviewing all papers, T.-H.M.; English editing, W.Z.; checking and correcting the manuscript, C.-F.Y. All authors have read and agreed to the published version of the manuscript.

Acknowledgments: The guest editors would like to thank the authors for their contributions to this special issue and all the reviewers for their constructive reviews. We are also grateful to Michelle Zhou, the assistant editor of Electronics, for her time and efforts on the publication of this special issue for Electronics.

Conflicts of Interest: The authors declare no conflict of interest.

References

1. Lei, Q.; Zhang, H.B.; Du, J.X.; Hsiao, T.C.; Chen, C.C. Learning effective skeletal representations on RGB video for fine-grained human action quality assessment. *Electronics* **2020**, *9*, 568. [CrossRef]
2. Lee, M.T.; Chen, B.Y.; Lai, Y.C. A hybrid tabu search and 2-opt path programming for mission route planning of multiple robots under range limitations. *Electronics* **2020**, *9*, 534. [CrossRef]
3. Chien, W.; Cheng, Y.T.; Hsiao, C.F.; Han, K.X.; Chiu, C.C. Research on anti-radiation noise interference of high definition multimedia interface circuit layout of a laptop. *Electronics* **2020**, *9*, 426. [CrossRef]
4. Du, Y.C.; Muslikhin, M.; Hsieh, T.H.; Wang, M.S. Stereo vision-based object recognition and manipulation by regions with convolutional neural network. *Electronics* **2020**, *9*, 210. [CrossRef]
5. Perng, J.W.; Hsieh, T.L. An electromagnetic lock actuated by a mobile phone equipped with a self-made laser pointer. *Electronics* **2019**, *8*, 1524. [CrossRef]
6. Lin, P.L.; Chen, S.L.; Fan, S.K. Electrostatic-discharge-immunity impacts in 300 V nLDMOS by comprehensive drift-region engineering. *Electronics* **2019**, *8*, 1469. [CrossRef]
7. Lam, D.N.; Du, Y.C. Generative noise reduction in dental cone-beam CT by a selective anatomy analytic iteration reconstruction algorithm. *Electronics* **2019**, *8*, 1381. [CrossRef]
8. Shen, C.L.; Chen, L.Z.; Chen, H.Y. Dual-input isolated DC-DC converter with ultra-high step-up ability based on sheppard taylor circuit. *Electronics* **2019**, *8*, 1125. [CrossRef]
9. Tseng, W.H.; Hsiao, W.C.; Juan, D.; Chan, C.H.; Hsiao, W.S.; Ma, H.Y.; Lee, H.Y. secondary freeform lens device design with stearic acid for a low-glare mosquito-trapping system with ultraviolet light-emitting diodes. *Electronics* **2019**, *8*, 624. [CrossRef]
10. Wei, Y.F.; Chung, W.Y.; Yang, C.F.; Shen, J.R.; Chen, C.C. Using different ions in the hydrothermal method to enhance the photoluminescence properties of synthesized ZnO-based nanowires. *Electronics* **2019**, *8*, 446. [CrossRef]
11. Ulansky, V.; Raza, A.; Oun, H. Electronic circuit with controllable negative differential resistance and its applications. *Electronics* **2019**, *8*, 409. [CrossRef]
12. Zhang, G.; Yuan, Y.; Liang, F.; Wei, S.; Yang, C.F. Low Cost Test Pattern Generation in Scan-Based BIST Schemes. *Electronics* **2019**, *8*, 314. [CrossRef]
13. Pamungkas, D.S.; Caesarendra, W.; Soebakti, H.; Analia, R.; Susanto, S. Overview: Types of lower limb exoskeletons. *Electronics* **2019**, *8*, 1283. [CrossRef]

 *electronics*

Article

Learning Effective Skeletal Representations on RGB Video for Fine-Grained Human Action Quality Assessment

Qing Lei [1,2,3], Hong-Bo Zhang [1,2,3], Ji-Xiang Du [1,2,3], Tsung-Chih Hsiao [1,3,*] and Chih-Cheng Chen [4,*]

[1] College of Computer Science and Technology, University of Huaqiao, Xiamen 361021, China; leiqing@hqu.edu.cn (Q.L.); zhanghongbo@hqu.edu.cn (H.-B.Z.); jxdu@hqu.edu.cn (J.-X.D.)
[2] Fujian Key Laboratory of Big Data Intelligence and Security, Huaqiao University, Xiamen 361021, China
[3] Xiamen Key Laboratory of Computer Vision and Pattern Recognition, University of Huaqiao, Xiamen 361021, China
[4] School of Information Engineering, Jimei University, Xiamen 361021, China
* Correspondence: hsiaotc@hqu.edu.cn (T.-C.H.); 201761000018@jmu.edu.cn (C.-C.C.)

Received: 23 January 2020; Accepted: 26 March 2020; Published: 28 March 2020

Abstract: In this paper, we propose an integrated action classification and regression learning framework for the fine-grained human action quality assessment of RGB videos. On the basis of 2D skeleton data obtained per frame of RGB video sequences, we present an effective representation of joint trajectories to train action classifiers and a class-specific regression model for a fine-grained assessment of the quality of human actions. To manage the challenge of view changes due to camera motion, we develop a self-similarity feature descriptor extracted from joint trajectories and a joint displacement sequence to represent dynamic patterns of the movement and posture of the human body. To weigh the impact of joints for different action categories, a class-specific regression model is developed to obtain effective fine-grained assessment functions. In the testing stage, with the supervision of the action classifier's output, the regression model of a specific action category is selected to assess the quality of skeleton motion extracted from the action video. We take advantage of the discrimination of the action classifier and the viewpoint invariance of the self-similarity feature to boost the performance of the learning-based quality assessment method in a realistic scene. We evaluate our proposed method using diving and figure skating videos of the publicly available MIT Olympic Scoring dataset, and gymnastic vaulting videos of the recent benchmark University of Nevada Las Vegas (UNLV) Olympic Scoring dataset. The experimental results show that the proposed method achieved an improved performance, which is measured by the mean rank correlation coefficient between the predicted regression scores and the ground truths.

Keywords: action quality assessment; human activity analysis; skeletal feature representation

1. Introduction

Human action evaluation (HAE) aims to tackle the challenging problem of making computers automatically quantify how well people perform actions. It has been largely unexplored in past decades [1,2], and has been involved in a wide range of applications, such as sport activity scoring and training systems [1,3], physical therapy and rehabilitation [4–7], interactive entertainment [8–10], skill training for expertise learners, and video retrieval [11–13]. With the rapid progress in human activity understanding in the research area of computer vision, research efforts have recently been devoted to human action quality assessment [14,15].

Since the traditional manual assessment of human motion quality needs a great deal of expertise from specialized fields, longtime learning, and training processes are required to summarize the experience and evaluation rules for automatic scoring sport activity in a specialized field. This requires a great amount of time and high labor cost. Apart from traditional action recognition research, human action evaluation aims to design computation models for automatically assessing the quality score of human actions or activities and further give interpretable feedback to improve human body movement. It relies on accurate human motion detection and segmentation, action feature extraction and representation, and effective evaluation methods for measuring the quality of action performance. Severe challenges have to be dealt with when the action evaluation learning method is applied in realistic scenes such as intra-class variations in the scale, appearance, illumination, view, and inter-class ambiguity.

Most of the published research on human action evaluation directly employed advanced action recognition approaches to segment an action video into several action fragments, extract local or holistic motion features for video fragments, and aggregate fragmented features into a final feature representation. Then, regression models [14,15] or Hidden Markov Models (HMM) [16] were trained to estimate the quality score of featured actions. Furthermore, interpretable feedback was provided for improving the action performance. Some of these studies employed an optimization framework to formalize the action quality assessment problem [1,5,15]. They commonly employed this framework to develop a unified regression model for all action categories. One of the disadvantages of using a unified regression model is that large approximation errors caused by in-class variation lead to poor fitting in regression analysis. The second disadvantage is that a common evaluation function shared among all action categories can be fragile when dealing with unbalanced distributions of training data. Third, similar postures shared by different action categories significantly decrease the performance of action evaluation methods, such as the swing in a tennis serve and badminton smash, the spin in figure skating, and the floor exercise. Consequently, single assessment function learning for all action categories tends to generate an inaccurate assessment score and even provide the wrong feedback information. With significant progress in pose estimation methods, skeleton data of the human body can be estimated from an RGB video to facilitate detailed motion quality analysis. Most of the existing research introduced alignment and normalization methods developed for action recognition in action quality analysis to preprocess the skeleton data [17–20]. However, view variation due to a change of the camera position has not yet been addressed, despite the fact that it cannot be trivial for human action evaluation of a realistic dataset.

In this paper, we attempt to extract effective skeletal feature representation of human motion from an RGB video for a fine-grained human action quality assessment and develop a learning framework for assessing the action quality of sport activity. First, it is reasonable to assume that the action quality directly depends on the dynamic changes in human body movements. Therefore, in this study, we employed the OpenPose estimation method [21] to extract skeleton data from RGB videos for action quality analysis. Then, to determine the action category of test videos, we extracted the local motion pattern from each joint movement volume and aggregated all volumes to form a feature description for action classification. To accurately predict the quality score of an action video, we developed effective self-similarity feature descriptors extracted from the self-similarity matrices (SSMs) of joint trajectories and a joint displacement sequence that has been proven to alleviate the impact of camera motion in diving, figure skating, and vaulting videos. Lastly, a class-specific regression model learning strategy was employed to weigh the impact of joints on different action category evaluations. In the testing stage, with the supervision of the action classifier's output, the target regression model was determined to predict the quality score of the testing action video. The framework of our approach is illustrated in Figure 1.

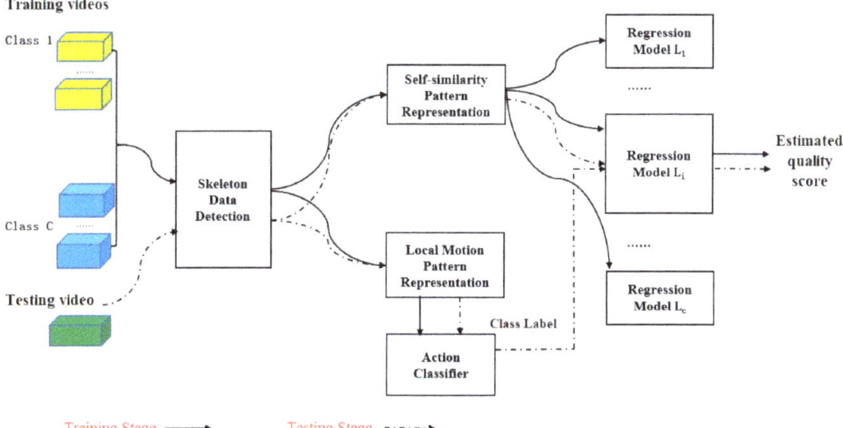

Figure 1. The framework of our action quality assessment method. (The input data were RGB videos of diverse action categories. Skeleton data detection: OpenPose algorithms provided by the work of Reference [21] were performed to capture skeleton data from the RGB video. Local motion pattern representation: the local motion pattern was extracted from each joint movement volume and aggregated to train the action classifier. Action classifier: the Support Vector Machine (SVM) classifier was developed from labeled training samples to determine the class label of the action video. Self-similarity pattern representation: we developed self-similarity feature descriptors extracted from joint trajectories and joint displacement sequences to represent the periodic property of sport activities. Regression models: we employed a class-specific learning strategy, and trained multiple regression models specific to different action categories for action quality score estimation.).

The contributions of this paper can be summarized as follows.

(1) First, in this paper, we propose an integrated action classification and action quality regression learning framework for the human action evaluation of RGB videos. An action classifier and assessment regression models are utilized, respectively, in different components. The former is used to predict the class label of a testing video, while the latter is employed to estimate the quality score with the supervision of a class label.

(2) Second, taking advantage of the view invariant property provided by self-similarity, in this paper, we develop self-similarity feature representation extracted from joint trajectories and joint displacement sequences to describe motion patterns of joints and posture changes, respectively. This encodes not only the dynamic changes of individual joints, but also the layout changes of all body joints. The experimental results prove that it alleviates the impact of camera motion in realistic scenes, and improves the skeleton representation's performance for diving, figure skating, and vaulting videos of an Olympic sports event.

(3) Experiments on a benchmark dataset were carried out in this study. The results show that the proposed method improved the rank correlation coefficient of the predicted scores against the judge's scores when compared with the baseline and other handcrafted feature methods.

The remainder of this paper is organized as follows. Section 2 introduces the related research works. Section 3 describes the algorithms developed to implement our action quality assessment method. After showing experiments in Section 4, we conclude the study in Section 5.

2. Related Works

On the basis of skeleton data analysis, several video-based human action evaluation research studies have been published in the last decade. There are two major issues focused on in these studies. The first relates to capturing useful features from an RGB video to represent human actions for video-based human action evaluation, and the second relates to developing a robust evaluation method for accurately assessing the similarity between action features on account of complex environments. Some of the reviewed published works regarded the task of human action evaluation as a video sequence recognition problem. They tackled the challenge with the help of a machine learning method. In summary, the reviewed action quality assessment models proposed for human action evaluation research can be divided into three categories. These three categories include linear regression models (LR) [1,5,15], Hidden Markov Models (HMM) [4,7,10], and other statistic-based learning models [3,6,11,12,14].

In the early work of Reference [11], linear regression was employed to reduce the raw Motion Capture data for online action recognition. They designed a graph-based action model embodied with recurrent transitions for motion retrieval. The algorithm was linear and incremental, which makes it convenient for adding new actions and suitable for real-time application. Pirsiavash et al. [1] first proposed a two-layer processing framework for video-based human action quality assessment. In the first layer, they extracted spatio-temporal interest point features from regions returned by a human detection algorithm and computed discrete cosine transformation (DCT) features of joint displacement vectors from body joints' trajectories to represent human actions. Then, in the second layer, they developed a regression estimator to predict the quality score of a sports activity. Venkataraman et al. [15] tried to encode human actions through the dynamic changes of each body joint and the relationship between body joints. They developed two kinds of entropy-based features extracted from human skeleton data to represent these two clues and used them to realize human action segmentation from long video sequences and assess the quality score of diving in sports competition videos. Antunes et al. [5] presented a visual and human-interpretable feedback system for assisting with the physical activities of patients or athletes suffering from sport injuries. They also used skeleton data extracted from videos to quantify the action quality of human movements. First, the pre-processing transformation steps were conducted in both spatial and temporal dimensions to align a testing sequence with the template. Then, the matching error between the testing sequence and the template was computed based on the Euclidean distance of joints' coordinates in order to quantify the similarity between a testing sequence and a normal one. Furthermore, feedback for guiding how to perform properly was computed by minimizing the skeleton matching error and returned to the users.

Paiement et al. [4] used 3D skeleton data captured by two kinds of depth sensors and pre-processed the coordinates of joint sequences for online estimation of the quality of human movements. They proposed two statistical models: one was the probability density function (PDF) of each individual pose, and the other was the conditional PDF of a pose sequence in order to represent the features of normal movements. Then, log-likelihoods of observations compared with the model of normal ones were computed for quality assessment. They evaluated their methods using a gait on the stairs' dataset. In their further work of Reference [7], they studied four low-level pose features such as joint positions, joint velocities, pairwise joint distances, and pairwise joint angles. They also compared three kinds of discrete-state HMM and one continuous-state HMM to represent pose features and temporal dynamics of motion. They tested these features and models using periodic and non-periodic motions, including walking on a flat surface, gait on stairs, sitting, and standing. The experimental results demonstrated that continuous-state HMM performed better when describing motion dynamics for these action categories than other models for a frame-by-frame analysis of a motion quality assessment.

To concurrently evaluate the relevant spatial and temporal information of motion, Morel et al. [3] proposed an automatic morphology-independent and sport-independent method for assessing the motion of a player by comparing the features with the model learned from experts' motions. To deal with the different motion durations, they employed Dynamic Time Warping (DTW) to temporally align the skeleton data of joint trajectories. Considering the limitation of DTW—that the first and last frame of two aligned sequences are required to be in correspondence—Baptista et al. [6] investigated adapting Subsequence Dynamic Time Warping (SS-DTW) and Temporal Commonality Discovery

(TCD) to provide feedback proposals for improving the action performance of stroke survivors in a video-based rehabilitation system. Hu et al. [12] presented an action tutor system, which aimed to achieve a high-level evaluation of human action movements with the aid of Kinect. The body-joint configuration and shape/depth distribution of the human silhouette were encoded as pose descriptors to reflect the difference between various postures. Modified DTW and approximate string matching were further proposed to measure the similarity of actions. In the work of Reference [14], a novel framework for motion analysis, including the real-time action detection, recognition, and evaluation of motion capture data, was presented. The descriptor named Gesturelet was calculated for 3D skeleton joint positions, and combined the Moving Pose [22] and the angle descriptor [23] with appropriate weighting. Kinetic energy features were employed to construct Bag-of-Words data representation for action segmentation. In the evaluation component, 3D joint position and linear velocity errors were calculated and normalized, and then fed into a fuzzy logic engine to produce semantic feedback interpretations.

In this work, we investigated the effectiveness of using an integrated learning framework combining action quality assessment with action classification to boost the performance of action evaluation in realistic scenes. In this framework, a novel feature descriptor extracted from the self-similarity matrix of joint trajectories and joint displacement sequences was developed to alleviate the impact of camera motion. Then, class-specific regression models were established to predict the quality scores of action videos. Additionally, we trained the action classifier to supervise the determination of the regression model to assess the quality score of a testing action video. The experimental results proved that this approach is helpful for alleviating approximation errors caused by inter-class confusion.

3. Proposed Method

3.1. Skeleton Data Extraction

It is a reasonable assumption that the quality of human motion directly depends on the changing process of human body movement, which can be represented by the motion trajectory and relative location relationship between joints or body parts. Therefore, learning effective action features from motion trajectories of action videos plays an important role in developing reliable quality assessment algorithms for action evaluation. Most recent works employed skeleton data that was captured or detected from a depth or color camera for action evaluation research. With significant progress in recent pose estimation techniques and methodologies, skeleton data can now be estimated from RGB image data. Traditional skeletonization models, such as the deformable part model and flexible mixtures of parts model, have been replaced by deep neural network-based approaches. OpenPose [21] is an effective pose estimation method developed by the perceptual computing lab of Carnegie Mellon University. It is the first real-time multi-person skeleton detection system and works well when applied to RGB videos. Therefore, to obtain skeleton data of an action performer in an RGB video, we employed the state-of-the-art OpenPose algorithm to detect joints' position for each frame and extracted the trajectories of joints to represent the action video.

OpenPose provides the functionality of 2D real-time multi-person key point detection (15-, 18-, or 25-key point body/foot key point estimation). The 18-key point skeleton model is composed of 18 human body joints, as illustrated in Figure 2a, including the nose, neck, right shoulder, right elbow, right wrist, left shoulder, left elbow, left wrist, right hip, right knee, right ankle, left hip, left knee, left ankle, right eye, left eye, right ear, and left ear. Since action quality assessment is highly dependent on the changing positions and configuration of human body parts, the motion changes of key points on eyes, ears, and feet are not clear. Consequently, we only used the motion information of 14 body joints ($N = 0 \sim 13$) except for the eyes and ears for analysis. After the multi-person's skeleton detection results were returned by OpenPose, we carried out scale computation of the human body and a key point confidence comparison to extract the target performer's skeleton data and filter out noise data, which results from a cluttered background. In the case of occlusion or self-occlusion, zero values of the joints' coordinate were obtained due to the failure detection of the human body. Then, linear interpolation was employed to capture the missing skeleton data from the pose estimation results of the previous frame and the next frame. In this way, the joint trajectories of the target performer were obtained. Some detection examples of diving and figure skating videos from the MIT Olympic Scoring Dataset [1] are illustrated in Figure 2b,c.

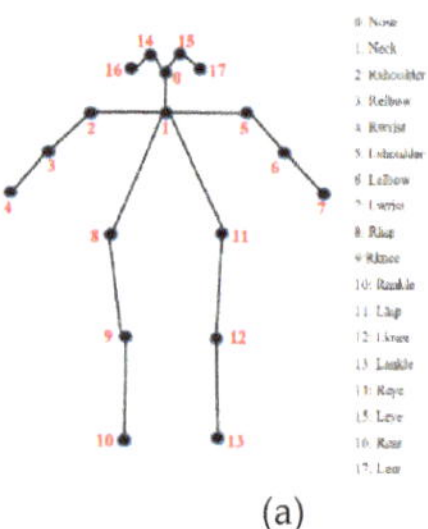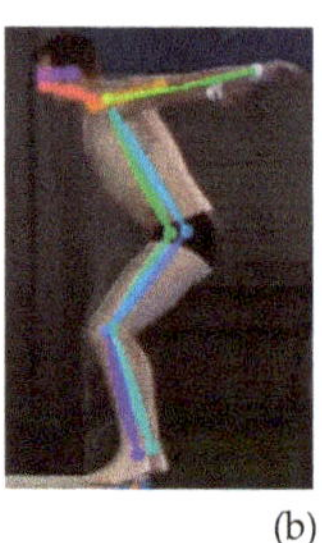

(a)   (b)   (c)

Figure 2. The 18-key point skeleton model and detection examples of OpenPose [21]. (**a**) Skeleton model of 18 joints, (**b**) example of diving, and (**c**) example of figure skating.

3.2. Action Classification Component

We considered local feature representations' advantage in terms of their robustness in dealing with intra-class variation, and the fact that holistic feature representations provide a comprehensive description of human body movement for a time sequence. Combining both of their strengths, we proposed a joint movement feature to acquire discriminative information for action classification. It is believed that the semantics of human action are related to the movement pattern of joints and the relationship of the human body interacting with its surrounding environment. Therefore, building effective features that capture both the discriminative dynamics of body joints and the descriptive spatial context for class label determination is investigated in this paper. The pipeline of our action classification component is illustrated in Figure 3.

Let $\left\{S^1, S^2, \ldots, S^N\right\}$ denote a set of joint trajectories obtained by skeleton detection for video V, where N is the number of human body joints, $S^k = \left[s_1^k, s_2^k, \ldots, s_T^k\right]$ represents the trajectory of the kth joint, and T is the number of frames of V. Each joint position is located by its coordinates $s = \left[s_x, s_y, s_t\right]$ in discrete (x, y, t)-space. To capture the spatial context of joints, a n × n dimensional local patch centered at each joint position $s^k = \left[s_x^k, s_y^k, s_t^k\right]$ is extracted from video frames where $k = 1 \sim N$. All patches extracted over a temporal duration ($t = 1 \sim T$) are assembled into a motion volume of joint k, denoted by $v^k (k = 1 \sim N)$, which is illustrated by the red cuboids in Figure 3.

To filter out noise data resulting from failure or false detection of the joint position, 2D Gaussian smoothing is first performed for the joint motion volume v^k. Then, the central moment features $m_{i,j}^r (i, j = 1 \sim n, r = 1, 2)$ for each super-pixel $v_{i,j}^k$ of joint motion volume v^k are calculated according to Equation (1).

$$m_{i,j}^r = \frac{1}{T}\Sigma_{t=1}^{T}\left(G_{i,j,t} - \overline{G}\right)^r, \ \overline{G} = \frac{1}{n \times n \times T}\Sigma_{t=1}^{T}\Sigma_{i=1}^{n}\Sigma_{j=1}^{n}G_{i,j,t} \tag{1}$$

where $G_{i,j,t}$ is the value of the pixel located at the (i, j, t)-coordinate in filtered volume v^k. Features of all the super pixels contained in a joint motion volume v^k are assembled to form the motion feature m^k, and to represent the motion pattern of the kth joint. Then, all the features of N joint volumes are concatenated to form the final spatio-temporal feature description to represent the action instance occurring in video V.

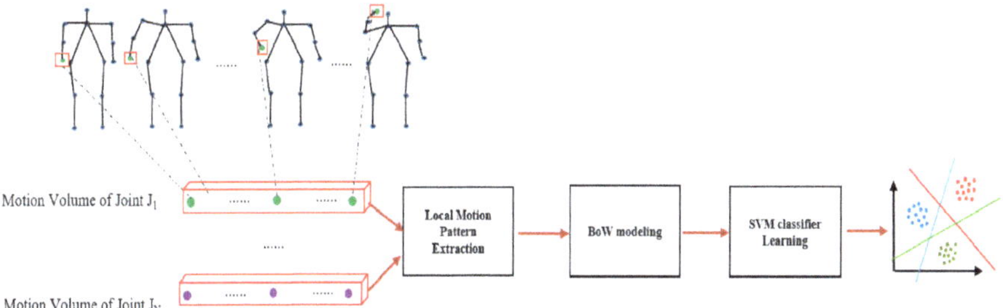

Figure 3. The pipeline of the action classification component. (The input of the action classification component was joint trajectories obtained by skeleton detection from an RGB video. Joint motion volume: A joint motion volume centered at each of the joint positions and comprised of the $n \times n$ spatial context was extracted to represent the movement of each joint. Local motion pattern representation: We employed the central moment features computed by the noise filtered joint motion volume to describe the motion patterns of each corresponding joint, and concatenated all joint features into the final motion descriptor. Bag-of-words (BoW) modeling: For the sake of alleviating intra-class variation, we employed bag-of-words feature modeling, which utilizes K-means feature clustering and cluster frequency statistics to form the final action representation of a video. Action classifier: The SVM classifier was developed from features of labeled training samples to determine the class label of action instance.).

We propose applying the Bag-of-Words (BoW) [24] model for action representation. Specifically, the unsupervised K-means algorithm is first performed on all joint motion volume features extracted from training videos to obtain K clusters for constructing the action codebook. The center of each cluster is called a visual word, and all centers of K clusters form a visual codebook for action modeling. Then, the original feature is projected onto the closest visual word in the action codebook. All features of an action video are projected and aggregated into an occurrence frequency histogram of visual words. This forms the final BoW feature representation of the action video.

Lastly, BoW features with class labels are fed into maximum margin classifier learning, as formulated in Equation (2).

$$min_{w,b} \frac{\|w\|^2}{2}, s.t.\ Y\left(w^T H_C + b\right) - 1 \geq 0, \tag{2}$$

where H_C denotes the feature vector of a training video for classification and Y is the ground-truth action class label of each training video provided by manual annotation in the benchmark dataset. w and b represent the normal vector and bias of the classification hyperplane, respectively.

3.3. Quality Assessment Component

As stated above, skeleton-based pose feature representations intuitively reflect the changing process of human body movement and provide significant information for action quality assessment. Therefore, when developing an accurate action quality assessment method, it is preferable to encode the dynamic changes of joints or body parts into feature representation for action analysis. However, the fine-grained quality assessment of human action is confronted by the challenges of intra-class variations, such as different scales of the human body, variation in motion velocity, individual style, and changes due to camera motion. We performed skeleton data preprocessing, including noise filtering, scale normalization, and spatial alignment, to tackle the problem of intra-class variations. Furthermore, self-similarity patterns were extracted from joint trajectories and joint displacement sequences, respectively, to model the invariant dynamics of human body motion and to deal with view changes in realistic scenes.

Another motivation is the observation that the movement ranges of joints are different for various action categories. Therefore, the importance of different joint movements cannot be considered identical for action quality evaluation. For example, as far as serves, smashes, and swings in tennis and badminton activity are concerned, the movement of upper limbs ranges more significantly than that of lower limbs. On the contrary, the remarkable change that occurs in lower limb motion should be addressed in pommel horse riding, parallel bars, and football. Furthermore, all limb movement may be considered equal for figure skating and diving quality assessments. Therefore, it should be noted that, for different action categories, the similarity measurement of quality features should address weight assignment for the impact of different joints. To address this problem, a class-specific regression model was developed to weigh different impacts of joint movements and to obtain more accurate evaluation scores for fine-grained human action quality assessment in this work.

The quality assessment component of our proposed learning framework consists of the preprocessing of joint trajectories, joint displacement sequence extraction, feature extraction from the self-similarity matrix (SSM) of joint trajectories and joint displacement sequence, and class-specific regression learning. The training process of our class-specific action quality assessment component is illustrated in Figure 4.

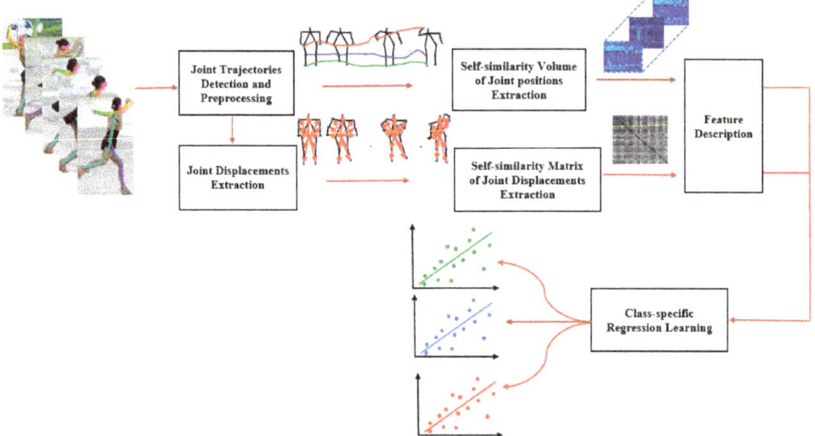

Figure 4. The training process of our quality assessment component. (The input of the action classification component was joint trajectories obtained by skeleton detection from an RGB video. Joint trajectory detection and preprocessing: We performed noise filtering, normalization, and spatial alignment processes for detected joint trajectories to deal with failure or false detection, scale variation, and spatial transformation of human motion. Joint displacement extraction: As the middle of the left hip and right hip was selected as the central point, the displacement of each joint relative to the central point was computed to represent the layout relationship between human body parts. Self-similarity volume of joint position extraction: Temporal self-similarity matrix of each joint trajectory was computed to capture view invariant patterns of intra-joint dynamic changes. Self-similarity matrix of joint displacement extraction: Temporal self-similarity matrix of all joint displacements was computed to capture view invariant patterns of inter-joint dynamic changes. Feature description: the Histogram of Gradient (HOG) descriptor was employed to describe the pattern structure of two kinds of self-similarity matrices. Class-specific regression learning: We utilized two types of regression strategy—support vector regression and ridge regression—to develop class-specific regression models for an action quality assessment.).

3.3.1. Pre-Processing of the Joint Trajectory

The original skeleton data detected from RGB videos often contain noise in cases of occlusion and cluttered environments in realistic applications. To obtain robust similarity quantization for fine-grained assessment, the first processing step of skeleton data is the noise filtering of incorrect joint coordinates resulting from incorrect human detection. Then, normalization and alignment processing, including the scale, transition, rotation, and appearances, are subsequently conducted to deal with various intra-class variations.

In the case of occlusion or self-occlusion, the failure or false detection of the human body leads to outliers of joint coordinates in skeleton estimation that lead to an unreliable representation of human motions. In this study, discrete cosine transformation was performed for discrete coordinates of each joint trajectory to filter out zero values or sharply changing coordinates resulting from failure or false pose estimation. Low-frequency components were preserved for the reliable detection of human body positions.

On one hand, the scale of captured skeleton data can be quite diverse due to different distances between the subject and camera. The original joint coordinates are required to be normalized to a prototypical range for comparison. The scale normalization process is formulized as follows. Suppose that you are given skeleton data $\{S^1, S^2, \ldots, S^N\}$ detected from an action video I, where N denotes the number of joints and $S^k = \left[s_1^k, s_2^k, \ldots, s_T^k\right]$ represents the motion trajectory of the kth joint through T consecutive frames of I. First, the middle of Rhip (Joint 8) and Lhip (Joint 11) of the first frame is defined as the central point, and the distance from Neck (Joint 1) to the central point is defined as the normalized length. Then, the joints' coordinates are scaled by the normalized length. The normalization process can be formulated by the following equation.

$$s'^k_t = \frac{s_t^k}{\left\| s_1^1 - \frac{s_1^8 + s_1^{11}}{2} \right\|},$$

(3)

where $s_t^k = \left[s_x, s_y, t\right]$ denotes the origin coordinate of joint k in frame t in the (x, y, t)-space.

Then, spatial alignment is conducted by performing rotation transformation. The rotation angle θ is determined by projecting the vector from Lhip to Rhip onto the x-axis. Then, each joint's coordinate is transformed by rotating θ-degree, as formulated by Equation (4).

$$\theta = acos\left(\frac{\overrightarrow{\left(s^{11} - s^8\right)} \cdot \overrightarrow{ox}}{\left\|s^{11} - s^8\right\| \left\| \overrightarrow{ox} \right\|}\right), \quad \overrightarrow{s}' = \begin{bmatrix} cos\theta & sin\theta \\ cos\theta & -sin\theta \end{bmatrix} \overrightarrow{s},$$

(4)

where, $\overrightarrow{s} = \left[s_x, s_y\right]$ and $\overrightarrow{ox} = [1, 0]$.

3.3.2. Self-Similarity Feature Description

The motion pattern for different viewpoints varies significantly for an action, as shown in Figure 5. It illustrates two different joint trajectories of the same person performing a diving action in the side and front views. The first row presents the skeleton detection results on the side view and the second row shows the skeleton detection results on the front view. It should be noted that a reliable feature representation for fine-grained quality assessment has to be robust for different camera positions.

Junejo et al. [25] introduced the use of trajectory-based self-similarity matrices (SSMs) to encode humans observed from different views for classifying human actions. They proved that the Histogram of Gradient (HOG) and Optical Flow (OF) features extracted from a self-similarity matrix are stable under view changes of an action. In Reference [25], for calculating the trajectory-based SSM of $\{S^1, S^2, \ldots, S^N\}$, each element of SSM is computed by the equation below.

$$d_{ij} = \frac{1}{N} \sum_{k=1}^{N} \| s_i^k - s_j^k \|_2,$$

(5)

where $s_t^k = \left[s_x, s_y, t \right]$ denotes the original coordinate of joint k in frame t in the (x, y, t)-space, and $\| \cdot \|_2$ represents the Euclidean distance.

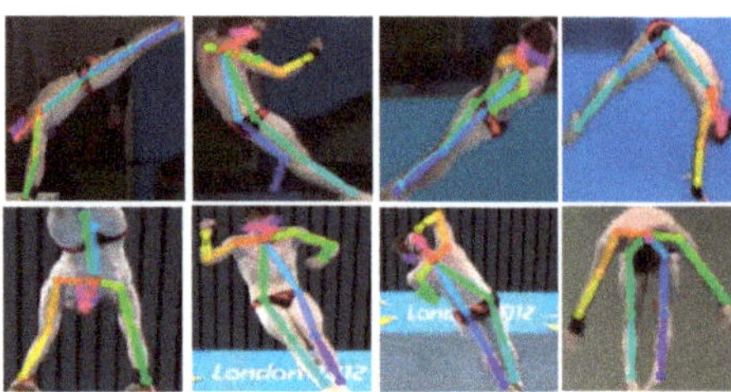

Figure 5. Joint trajectories of two different views for the same person performing diving. (The first row shows results on the side view and the second row presents results on the front view.).

It is worth noting that each element of the temporal self-similarity matrix in Reference [25] represents the accumulated coordinate's offset over all body joints for a frame of t, as formulated in Equation (5). However, it completely neglects the individual motion dynamics of each joint and the relationship between body joints' relative positions. We believe that both of these factors can be essential for an action similarity measurement and should be addressed in feature representation. Different from the SSM feature in Reference [25], we represent the trajectories of each joint of the human body independently, and further analyze the displacement sequence of inter-joints to build the temporal self-similarity matrices. As a result, temporal self-similarity matrices of joint trajectories and displacement sequences are computed, respectively, to capture view invariant patterns of intra-joint and inter-joint dynamics.

The calculation process is formulized as follows. Suppose that you are given preprocessed skeleton data $\left\{ S^1, S^2, \ldots, S^N \right\}$ through the procedures outlined in Section 3.3.1, where N denotes the number of joints and $S^k = \left[s_1^k, s_2^k, \ldots, s_T^k \right]$ represents the motion trajectory of the kth joint through T frames of the image sequence. The temporal self-similarity matrices of joint trajectories can be denoted by $SSM_{Js} = \left[SSM_{J_1}, SSM_{J_2}, \ldots, SSM_{J_N} \right]$, where $SSM_{J_k} = \left[d_{ij}^k \right]_{T \times T}$ and d_{ij}^k denotes the Euclidean distance between the position coordinates of the kth joint in frame i and j, as formulated by Equation (6).

$$d_{ij}^k = \| s_i^k - s_j^k \|_2 \tag{6}$$

The middle of Rhip (Joint 8) and Lhip (Joint 11) in each frame is regarded as the central point, and the displacement sequence of skeleton data can be represented by $P = [p_1, p_2, \ldots, p_T]$, where $\vec{p_t} = \left[p_t^1, p_t^2, \ldots, p_t^N \right]$ and p_t^k denotes the displacement of the kth joint relative to the central point in frame t. It is computed by the equation below.

$$p_t^k = |s_t^k - \frac{s_t^8 + s_t^{11}}{2}|, \tag{7}$$

where $| \cdot |$ denotes the norm of the vector. Then, the temporal self-similarity matrix of the joint displacement sequence can be denoted by $SSM_D = \left[d_{ij}' \right]_{T \times T}$, where d_{ij}' denotes the Euclidean distance between two displacement vectors $\vec{p_i}$ and $\vec{p_j}$ belonging to frame i and j. It is computed by Equation (8).

$$d_{ij}' = \| \vec{p_i} - \vec{p_j} \|_2 = \frac{1}{N} \sum_{k=1}^{N} \| p_i^k - p_j^k \|_2 \tag{8}$$

The Histogram of Gradient (HOG) feature descriptor extracted from a log-polar semicircle centered on the main diagonal of the matrix is used to describe the pattern structure of the self-similarity matrix of joint trajectories and joint displacement sequences. For a given semicircle area, $C = \{r_i, \theta_i\}(r_i = 0 \sim r, \theta_i = 0 \sim \pi)$, where r is the radius of the semicircle, which denotes the temporal window extent to be considered, and the origin of the pole coordinate is located at the jth element of SSM's main diagonal. C is segmented into 11 blocks equally divided into three sections of the polar axis and five bins of the polar angle. One block near the origin of the pole ordinate is kept within one-third of the radius and without division of the polar angle. For each block of C, the normalized eight-bin HOG feature vector $h_j^a = [h_{j,b}^a]'_{b=1\sim8}$ is calculated, and the vectors of all 11 blocks are concatenated to form a feature vector $h_j = [h_j^a]'_{a=1\sim11}$ of the jth element of SSM's main diagonal. $h = (h_1, h_2, \ldots, h_T)$ is computed for all elements of SSM's main diagonal to form a feature description of SSM. Lastly, HOG feature vectors of all joint trajectories and joint displacement sequences are concatenated to generate self-similarity feature representation of an action instance $H = [h_{SSM_{J1}}, \ldots, h_{SSM_{JN}}, h_{SSM_D}]$.

3.3.3. Class-Specific Regression Model

Not all joints are equally involved in the motion of the human body, and, for different action categories, the participating joints show different levels of significance. Therefore, different impact weights should be assigned to distinguish the importance of body joints. We propose a class-specific regression learning strategy to address this problem. In this strategy, training samples annotated with class labels and quality scores were used to train regression weight vectors for specific action categories, which resulted in a more effective evaluation of different categories that shared postures. This alleviated the confusion. Correspondingly, a more accurate score can be estimated to boost the performance of a fine-grained quality assessment. The learning process is formulated as follows.

The training set $D = \{(H_1^1, y_1^1), (H_2^1, y_2^1), \ldots, (H_{n_1}^1, y_{n_1}^1), \ldots, (H_1^c, y_1^c), (H_2^c, y_2^c), \ldots, (H_{n_c}^c, y_{n_c}^c)\}$ consists of action videos annotated with action class labels and quality assessment scores. $H_j^i \in R^K$ is the self-similarity feature vector of the jth action instance that belongs to action class i, which is computed through the feature description procedure described in the previous subsection. $y_j^i \in R$ denotes the ground-truth quality score and n_i is the number of training samples belonging to action class i. The regression model w^i of the ith action category is trained for all videos of the same action category by finding a real-valued linear function $w^{i^T} H_j^i$, which acquires the minimized approximation errors between the ground truths and estimated scores. The estimated score is obtained by projecting the original features onto the transformation space. The learning process is formulated as the following.

$$\text{argmin}_{w^i} \sum_{j=1}^{n_i} \| y_j^i - w^{i^T} H_j^i \|_2, \tag{9}$$

where $w^i \in R^K$, y_j^i is the ground-truth quality score of the jth action instance belonging to class i and $w^{i^T} H_j^i$ is the corresponding predicted score.

The algorithm is performed individually for each action category to obtain a specific regression model. In the proposed method, two types of regression strategies—Support Vector Regression (SVR) and Ridge Regression (RR)—are employed to implement this training process. During testing, with the supervision of the action classifier's output, the specific regression model is determined to predict the quality score of the testing video.

4. Experiments

The construction of an action evaluation dataset requires professional experts to annotate training videos according to their knowledge and experience of relevant fields. The annotation is highly dependent on the subjective judgement of professionals. It is difficult to collect training samples from massive action categories, and accurate annotation acquisition requires a great deal of manual effort when constructing a large-scale dataset of human action evaluation. Therefore, several limitations are shared with the published dataset of action quality assessment, such as insufficient training data, a limited number of action categories, a fixed position of the RGB camera, and identical scenes. In this study, we investigated the performance of our proposed method for the diving and figure skating videos of the MIT Olympic Scoring dataset [26], and the gymnastic vaulting videos of the UNLV Olympic Scoring dataset [27,28].

4.1. Introduction of Datasets and Experimental Setups

The MIT Olympic Scoring dataset [26] contains two action category sport videos: diving and figure skating. The diving dataset contains 159 videos that include 25,000 frames with scores varying between 20 and 100. The frame rate of diving videos is 60 fps and each diving instance is about 150 frames. The figure skating dataset consists of 150 videos captured at 24 fps. There is a total of 630,000 frames, and each action instance is almost 4200 frames. The ground-truth score of each action video is freely obtained by extracting the judge's score that is released publicly in sports footage. The quality assessment score of each action instance varies between 0 and 100. The figure skating assessment is more suitable for activity evaluation since several action components, such as jumps, spins, and steps, are included and repeated in each of the videos. Compulsory routines are required in the performances of the Olympic games, including a spin, axel, spiral, and transition. However, the sequential order of routines is different in action videos. Therefore, figure skating is more challenging than diving because of the complexity of activity analysis rather than simple actions. Each video is annotated with the start frame, end frame, and judgement score of action occurrence. The score is extracted from the frame by displaying the referee's decision from the video. Figure 6a,b show some examples of diving and figure skating frames performed by different people with different views.

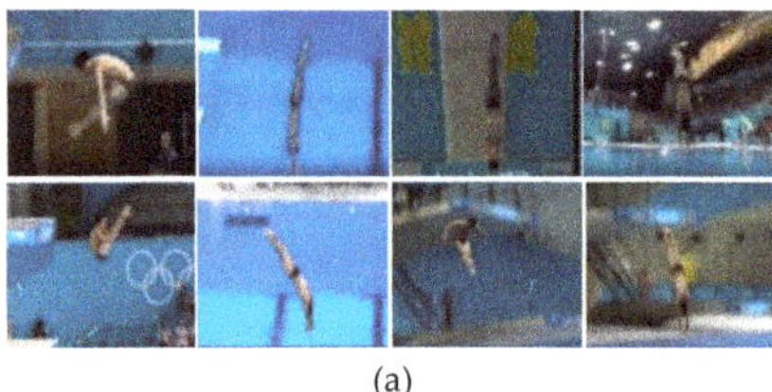 

(a)   (b)

Figure 6. Example frames of the MIT Olympic Scoring dataset [26]. (**a**) Example frames of diving. (**b**) Example frames of figure skating.

The UNLV Olympic Scoring dataset [27,28] includes sport videos from Summer and Winter Olympics events on YouTube. It comprises 1189 videos from seven action categories, including 370 videos of single-person diving from a ten-meter platform, 88 videos of synchronized diving from a three-meter springboard, 91 videos of synchronized diving from a ten-meter platform, 176 videos of a gymnastic vault, 175 videos of a skiing sport, 206 videos of snowboarding, and 83 videos of trampolining in Olympic events. In the action categories of vaulting, skiing, and snowboarding, severe view changes existed in the action videos. Since the motion patterns of a single person's action are addressed in our feature representation, synchronized events of more than one person are not taken into account. Additionally, a long-range shot distance is commonly adopted in filming videos of skiing, snowboarding, and trampolining. The accuracy of skeleton detection is severely reduced with the influence of poor pose estimation results. Therefore, we only evaluated our proposed method for vault videos of this dataset. The vault dataset of UNLV Olympic Scoring includes 176 videos with an average length of about 75 frames. The frame size is 320 × 240. The ground-truth score of a vaulting video that ranges from 12.30 to 16.87 is determined by the sum of the "execution" score and "difficulty" score.

In the action classification component, we propose extracting local motion patterns from the joint motion volume and training the linear kernel-based SVM classifiers to determine the action label of the testing action video. The estimation action class of the testing video can be obtained by comparing the output of all classifiers and finding the highest category confidence. For the evaluation of our classification component, we adopted leave-one-video-out validation for 159 diving videos, 150 figure skating videos, and 176 vaulting videos. The performance was measured by the average accuracy over all videos. The proposed classification method achieved the performance of 92.6% for all action videos. The average classification accuracies of diving, figure skating, and vaulting were 93%, 89.3%, and 94.8%, respectively.

For validating the proposed action quality assessment method, the rank correlation coefficient between the predicted scores and the ground truths was computed to evaluate the performance. We employed the rank correlation coefficient measurement of two vectors' similarity to evaluate our proposed method. It was computed by Equation (10).

$$\rho_{y,\hat{y}} = 1 - \frac{6\sum_{i=1}^{N} d_i^2}{N(N^2-1)},$$ (10)

where y is the ground truth score vector and $\hat{y}$ is the predicted score vector. d_i denotes the ranking difference between y_i and $\hat{y}_i$ of the ith video. A higher ρ value denotes a better estimation performance. In experiments using the MIT diving and figure skating dataset, we followed the testing schema introduced in Reference [1]. A random split of action videos was adopted, resulting in 100 videos being selected as training samples and the rest as test samples. The average rank correlation coefficient of 200 rounds' testing results was computed as the final evaluation performance of the dataset. Additionally, in experiments using the UNLV vaulting dataset, the testing schema introduced in Reference [27] employing a fixed split of the dataset, namely 120 videos for training and the remaining 56 videos for testing, was followed. The detailed results are presented in the next section.

4.2. Results of the MIT Diving Dataset

According to the evaluation protocol introduced in Reference [1], 100 labeled action videos were selected randomly as training samples from the diving dataset and the remaining 59 videos were used as test samples. The rank correlation coefficient between the predicted scores and the ground truths of this round split was computed. Then, 200 rounds of experiments with different random splits were conducted according to the same strategy. The evaluation performance was obtained by averaging all rounds' rank correlation coefficients. We compared the performance of our proposed method with four state-of-the-art action feature methods for human action recognition, and several evaluation learning methods, including the all-action regression model, single-action regression models, support vector regression of different kernels, and ridge regression.

We compared three feature extraction methods, including spatio-temporal interest points (STIP) [29], dense sampling (Dense) [30], and skeleton data (Skeleton). The benchmark MIT Olympic Scoring dataset that we employed to evaluate the proposed method was published in Reference [1]. Additionally, a similar solution strategy was employed both in Reference [1] and our proposed learning framework. A handcrafted feature engine was first built for feature representation and a regression model was then developed from the action features for quality assessment. Therefore, we chose Reference [1] as the baseline method. In Reference [1], the researchers extracted both low-level spatial and temporal filtering features that captured edges and velocities, as well as high-level pose features obtained from the discrete cosine transformation of joint displacement vectors, and estimated a regression model that predicted the scores of actions. They compared their performance with the space-time interest points (STIP) method [29] and Discrete Fourier Transform (DFT) pose features. STIP is the abbreviation for space-time interest points and was developed for feature detection in traditional action recognition research. The method was presented based on the observation that actions frequently occurred in positions with sharp changes in both the spatial and temporal domains. It employed a space-time extension of the Harris corner detector to extract the prominent positions of significant changes in spatial and temporal dimensions from the action video. Then, histograms of oriented gradients (HOG) and the optical flow (HOF) were calculated and concatenated for each local spatial and temporal volume centered at the prominent position. All the local features were aggregated to form the feature descriptor for action representation. However, it has been proven in Reference [30] that dense sampling has demonstrated a better classification performance than original STIP for human action recognition in realistic scenes. Dense sampling extracted video blocks at regular intervals and scales in space and time by a sliding window moving throughout the whole video. The HOG and HOF descriptors of each video block were computed and concatenated to represent the local feature of each spatial and temporal position. All local feature descriptors were aggregated to form the final feature representation of the whole video. Skeleton data can be obtained from RGB videos using the pose estimation algorithm. Pose features extracted by transformation and projection of the original skeleton data were regarded as being encoded with high-level semantics of human actions. Therefore, we compared the performance of our skeleton data extraction method with that of STIP [29] and dense sampling [30] using the benchmark MIT Olympic Scoring dataset.

On the other hand, we developed self-similarity feature representation extracted from joint trajectories and joint displacement sequences to describe motion patterns of joints and posture changes. The self-similarity matrices employed to encode human actions were initially proposed in Reference [25], in which only the coordinate's offset over all body joints was accumulated for a single frame, and the individual motion dynamics of each joint and the relationship of body joints' relative positions were neglected. In contrast to Reference [25], we encoded the motion dynamics of each body joint independently, as well as the displacement sequence between body joints, to build temporal self-similarity matrices. On the basis of skeleton data representation for human actions, we compared our proposed feature method with the baseline self-similarity matrix feature [25] and the pose feature captured from discrete cosine transformation (DCT) [1] for the original coordinates of human body joints. The performance comparison of our proposed feature and the reviewed features of the MIT diving dataset is presented in Table 1.

Table 1. Quality assessment results on the Massachusetts Institute of Technology (MIT) diving dataset. The average rank correlation coefficients between the predicted results and the ground truth of different feature representation and evaluation methods are presented (The higher the value, the better the performance.).

	STIP *	Dense	Skeleton + DCT	Skeleton + SSM *	Our Method
All-action SVR-Linear	0.07	0.09	0.19	0.13	**0.20**
Single-action SVR-Linear *	0.18	0.16	0.45	0.35	**0.52**
Single-action Ridge Regression	0.07	0.10	0.32	0.30	**0.4** [1]

[1] The bolded value indicated the best performance of all compared feature methods. * STIP means space-time interest points feature method. SSM indicates self-similarity matrices feature method. SVR means support vector regression method.

As shown in Table 1, skeleton-based feature representations exhibited a significantly greater strength than low-level features of STIP or dense sampling under different evaluation strategies for the diving video assessments. On the basis of the same skeleton data detection results, the DCT feature method captured from the discrete cosine transform on the original coordinates of joints achieved better performances at 0.19, 0.45, and 0.32 under different evaluation strategies, in comparison to the corresponding results of 0.13, 0.35, and 0.30 obtained by the SSM feature method. This is likely because the SSM feature accumulated the position offsets of all joints between each frame of the video sequences. Therefore, it completely discarded the motion information of individual joints of the human body and the relationship between body joints. The DCT feature captured the dynamic changes of human body movement. However, it simply considered the change of joints' position relative to the centroid of the human body along the time dimension, and neglected the motion patterns of each joint trajectory and the changes of the joints' correlation. The proposed feature method extracted patterns from the self-similarity matrix of each joint trajectory and joint displacement sequence. It encoded not only the dynamic changes of individual joints, but also the pose feature described by the layout changes of all body joints. It achieved the best rank correlation coefficients of 0.20, 0.52, and 0.4 among all the experienced feature methods for the MIT diving dataset.

Furthermore, we compared the performance of different evaluation methods, namely all-action support vector regression (all-action SVR) evaluation, single-action support vector regression (single-action SVR) evaluation, and single-action evaluation ridge regression (single-action ridge regression) evaluation for this dataset. All-action evaluation trained a unified regression function to assess all action categories, and single-action evaluation employed the strategy of specific-action training to learn a specific assessment function for each action category from annotated action features. As shown in Table 1, the performance difference is not clear between single-action ridge regression and all-action evaluation under low-level STIP (single-action, 0.18, all-action, 0.07) and dense sampling features (single-action, 0.16, all-action, 0.09). However, the estimation results were significantly improved under the same skeleton-based feature representation, and the performance of single-action evaluation was significantly superior to that of all-action evaluation. This indicated that a dedicated quality assessment model for specific action categories is suitable for sport activity scoring. Therefore, it is less effective to design one unified feature evaluation function to assess various kinds of feature patterns' similarities for all action categories. We also attempted different kernel functions of support vector regression evaluation and compared the evaluation methods of SVR and Ridge Regression (RR). It was found that SVR with a linear kernel achieved a better performance of 0.52 than the ridge regression method, which displayed a value of 0.4. In addition, the linear kernel always achieved better results than the Radial Basis Function (RBF) kernel and sigmoid kernel in SVR regression.

The predicted score of each testing video for the MIT diving dataset from our best rank correlation coefficient performance is presented in Figure 7. The horizontal axis denotes the index of the action video for testing, and the vertical axis represents the predicted quality score. The data series of the GT_test indicate the ground truth scores of test videos, and the pred test corresponds to the estimated scores of the proposed method.

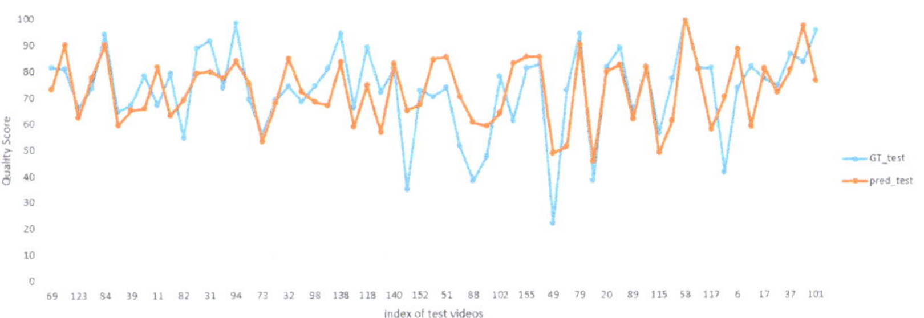

Figure 7. The predicted scores of test videos from our best rank correlation coefficient for the MIT diving dataset.

4.3. Results on the MIT Figure Skating Dataset

The figure skating activity of the MIT Olympic Scoring dataset is more challenging than the diving action. The frame rate of figure skating videos is 24 fps and the length of each video is about 4200 frames. Therefore, complex activity, rather than a single action, is involved in a figure skating video. Other challenges are the irregular camera motion and cluttered background that commonly exist in long-duration recorded videos. Compulsory routines are required and performed successively in each video, such as jumps, spins, turns, steps, and moves. However, the order of execution for custom actions is not strictly restricted. Moreover, significant pose variations seriously affected the performance of action quality assessment due to the appearance of athletes, irregular changing of the camera position, and zooming in and out for the sake of capturing the best recording effect.

To obtain effective feature representation of complex activities in long-duration videos, we divided each video into 10 segments with an equal length. Feature extraction was performed for each segment, and all segments' features were concatenated to represent the action feature of the whole video. The performance using the figure skating dataset was also evaluated according to the testing protocol in Reference [1] where the average rank correlation coefficient of 200 experiments was computed with random splitting. In this case, 100 action videos were selected as training samples and the rest were selected as tests. The performance comparison of our feature method and the baseline method and other state-of-the-art feature methods was investigated. The details are presented in Table 2.

Table 2. Quality assessment results on the MIT figure skating dataset. The average rank correlation coefficients between the predicted results and the ground truth of different feature representations and evaluation methods are presented.

	STIP *	Dense	Skeleton + DCT	Skeleton + SSM *	Our Method
All-action SVR-Linear *	0.13	0.15	0.21	0.15	**0.25**
Single-action SVR-Linear	0.21	0.23	0.37	0.19	**0.41**
Single-action Ridge Regression	0.20	0.21	0.25	0.17	**0.28**

* STIP means space-time interest points feature method. SSM indicates self-similarity matrices feature method. SVR means support vector regression method.

From Table 2, it can be observed that most of the skeleton-based feature representations achieved a superior performance in comparison to low-level STIP or dense sampling feature methods. In skeleton data-based feature representation, the best mean rank correlation coefficients of the Discrete Cosine Transform (DCT) transformation feature and our proposed feature were 0.37 and 0.41, respectively, which outperformed STIP (0.21) and dense sampling (0.23) with single-action Support Vector Regression (SVR) employing a linear kernel. The proposed feature representation slightly improved the result of DCT and achieved the best value of 0.41. The promotion was limited, likely due to the simple segmentation of videos according to the equal length strategy, and the synchronization of segments between different action instances that were not considered. It is noted that the performance of the original SSM feature method was clearly decreased for this dataset, and the best result obtained was 0.19, which was inferior to the low-level feature of STIP and dense sampling. This indicated that the original SSM feature is unsuitable for fine-grained feature representation of long-duration video sequences, likely because it accumulated all joints' relative position offsets between each two successive frames of a video sequence. However, the motion data of each joint and joint relationship important for describing the intrinsic characteristic of different contained actions were completely ignored. We also compared the all-action evaluation and single-action evaluation methods, SVR regression, and the RR regression method using this dataset. The comparison of different evaluation methods is presented in Table 2. In terms of the best rank correlation coefficient obtained, the predicted score of each testing video for the MIT figure skating dataset is illustrated in Figure 8.

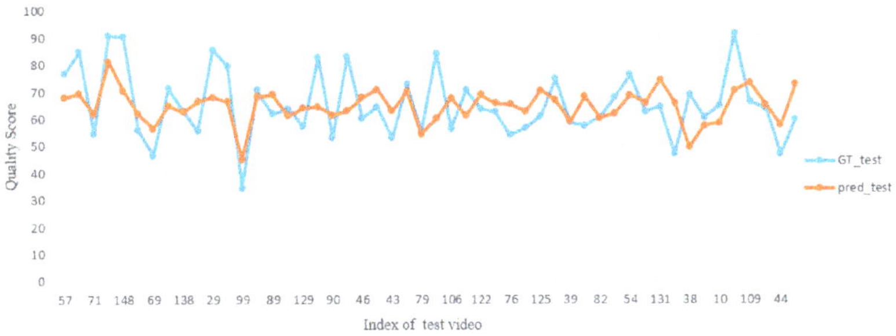

Figure 8. The predicted scores of test videos in terms of the best rank correlation coefficient obtained for the MIT figure skating dataset.

4.4. Results on the UNLV Vault Dataset

The UNLV vault dataset comprises 176 videos of gymnastic vaulting captured from five international competitions of the Summer and Winter Olympics on YouTube. The average length is 75 frames per vault video. The frame size is 320×240. The ground-truth score of each vaulting video is freely obtained by extracting the judge's score that is publicly released in sports footage. It ranges from 12.30 to 16.87, and is determined by the sum of the "Execution" score and "Difficulty" score. Although a short average length and relatively simple actions are contained in vault videos compared with those of diving and figure skating, severe view changes exist in this dataset due to the different broadcast configurations employed for different events. Additionally, the low-resolution of some broadcasting videos make them more difficult to employ for action quality scoring.

We followed the evaluation protocol in Reference [27], which consists of a fixed split of the dataset, where 120 videos were chosen as training samples and the rest of the 56 videos were selected as testing samples. The comparison of our feature method with other reviewed state-of-the-art feature methods and different evaluation methods is presented in Table 3.

Table 3. Quality assessment results for the UNLV vault dataset. The rank correlation coefficient between the predicted results and the ground truth is presented.

	STIP	Dense	Skeleton + DCT	Skeleton + SSM	Our Method
All-action SVR-Linear	0.05	0.10	0.15	0.09	**0.17**
Single-action SVR-Linear	0.13	0.16	0.41	0.33	**0.47**
Single-action Ridge Regression	0.11	0.12	0.35	0.31	**0.39**

From Table 3, it can be concluded that skeleton-based feature representations exhibited a significantly greater strength than low-level features of STIP and dense sampling under most of the different evaluation strategies used for vault video assessment. In skeleton data-based feature representation, the best mean rank correlation coefficients of the DCT transformation feature, SSM feature, and our proposed feature were 0.41, 0.33, and 0.47, respectively. It outperformed STIP (0.13) and dense sampling (0.16) with single-action SVR using a linear kernel. The proposed feature representation improved the performance and achieved the best result of 0.47. In the comparison of different evaluation methods, as is shown in Table 3, it could also be found that, even though the performance difference between single-action evaluation and all-action evaluation under low-level STIP and dense sampling features is not clear, singe-action evaluation significantly improved the performance of skeleton-based feature representations. It is noted that the ridge regression obtained better results for this dataset than for the diving and figure skating datasets. The predicted score of each testing video for this dataset is illustrated in Figure 9.

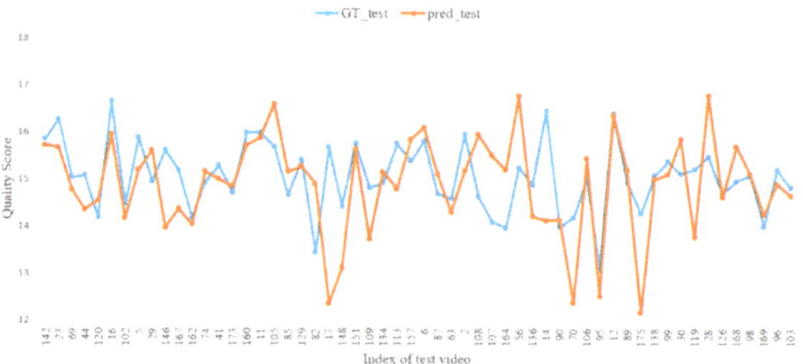

Figure 9. The predicted scores of test videos for the UNLV vault dataset.

4.5. Comparison with State-of-the-Art Feature Methods

We also compared our method with other state-of-the-art feature methods for the MIT diving, MIT figure skating, and UNLV vault dataset, as shown in Table 4.

Table 4. Comparison results of the proposed method with the baseline and other relevant methods.

	MIT Diving Dataset	MIT Figure Skating Dataset	UNLV Vaul Dataset
STIP feature method [1]	0.10	0.21	–
Pose+DFT method [1]	0.27	0.31	–
Pose+DCT method [1]	0.41	0.35	–
ApEn method [15]	0.45	–	–
Our approach	0.52	0.41	0.47

As a baseline method, Pirsiavash et al. [1] proposed a general learning-based framework to assess the quality of human-based actions from videos. They proposed using DFT and DCT transformation of original coordinates of human body joints detected by the Flexible Parts Model [31], and selected k low frequency coefficients to represent human actions. They evaluated the proposed feature method using the MIT diving and figure skating datasets and compared the performance with two kinds of low-level features, namely the STIP feature and DFT pose feature. For the MIT diving dataset, the best results of the STIP feature were 0.07 with the support vector regression and 0.10 with ridge regression. The DCT pose feature achieved the best result of 0.41, which was superior to the STIP feature value of 0.10 and DFT pose feature value of 0.27. For the MIT figure skating dataset, the DCT transformation-based pose feature obtained the result of 0.35, which was better than the STIP feature value of 0.21 and DFT transformation pose feature value of 0.31. Venkataraman et al. [15] investigated approximate entropy-based (ApEn) feature representation for segmenting untrimmed motion capture data, and only evaluated their method for assessing the quality of diving actions included in the MIT Olympic Scoring dataset. They reported the best result of 0.45 for the diving dataset, which showed that a 10% improvement was achieved in the rank correlation coefficient when compared to Reference [1].

As shown in Table 4, from the results on MIT diving and figure skating of the above-mentioned handcrafted features, we can conclude that our proposed feature method achieved the best rank correlation coefficients of 0.52 and 0.41, respectively, for the two actions. It is believed that the proposed feature representation can better encode the dynamics of human motion for the fine-grained quality assessment of human actions than the reviewed state-of-the-art feature methods by capturing the self-similarity patterns from joint trajectories and joint displacement sequences.

5. Conclusions

In this paper, we have proposed an integrated category classification and regression-based evaluation framework for fine-grained human action quality assessment. In this framework, for action classification, the local motion patterns of body joint-based feature representation are extracted to train the discriminative classifier. The output of the classifier is used to supervise the quality assessment process in the testing stage. To deal with intra-class variations and acquire effective dynamic representation, the semantic pose feature captured from the self-similarity matrix of joint trajectories and joint displacement sequences is developed. A class-specific learning algorithm is employed to build an evaluation function for each action category. The experimental results show improvements for both the diving and figure skating datasets in comparison with other handcrafted feature methods.

The limitations of our proposed method include the fact that the segmentation of long videos has simply been considered and the synchronization of segments has not been researched. Our method is more suitable for assessing well-segmented action instances. When complex activities rather than actions are contained in videos with a long-time duration, the quality score is strongly affected. In future studies, semantic segmentation and alignment methods will be addressed to promote the practical application of the proposed framework.

Author Contributions: Conceptualization, Q.L. Methodology, Q.L. Software, Q.L. Validation, Q.L., H.-B.Z., and J.-X.D. Formal analysis, Q.L. Investigation, Q.L. Resources, Q.L. Data curation, T.-C.H. Writing—original draft preparation, Q.L. Writing—review and editing, Q.L. Visualization, T.-C.H. Supervision, J.-X.D. Project administration, C.-C.C. Funding acquisition, H.-B.Z. All authors have read and agreed to the published version of the manuscript.

Funding: The National Nature Science Foundation of China (Grant no. 61673186, 61871196), the Natural Science Foundation of Fujian Province, China (Grant no. 2019J01082, 2017J01110), and the Scientific Research Funds of Huaqiao University, China (16BS812), supported this work.

Acknowledgments: The authors would like to thank the anonymous reviewers for their valuable and insightful comments on an earlier version of this manuscript.

Conflicts of Interest: The authors declare no conflicts of interest.

References

1. Pirsiavash, H.; Vondrick, C.; Torralba, A. Assessing the Quality of Actions. In Proceedings of the European Conference on Computer Vision 2014, Zurich, Switzerland, 6–12 September 2014; Springer: Berlin/Heidelberg, Germany, 2014; pp. 556–571.
2. Lei, Q.; Du, J.-X.; Zhang, H.-B.; Ye, S.; Chen, D.-S. A Survey of Vision-Based Human Action Evaluation Methods. *Sensors* **2019**, *19*, 4129. [CrossRef] [PubMed]
3. Morel, M.; Kulpa, R.; Sorel, A. Automatic and Generic Evaluation of Spatial and Temporal Errors in Sport Motions. In Proceedings of the International Conference on Computer Vision Theory and Applications, Rome, Italy, 27–29 February 2016; pp. 542–551.
4. Paiement, A.; Tao, L.; Hannuna, S. Online quality assessment of human movement from skeleton data. In Proceedings of the British Machine Vision Conference (BMVC 2014), Nottingham, UK, 1–5 September 2014; pp. 153–166.
5. Antunes, M.; Baptista, R.; Demisse, G.; Aouada, D.; Ottersten, B. Visual and Human-Interpretable Feedback for Assisting Physical Activity. In Proceedings of the European Conference on Computer Vision (ECCV), Amsterdam, The Netherlands, 11–14 October 2016; pp. 115–129.
6. Baptista, R.; Antunes, M.; Aouada, D. Video-Based Feedback for Assisting Physical Activity. In Proceedings of the International Joint Conference on Computer Vision, Imaging and Computer Graphics Theory and Applications (VISAPP), Rome, Italy, 27 February–1 March 2017.
7. Tao, L.; Paiement, A.; Damen, D. A comparative study of pose representation and dynamics modelling for online motion quality assessment. *Comput. Vis. Image Underst.* **2016**, *148*, 136–152. [CrossRef]
8. Meng, M.; Drira, H.; Boonaert, J. Distances evolution analysis for online and off-line human object interaction recognition. *Image Vis. Comput.* **2018**, *70*, 32–45. [CrossRef]
9. Zhang, W.; Liu, Z.; Zhou, L.; Leung, H.; Chan, A.B. Martial arts, dancing and sports dataset: A challenging stereo and multi-view dataset for 3d human pose estimation. *Image Vis. Comput.* **2017**, *61*, 22–39. [CrossRef]

10. Laraba, S.; Tilmanne, J. Dance performance evaluation using hidden markov models. *Comput. Animat. Virtual Worlds* **2016**, *27*, 321–329. [CrossRef]

11. Barnachon, M.; Boufama, B.; Guillou, E. A real-time system for motion retrieval and interpretation. *Pattern Recognit. Lett.* **2013**, *34*, 1789–1798. [CrossRef]

12. Hu, M.C.; Chen, C.W.; Cheng, W.H.; Chang, C.H.; Lai, J.H.; Wu, J.L. Real-time human movement retrieval and assessment with kinect sensor. *IEEE Trans. Cybern.* **2014**, *45*, 742–753. [CrossRef] [PubMed]

13. Liu, X.; He, G.F.; Peng, S.J.; Cheung, Y.M.; Tang, Y.Y. Efficient human motion retrieval via temporal adjacent bag of words and discriminative neighborhood preserving dictionary learning. *IEEE Trans. Hum. Mach. Syst.* **2017**, *47*, 763–776. [CrossRef]

14. Patrona, F.; Chatzitofis, A.; Zarpalas, D.; Daras, P. Motion analysis: Action detection, recognition and evaluation based on motion capture data. *Pattern Recognit.* **2018**, *76*, 612–622. [CrossRef]

15. Venkataraman, V.; Vlachos, I.; Turaga, P. Dynamical Regularity for Action Analysis. In Proceedings of the 26th British Machine Vision Conference, Swansea, UK, 7–10 September 2015; British Machine Vision Association: Swansea, Wales, 2015; pp. 67.1–67.12.

16. Vicente, I.S.; Kyrki, V.; Kragic, D.; Larsson, M. Action recognition and understanding through motor primitives. *Adv. Robot.* **2007**, *21*, 1687–1707. [CrossRef]

17. Han, F.; Reily, B.; Hoff, W.; Zhang, H. Space-time representation of people based on 3d skeletal data: A review. *Comput. Vis. Image Underst.* **2017**, *158*, 85–105. [CrossRef]

18. Pazhoumand-Dar, H.; Lam, C.P.; Masek, M. Joint movement similarities for robust 3d action recognition using skeletal data. *J. Vis. Commun. Image Represent.* **2015**, *30*, 10–21. [CrossRef]

19. Ofli, F.; Chaudhry, R.; Kurillo, G.; Vidal, R.; Bajcsy, R. Sequence of the Most Informative Joints (SMIJ): A new representation for human skeletal action recognition. *J. Vis. Commun. Image Represent.* **2014**, *25*, 24–38. [CrossRef]

20. Wang, P.; Li, W.; Li, C.; Hou, Y. Action recognition based on joint trajectory maps with convolutional neural networks. *Knowl.-Based Syst.* **2018**, *158*, 43–53. [CrossRef]

21. Cao, Z.; Simon, T.; Wei, S.E.; Sheikh, Y. Realtime multi-person 2D pose estimation using part affinity fields. In Proceedings of the 30th IEEE Conference Computer Vision and Pattern Recognition, Honolulu, HI, USA, 21–26 July 2017; pp. 1302–1310.

22. Zanfir, M.; Leordeanu, M.; Sminchisescu, C. The moving pose: An efficient 3D kinematics descriptor for low-latency action recognition and detection. In Proceedings of the IEEE International Conference on Computer Vision, Sydney, NSW, Australia, 1–8 December 2013; pp. 2752–2759.

23. Nowozin, S.; Shotton, J. Action points: A representation for low-latency online human action recognition. *Mark. Health Serv.* **2013**, *32*, 3–5.

24. Schuldt, C.; Laptev, I.; Caputo, B. Recognizing human actions: A local SVM approach. In Proceedings of the International Conference on Pattern Recognition, Cambridge, UK, 23–26 August 2004; pp. 32–36.

25. Junejo, I.N.; Dexter, E.; Laptev, I. View-independent action recognition from temporal self-similarities. *IEEE Trans. Pattern Anal. Mach. Intell.* **2010**, *33*, 172–185. [CrossRef] [PubMed]

26. MIT Olympic Scoring Dataset. Available online: https://www.csee.umbc.edu/~{}hpirsiav/quality.html (accessed on 23 January 2020).

27. UNLV Olympic Scoring Dataset. Available online: http://rtis.oit.unlv.edu/datasets.html (accessed on 23 January 2020).

28. Parmar, P.; Morris, B.T. Learning to score olympic events. In Proceedings of the IEEE Conference on Computer Vision and Pattern Recognition Workshops 2017, Honolulu, HI, USA, 21–26 July 2017; pp. 20–28.

29. Laptev, I.; Lindeberg, T. On Space-time interest points. In Proceedings of the International Conference on Computer Vision 2003, Nice, France, 14–17 October 2003; pp. 432–439.

30. Wang, H.; Ullah, M.M.; Klaser, A.; Laptev, I.; Schmid, C. Evaluation of Local Spatio-temporal Features for Action Recognition. In Proceedings of the British Machine Vision Conference, London, UK, 7–10 September 2009; pp. 1–10.

31. Yang, Y.; Ramanan, D. Articulated pose estimation with flexible mixtures-of-parts. In Proceedings of the IEEE Conference on Computer Vision and Pattern Recognition, Colorado Springs, CO, USA, 20–25 June 2011; pp. 1385–1392.

 *electronics*

Article

A Hybrid Tabu Search and 2-opt Path Programming for Mission Route Planning of Multiple Robots under Range Limitations

Meng-Tse Lee [1,*], Bo-Yu Chen [1] and Ying-Chih Lai [2,*]

[1] Department of Automation Engineering, National Formosa University, Yunlin 632, Taiwan;
    40127227@gm.nfu.edu.tw
[2] Department of Aeronautics and Astronautics Engineering, National Cheng-Kung University,
    Tainan 701, Taiwan
* Correspondence: mtlee@nfu.edu.tw (M.-T.L.); yingclai@mail.ncku.edu.tw (Y.-C.L.);
    Tel.: +886-5-631-5388 (M.-T.L.); +886-6-2757575 (ext. 63648) (Y.-C.L.)

Received: 28 February 2020; Accepted: 23 March 2020; Published: 24 March 2020

Abstract: The application of an unmanned vehicle system allows for accelerating the performance of various tasks. Due to limited capacities, such as battery power, it is almost impossible for a single unmanned vehicle to complete a large-scale mission area. An unmanned vehicle swarm has the potential to distribute tasks and coordinate the operations of many robots/drones with very little operator intervention. Therefore, multiple unmanned vehicles are required to execute a set of well-planned mission routes, in order to minimize time and energy consumption. A two-phase heuristic algorithm was used to pursue this goal. In the first phase, a tabu search and the 2-opt node exchange method were used to generate a single optimal path for all target nodes; the solution was then split into multiple clusters according to vehicle numbers as an initial solution for each. In the second phase, a tabu algorithm combined with a 2-opt path exchange was used to further improve the in-route and cross-route solutions for each route. This diversification strategy allowed for approaching the global optimal solution, rather than a regional one with less CPU time. After these algorithms were coded, a group of three robot cars was used to validate this hybrid path programming algorithm.

Keywords: multi-robots; path programming; tabu search

---

1. Introduction

Mainstream applications are currently focused on unmanned vehicle robots used in manufacturing; unmanned air vehicles (UAVs) in monitoring the earth's surface; emergency aid and disaster control and prevention efforts; commercial aerial photography; logistics; and unmanned combat air vehicle operations (UCAVs) [1,2]. When the scope of tasks and the areas involved are expanding, a system consisting of multiple unmanned vehicles agents is required to complete a mission with a very wide area. To complete the tasks more efficiently, well-planned path programming is a must, so as to minimize time and energy consumption by shortening the overall distances of the routes.

Facing this problem of a large area multi-waypoints mission dealing with a multi-agent system, we designed a hybrid dynamic path programming algorithm to help us achieve the goals of saving time and energy, with shorter and more efficient routes so that the robot cars were not running redundant paths. The maximum travelable distance (limited by the battery energy capacity) was used as one of the constraints during the algorithm iterations.

Unlike other path-programming works, this study contributes to the field with a hybrid path programming algorithm involving a combined tabu search and a 2-opt swap under the maximum travelable range consideration. A set of multiple robot cars was built to validate the tasks' completeness in the final phase.

2. Related Work

The idea of using a multi-agent robot system has been gradually adapted for wide-area searching in large-scale missions. This includes the use of multi-robot Simultaneous Localization and Mapping (SLAM), as explained by Atanasov et al. [3], along with hitchhiking robots as part of a collaborative approach for efficient multi-robot navigation, as explored by Ravankar et al. [4]. However, neither of these studies were focused on including an optimized mission path for each robot of the system. In fact, many scholars have been conducting research on path programming, as well as various factors regarding the number of vehicles dispatched. Most research is focused on solving the Vehicle Routing Problem (VRP) and the multiple traveling salesman problem (mTSP). This mTSP involves m salesmen who must visit a set of n cities, with each salesman starting and ending at the same place (city). Each city must be visited exactly once by one salesman, with the objective of finding the shortest total distance traveled by all of the salesmen. It could happen that we have one of the salesmen travel to all of the cities, while others visit only one. However, in practice, every salesman has similar abilities and limits. So the mTSP with ability constraints is more appropriate for real-world problems. Suppose the number of cities traveled to by each salesman is limited [5]. The multi-robot, multi-target exploration problem further extends the traveling salesman problem (TSP). This problem is called the Multiple Traveling Robot Problem (MTRP), and it involves a team of robots visiting target points at least once (ideally, no more than once). The overall solution quality is dependent upon both the quality of the solution constructed by the paths of the robots, and the efficient allocation of the targets to those robots [6].

The mTSP is a generalization of the well-known TSP [7], which is an obviously non-deterministic polynomial-time hardness (NP-hard) problem [8,9]. With NP-hard problems, the complexity of the solution increases as the number of target points expands. Since it is almost impossible to go through all of the feasible paths, and produce the best solution within an acceptable timeline, a heuristic algorithm was adopted to obtain an approximate solution.

Dorigo et al. [10] suggested a new computational paradigm called ant system (AS) for solving the TSP problem. This innovative contribution formed the foundation of today's Ant Colony Optimization (ACO) algorithm.

Yousefikhoshbakht et al. [11] used the New Modified Ant Colony Optimization (NMACO) that mixed the insert, swap and 2-opt algorithms, as well as an efficient Candidate List to improve the efficiency of the ACO. It is quite competitive with other meta-heuristic algorithms that solve a library of sample instances for the TSP (TSPLIB) instances. But the solution it yielded still has room to improve.

Necula et al. [12] used five modified Ant Colony Optimization (ACO) algorithms to solve mTSP. With this method, the maximum number l and the minimum number k of the city that each salesman should visit are defined and limited. The mTSP is then solved with TSPLIB instances.

Xu et al. [13] used a hybrid Genetic Algorithm (GA) and Simulated Annealing (SA) to solve mTSP. This enhanced the local search ability so as to achieve a better global optimal solution than general GA. But the solution it solved is not the best. In the previously mentioned literature, the authors used ACO and GA to solve the mTSP problem, but it does not produce the same solution with every execution, and must be tested continuously to ensure the best solution.

Côté et al. [14] used Tabu Search (TS) to solve the VRP, which they tested with benchmark instances. They not only achieved convergence in a reasonable time, but also surpassed many of the best solutions in benchmark instances.

Huang and Liao [15] used a two-phased method to solve the Dynamic Vehicle Routing Problem (DVRP). Different from mTSP, it limits its vehicle capacity. Fuzzy c-mean techniques were used to divide existing customers, while cost function was adopted to figure out the initial solution in the first phase. Then, in the second phase, a tabu search was combined with 2-opt and or-opt to improve cross-route and in-route respectively, aiming to solve the problem of multi-vehicle paths programming.

In fact, the tabu search algorithm not only adapted in path programming, but also in many other fields, such as those involving the economic dispatch of electric generators [16].

Sariel-Talay et al. [6] studied MTRP. They mentioned that their system was able to obtain real-time, optimal paths for traveling among multiple target points on their own platform with a robot swarm, and the assignment was completed successfully. However, a single robot car's maximum traveling capacity was not considered.

There are many published articles on multi-vehicle path programming, but only a few involve on-site experiments with cars to verify the effectiveness of the programming. In mTSP, the maximum and minimum number of cities that each salesman should visit are defined and limited, but the tasks may not get completed due to the limited energy capacity of one car. Therefore, we focused on the "maximum travelable distance" for this study, and also adopted a two-phased module as our solving module. In the first phase, a tabu search combined with a 2-opt swap method was used to program a single optimal path, and then to get the initial solution by splitting the path into multiple sub-paths. In the second phase, Huang and Liao's solving module was adopted—a tabu search combined with a 2-opt swap method to improve the initial paths, and to determine whether the result exceeded the maximum distance limit.

Additionally, a diversification strategy similar to GA was implemented for in-route path improvements. The use of this strategy resulted in the recording and selection of the optimal solution or second-best solution as the initial solution for the in-route path improvement later on, so that an unwanted regional optimal solution could be avoided. In this research, the 2-opt swap method was adopted to improve in-route paths. Once it finished, the calculation continued to improve cross-route paths until the conditions of termination were reached. In the last stage, this solution was verified by our experiments with real robot cars running the programmed paths. From the experiment, we looked into the distance designed for the path programming, and the trajectory difference between the actual paths and theoretical ones.

Research Highlights

The objective of the path programming problem of this research was to assign several series of target points to multi-robots to maximally reduce total energy consumption. The robots had limited onboard (fuel or battery) energy, so a maximum travelable distance constraint was expected to be satisfied. Highlights of this research speak to these issues:

1. We developed an optimal route planning algorithm for multi-robots' application by using an innovative, hybrid, two-phase tabu and 2-opt search.
2. In previous research involving salesman problems, constraints in VRP, mTSP and MTRP were mainly located on maximum reachable target points for single vehicle problems, or on maximum cargo capacity for multiple vehicles problems. In our study, we made this problem closer to an unmanned system's reality by building a new model with maximum range constraint.
3. Unlike most previous research in VRP and mTSP, in which problems end with a computational result, outdoor field tests were conducted to validate the algorithm developed in this project.

3. Design of the Two-Phase Path Programming

This research was focused on the dynamic path programming of multi-robots for achieving missions that lowered the energy costs in each group. The problem defined in this research is similar to the mTSP problems of visiting n targets with a shortest total route-distance by using m vehicles, with every target being visited only once.

However, the service range of unmanned robot vehicle systems is strictly limited by its onboard energy capacity (such as fuel or battery); hence the major difference in this research, as compared to traditional mTSP projects, is consideration of the range-limitation as a constraint during the programming loops. According to the previous statement, the problem definition for this research was modified from a standard module of SD-mTSP [17]. The object and subject were as follows:

$$x_{ijk} = \begin{cases} 1 & \text{the k path goes from city i to city j} \\ 0 & \text{otherwise} \end{cases} \tag{1}$$

$$\forall (i,j) \in A,\ k = 1, \ldots m \tag{2}$$

Object:

$$min \sum_{k=1}^{m} \sum_{(i,j) \in A} c_{ij} x_{ijk} \tag{3}$$

Subject to:

$$\sum_{(i,j) \in A} c_{ij} x_{ijk} \le D_{lmt},\ k = 1, \ldots m \tag{4}$$

$$\sum_{k=1}^{m} \sum_{j=2}^{n} x_{1jk} = m \tag{5}$$

$$\sum_{k=1}^{m} \sum_{j=2}^{n} x_{j1k} = m \tag{6}$$

$$\sum_{k=1}^{m} \sum_{i=1}^{n} x_{ijk} = 1,\ j = 2, \ldots, n \tag{7}$$

$$\sum_{k=1}^{m} \sum_{j=1}^{n} x_{ijk} = 1,\ i = 2, \ldots, n \tag{8}$$

$$\sum_{k=1}^{m} x_{i1k} + x_{1ik} \le 1,\ i = 2, \ldots, n \tag{9}$$

$$u_i \ge 1,\ i = 2, \ldots, n \tag{10}$$

where c_{ij} is the distance array of A; m is the number of robot cars; D_{lmt} is the value of maximum distance limit; n is the number of target points; and u_i is the number of target points visited on a car's path from the original point to target I. Equation (1) is variable integer, determining whether target i to j has been traveled by k car; with Equation (2), A in the formula is the set of all specified paths; Equation (3) is the object function of this problem; in Equation (4) the value of subject D_{lmt} should not be exceeded by all of the cars; Equations (5) and (6) ensure that cars start at the original point and come back to the same point; Equations (7) and (8) ensure that all of the targets have been entered and exited by the car; Equation (9) indicates that each car must visit at least one target; and Equation (10) prevents the problem of the sub-path not including the original point.

There are many kinds of heuristic algorithms to solve this problem, including ACO, Particle Swarm Optimization (PSO), GA and tabu search. In this research we mainly adopted the tabu search of a heuristic algorithm that can imitate human beings' memory, which keeps experiences in the past to prevent roundabout searches and to create a tabu list. It can learn from past solutions to avoid any regional optimal solution being seen as a global optimal solution. Tabu search is known for its ability to quickly converge iterations. In addition, its optimal solution and the number of iterations of convergence are more stable than what GA can achieve.

As GA randomly generates paths by mating, the solution could have turned out to be a suboptimal rather than an optimal solution, so we needed to repeatedly verify the result for the ultimate optimal solution. We knew that as long as the initial solution and tabu list for tabu search were well set, we could quickly complete a convergence and get the more stable, optimal solution; that being the main reason we adopted it for this study [2].

With tabu search as the core technology and involving the 2-opt swap method, a two-phase path programming algorithm model was established for this research. As shown in Figure 1, the initial solution established in the first phase was processed in two steps.

In the second phase, we needed to obtain the global optimal solution by improving the initial one we realized in the first phase. As in the first phase, three steps were used. Our first was the cross-route improvement by tabu search combined with 2-opt to exchange different target points within different paths. Then, a diversification strategy that recorded the optimal and suboptimal solutions from last computation was implemented to avoid the result falling into a regional optional solution. Step 3 was to improve each in-route path with tabu search combined with 2-opt. In other words, it was a refined solution exchanged from the first step, and we needed to repeat the process we carried out in step 2 until a termination condition was reached.

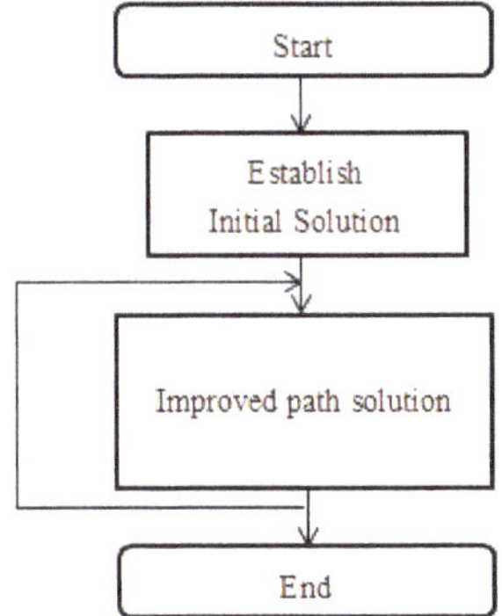

Figure 1. Main flow chart.

3.1. Tabu Search

Tabu search is a global search method. First, it establishes an initial solution, and then finds the neighborhood optimal solution, or accords the solution of aspiration criterion as the base for moving. That means, searching for solutions in the neighborhood domain of the current solution. Among them, the tabu list memory mechanism is noticeably important. It records the solutions which have been searched already to prevent any useless or redundant searching. Once the search on all neighborhood domains is completed, the optimal solution is selected. If any solution that be selected is found to be better than the current optimal solution, the optimal solution is updated until the termination condition is reached [18].

3.2. 2-opt Swap Method

The tabu search combined with the 2-opt nodal line swap method that was proposed by Lin (1965) [19] was adopted in this research. It is a method that we used to change the order of the path to expand the current solution. Initially it was designed on TSP, and now it is widely used in solving path problems (TSP, VRP, VRPTW, and so on). Its swap concept is shown in Figure 2. If (1, 3) and (2, 4) nodal lines are replaced, (1, 2) and (3, 4) could be connected to change its path.

The method for the cross-route path swap is different from the one we used for the single one path (see Figure 3 for the swap concept). If the nodal lines of (5, 6) and (1, 2) are exchanged, (5, 2) and (1, 6) as well as (5, 1) and (2, 6) are two possible paths that will be changed, respectively. Compared to the original path, the direction of the latter one will change. Thus, the use of 2-opt would be possible for reversing the directions of the paths.

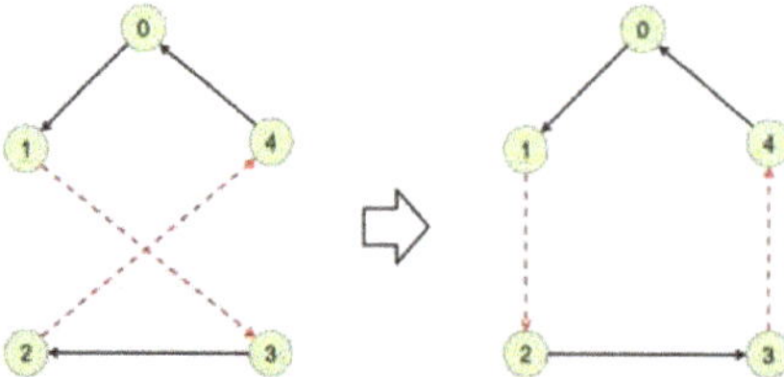

Figure 2. 2-opt Swap Method Concept (a).

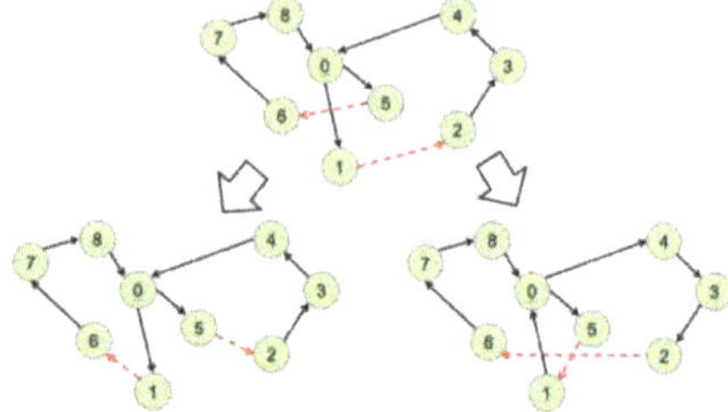

Figure 3. 2-opt Swap Method Concept (b).

3.3. Establishment of the Initial Solution

This phase was carried out in three steps, as shown in Figure 4. In step 1, tabu search and 2-opt were used to optimize the single path. In step 2, the single path was being divided into multiple sub-paths as the initial solution for our second phase.

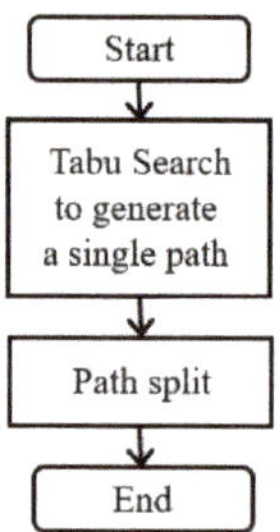

Figure 4. Establishment of the Initial Solution.

3.3.1. Use of Tabu Search to Program a Single Path

We used Nearest Distance Method to establish a rough single path as an initial solution for the tabu search. First, we established a distance array c_{ij}, and let the zeroth row of $c_{ij} = \infty$. Next, we set the point in which the row of c_{ij} had a minimum value as the first point, and let the value of this row of c_{ij} be infinity. Then, we repeated the same procedure to search and set the second point, the third point, and so on. The initial solution was generated until all of the points were searched. Following this, the tabu search and 2-opt were used to optimize the single path. The 2-opt was used as a move method to solve the problem of the single optimal path. We also set up the length of the tabu list, with the condition of stopping the search and using the solution of the Nearest Distance Method as the initial solution x_0 at that time. We then executed a tabu search to get the optimal single path.

3.3.2. Path Split

We roughly split the single path program into multiple sub-paths, and then split all the target points into m sub-paths. We also made the closed sub-paths link to the original point as much as possible, less than the maximum distance limit. At worst, no more than one route exceeded the maximum distance limit. The split path was not necessarily the optimal path, but it was good enough to be the initial solution for the next phase.

3.4. Improvement of Path to Obtain an Optimal Solution

In this phase, the rules of 2-opt were followed to swap the node lines in in-route and cross-route paths for moving. The solutions were gradually converged into the optimal solution. At this point, we decided to minimize the distance summation of all paths, and make sure the assigned distance of each robot car did not exceed the maximum distance limit. The flow chart is shown in Figure 5.

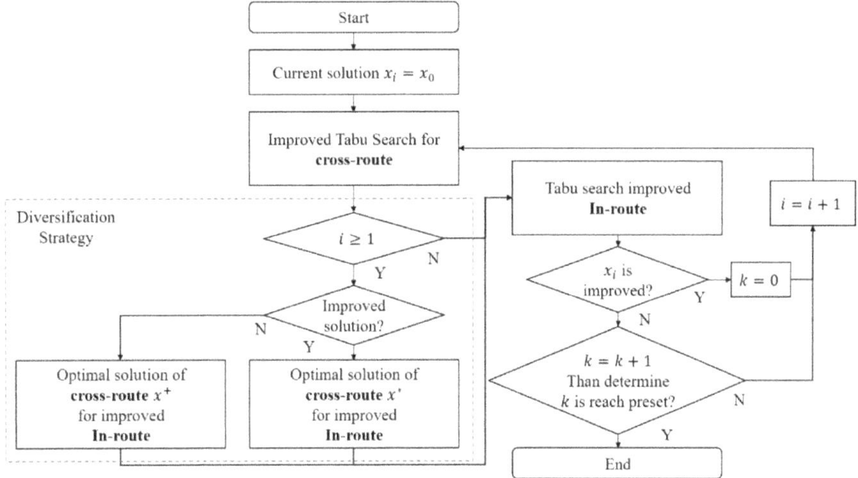

Figure 5. Path Improvement Flow Chart.

3.4.1. Modified Tabu Search for Improving Cross-Route Switch

Compared to a general tabu search, the major difference of this improved search is that S_{lmt}, which contains solutions exceeding the maximum distance limitation, were removed from the Candidate List to ensure that all path solution values were within range. However, this limitation resulted in an empty set for the Candidate List, thereby blocking all solutions from entering the next generation. When this occurred, an original tabu list was adopted with S_{lmt}, the solution set retaining all solutions, even with values exceeding the maximum distance limitation. With loosened criteria, the search also selected "the distance of the longest traveling path" as the solution of minimum value.

A tabu search combined with 2-opt was used in this step. By moving based on 2-opt, swapping the nodal lines among different paths was used to improve each of the current paths. So, 2-opt was applied to improve cross-route paths. At that time, we set up the length of the tabu list, the condition of stopping the search, and the initial solution as x_0. Then we executed the flow of the improved tabu search to get the optimal paths as follows (its pseudo code is shown in Algorithm 1):

Step 1.

First, an initial solution x_0 was set as the current optimal solution; the suboptimal solution was x^*, x^+. Then we set the current optimal and suboptimal distance function values as $d(x^*), d(x^+)$; with the initial generation of $i = 0$, the number of optimal solutions was not improved ($k = 0$) and tabu list T was the empty set ($T = \varnothing$).

Step 2.

According to characteristics of 2-opt, it extends all feasible neighborhood solutions of x_i as a neighborhood solution set $N(x_i)$. The candidate list included an $N(x_i)$ deduction T, and a solution of the maximum distance limit S_{lmt} $[N(x_i) - T - S_{lmt}]$. We selected the optimal neighborhood solution x_{i+1} which had an optimal total distance function value of $d(x_{i+1})$ in it. $(x_{i+1} = MinTotalDist(N(x_i) - T - S_{lmt}))$. If $N(x_i) - T - S_{lmt} = \varnothing$, we changed the Candidate List to $N(x_i) - T$ and selected an optimal neighborhood solution x_{i+1}, which had an optimal maximum distance function value of $t(x_{i+1})$ in it. (Note that the $x_{i+1} = MinMaxDist(N(x_i) - T)$)

Step 3.

We had to determine whether $d(x_{i+1})$ was less than $d(x^*)$. If it was, x^* had to be replaced by x_{i+1} ($x^* = x_i$). Then k = 0. Otherwise, the current optimal solution was preserved, and k = k + 1. Then, we determined whether $d(x_{i+1})$ was less than $d(x^+)$. If it was, x^+ was replaced with x_{i+1} ($x^+ = x_i$) Otherwise, we proceeded to the next step.

Step 4.

We determined whether the condition of stopping the search was reached (stopping the search and setting the current optimal solution as the global optimal solution when k reached the preset). Otherwise, we updated the tabu list. That is, we added x_{i+1} to T. If T was full, according to rule of first-in, first-out, to evict the solution which is the earliest one that enters T. Then we let $i = i + 1$ and returned to Step 2 to continue the operation.

Algorithm 1. Improved Tabu Search Pseudo Code

Begin
$i, k = 0$

$$x_i, x^* = x_0$$

$T = \emptyset$
Do
Use 2-opt method to expand x_i to get $N(x_i)$
Candidate List $= N(x_i) - T - S_{lmt}$
x_{i+1} = Neighborhood optimum have minimum total distance in Candidate List
If Candidate List $= \varnothing$ **then**

$$\text{Candidate List} = N(x_i) - T$$

x_{i+1} = Neighborhood optimum have minimum maximum-distance in Candidate List
end If
If $x_{i+1} < x^*$ **then**
$x^* = x_{i+1}$
$k = 0$
else
$k = k + 1$
If $x_{i+1} < x^+$ **then**
$x^+ = x_{i+1}$
end If
end If
add x_{i+1} into T
$i = i + 1$
   While $k <=$ preset
end While
end

3.4.2. Diversification Strategy

When this phase proceeded to the second generation, part of the cross-route improvements had solutions that were not improved. It was possible for a solution to be identical to the one obtained in the first generation. If this solution was sent to the calculation in the next step, it still came out as a useless local optimal solution without any further improvements. In order to avoid this situation, we had to determine whether the solution was improved before going to the second generation. If not, the suboptimal solution from the cross-route path exchange had to be implemented for improvement, to get rid of the circle of the local optimal solution and to obtain the global optimal solution.

3.4.3. Improvements for Each Single Path

In this part, the path solution given by the diversification strategy was adopted as the initial solution. We improved the single path of each group by using a tabu search in combination with 2-opt to be the same as in the first stage. Thus, we used each single path as an initial solution, and we used 2-opt as a move method to make improvements among different paths. Additionally, we set up the length of the tabu list with the condition of stop search. Then we executed the tabu search that was the same as that of the first stage. We did this until each group had been improved after the end of the sequence.

4. Experimental Vehicle and System Architecture

4.1. System Architecture

As we aimed to complete the task with multiple target points by multiple vehicles in the shortest possible time, a multi-agent system (MAS) under central control was set up for this research. The so-called "central control" was enlisted to allocate assignments to each individual robot, so that each of them could complete the tasks independently, rather than controlling the cars throughout the procedure. Through cooperation among multiple agents, the system was able to complete a task of a larger scale. From an MAS viewpoint on task allocation, the system was seen as having a Single-robot Task, Single-task Robots, and an Instantaneous Allocation of Task (ST-SR-IA) [20].

Various assignments at different levels of difficulty were randomly allocated by the system, meaning that each robot car may have received any assignment. Thus, each car in the robot swarm was equipped with identical specifications to ensure their performance and capacity to cope with various assignments.

From the viewpoint of MAS heterogeneity, the degree of similarity among individual robots within a collection Het(R) can be expressed as follows:

$$\mathrm{Het(R)} = -\sum_{i=1}^{Caste} p_i \log_2(p_i) \tag{11}$$

where *Caste* represents types of robots, and p_i is the decimal percent of robots belonging to any caste. Since all specifications of this system are the same (caste = 1, p_i = 1), Het(R) = 0 (Equation (11)) [21], which indicates a homogeneity system of a robot swarm.

Figure 6 illustrates the system architecture. This system included a Remote Monitoring Station, a Multi-Vehicle Paths Programming System, and Robot Cars. The Remote Monitoring Station acted as a central controller capable of programming the paths and distributing paths to the robot swarm, designed to complete the work together. From the viewpoint of network topology, the system was a sort of star topology. Each robot car was able to communicate with the Remote Monitoring Station, but they were not able to communicate with each other. The system mainly used LabVIEW to develop the Dynamic Remote Monitoring Station interface. By sending a URL to obtain a Google map web page as a display interface, it was used as a man–machine interface (MMI), in which each vehicle's location feedback and initial path programming could be displayed and managed. By clicking the target points displayed on the MMI, the location could be sent to the Remote Monitoring Station, where the algorithm was run for the path programming. Then, the planned paths were passed to each car via XBee. Once the cars received the data, they cross-compared their locations against the target points designated by the central monitoring station. Next they came out at an angle between the car and target point, and then headed out of the electronic compass. With these results, the robot car was able to drive the DC motor forward and control the servo motor differential to successfully complete the assignment.

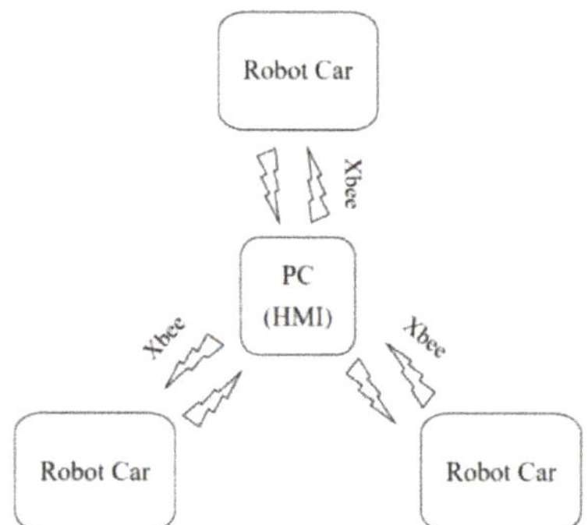

Figure 6. The Robot Cars System Architecture.

4.2. Experimental Vehicle

The robot car was used as an experimental vehicle (see Figure 7). Its function was to receive the data of target points from the Remote Monitoring Station and to establish a database for traveling. Once the car arrived at a designated target point, its real-time location was immediately sent to the Remote Monitoring Station.

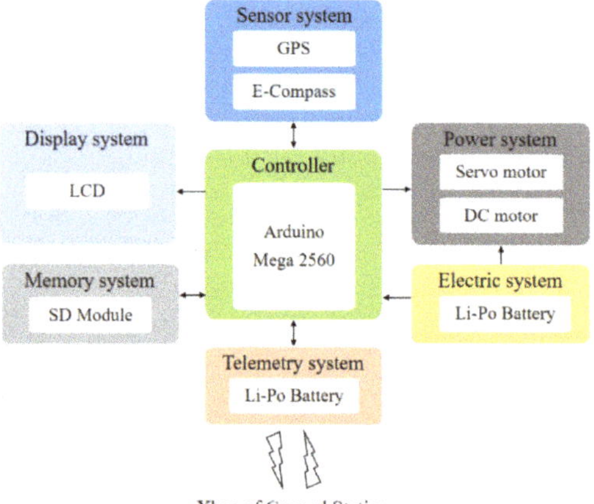

Figure 7. The Sub-System of Robot Car Architecture Diagram.

This experimental vehicle was modified from a 1/10 scale Shot Course Truck remote control car. We removed the car's shell and related remote control devices, then mounted additional off-the-shelf electronic components, including an Arduino Mega 2560 to act as an onboard computer; a U-blox NEO-7M Global Positioning System (GPS) module to provide position information for navigation; an HMC5883L Electronic Compass (E-Compass) to indicate heading; a 915 Mhz Xbee PRO S3B wireless module for duplex communication with the Remote Monitoring Station; a Liquid Crystal Display (LCD) to act as a health information indicator for field debugging; and a memory card (SD Module) to function as an integral data logger to include trajectory history. All the embedded automotive electronics are shown in Figure 8.

Figure 8. Experimental Robot Vehicle.

The robot car control principle was constructed from the latitude and longitude data of each target point stored in the database. Once the tasks assigned by the Remote Monitoring station were received by the robot car, the geographical data of each target point was extracted from the database for cars to travel among the points. On the other hand, a set of embedded steering strategies was also required for the car to automatically move along the planned paths and directions. In order to calculate the car's relative distance and angle against the target point, an electronic compass H was used. Also, the heading angle of the servo motor was set as θ_s. When $\theta_s = 0$, the robot car would go straight. When the θ_s value was positive, the robot car would turn right. When the θ_s became negative, the robot car would turn left. The maximum range of the angle was set at positive/negative 180°.

The input latitude and longitude data of the robot car (lat_r, lng_r) and the target point (lat_g, lng_g), which were received from the GPS, were used in Equation (12) to get $\varnothing$ (the angle of the target point against the exact north). Then we deducted H, the heading direction (see Equation (13)), to get the direction error θ_t, which was the angle between the current direction and the target point. We then used the proportional control θ_t to multiply K_P (the gain used for servo proportional control; $K_P = 0.161$) after adding θ_c (the center angle of the servo motor for the steering) to obtain θ_s (that is, the command sent to the servo motor for moving). In addition, the maximum range for the motor to move, ±15°, was set to prevent the cars from rolling over under high-speed turning. The $K_P\theta_t$ was determined with the understanding it should not exceed ±15° (Equation (14)).

$$\varnothing = \tan^{-1}\left(\frac{lng_g - lng_r}{lat_g - lat_r}\right) \tag{12}$$

$$\theta_t = \varnothing - H \tag{13}$$

$$\begin{cases} \theta_s = \theta_c + K_P\theta_t \\ K_P\theta_t = 15, \quad K_P\theta_t \geq 15 \\ K_P\theta_t = -15, \quad K_P\theta_t \leq -15 \end{cases} \tag{14}$$

In addition to steering, determining whether to reach the target point was equally important. Once the car reached the target point, only then was it allowed to move to the next one. The distance between the car and the target point was the key. We set the target point radius R. If the robot car was entering the range of R (the distance between the robot car and the target point < R), the target point hit would be determined. In this case, R was set with an appropriate value, understanding that it would be difficult for the car to hit the target points with too small or too large of an R value. The R value, then, was acknowledged as affecting the accuracy of the experiment [22].

5. Tests and Results

This section describes the tests for the proposed hybrid path programming method with maximum range constraint for the mission planning of multi-robot cars. Three types of tests were performed:

1. Convergence Test—This was conducted to confirm the convergence of the hybrid programming algorithm under range limitation.
2. Bench Test—This was performed to verify competitiveness by comparing with existing public TSPLIB instances.
3. Field Test—This was used to validate the practicality of the proposed algorithm by a group of three robot cars deployed on a field test.

In all tests, the proposed hybrid path programming computer code was processed on the remote control station based on a laptop PC with Intel Core i7 2.4 GHz CPU and 8 GB RAM.

5.1. Convergence Test

To see whether this hybrid path programming algorithm could successfully converge a solution set to optimize the route of each robot car group under their maximum travelable distance limitation, a series of tests were conducted. First, we ensured the algorithm convergence status by a simple 22-point test. As shown in Figure 9, a solution was converged at the 63rd generation. However, the solution was not immediately improved at the first generation. The solution exceeded the maximum distance limit as well as the Candidate List $= N(x_i) - T - S_{lmt} = \varnothing$, so we followed the improved tabu search and liberalized the Candidate List $= N(x_i) - T$. We compared the maximum distance of each of the paths and selected the solution with the minimum value in the Candidate List. Due to the maximum distance limit, there was a function to restrict the maximum distance of the feasible solution, which was not allowed to exceed this threshold. Starting from the maximum distance, we picked the minimum maximum distance of the neighborhood solution, and then made the current solution gradually close to the threshold (maximum distance limit). This eventually sufficed for meeting the threshold. Finally, the current solution was eligible (see the green line in Figure 9—the distance variation of the current solution). In this way, the current solution was improved and gradually got close to the feasible solution.

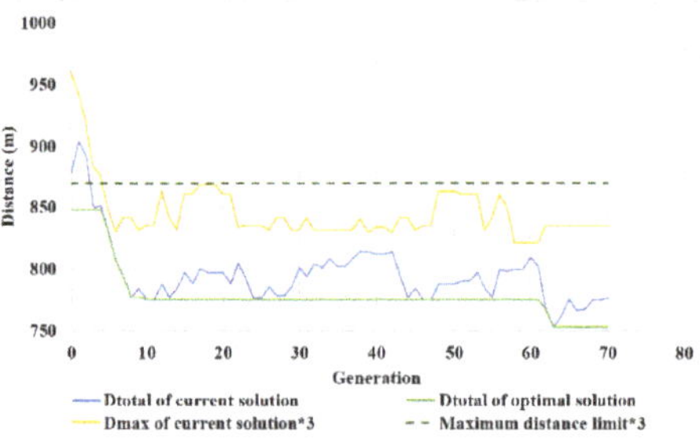

Figure 9. Solution Convergence Diagram.

5.2. Bench Test

The problem definition of this research is similar to mTSP. The difference, in comparison to mTSP, is that this research is limited to the travel distance of each robot car. Since mTSP instances are easy to obtain, we used the innovative algorithm developed in this research to solve the same mTSP problem for comparing.

Most scholars have modified TSPLIB instances to test mTSP, because mTSP does not have public instances. In this research, Pr76, Pr152, Pr226, Pr299 and Pr439 were tested. The mTSP rule that each salesman must visit more than two targets was used. We then compared with MGA [23], MACO [24], NMACO [11] and SA+EAS [25]. For establishing the initial solution, the tabu list length was set to 30, and the condition set for stopping was that the optimal solution was not updated, and it improved in 50 generations. The tabu list length of Improvement of Each Single Path was set to 50. The tabu list length of Improvement between Different Paths was 50 as well. As well as this, the condition set for stopping calculations for programming both paths was that the solution had neither been updated nor improved in the most recent 10 generations. The condition set for stopping the path improvement part was that the optimal solution had not been updated or improved in the second iteration.

Table 1 is the comparison result of our algorithm in TSPLIB instances with each other. The 2TS+2OPT was the hybrid algorithm developed in this research. The number of the target is expressed as n. The number of salesmen is shown as m. The maximum number of waypoints (cities) that each vehicle (salesman) could visit is denoted by u. As a result, the distance values of 2TS+2OPT are better than the others, and most of the CPU Times are less than the others as well.

Table 1. The Comparison Result of 2TS+2OPT Algorithm in the library of sample traveling salesman problem (TSPLIB) Instances.

Name	pr76	pr152	pr226	pr299	pr439
n	76	152	226	299	439
m	5	5	5	5	5
u	20	40	50	70	100
SA+EAS	157482	127755	167655	81922	161698
NMACO	157413	127781	167239	81261	160298
MACO CPU Time(s)	178597 51	130953 128	167646 143	82106 288	161955 563
MGA CPU Time(s)	178597 43	130953 91	167646 165	82106 363	173839 623
2TS+2OPT **CPU Time(s)**	**153840** **11.3**	**121165** **51.2**	**159831** **153.4**	**72813** **190.5**	**141526** **455.4**

5.3. Field Test with Multiple Robot Cars

In this study we set out to solve the path programming of a multi-target wide area. Subject to vehicle ability constraints, cars were not able to travel to all target points. Therefore, we sent multiple vehicles to respectively travel to target points and complete tasks. We selected a site in Yunlin, Taiwan, and set up several target points on it. With a maximum distance limit set up, the shortest total distance paths and the total distance limits for each robot car were set, so they did not exceed this limit. A hybrid tabu search combined with a 2-opt swap method was adopted to program the optimal path in the Remote Monitoring Station, which was a laptop with CI7, 2.4 GHz, and 8 GB RAM. For establishing the initial solution, the tabu list length was set to 30. The condition set for stopping was that the optimal solution had not been updated nor improved in 50 generations. The tabu list length of Improvement of Each Single Path was set to 30. The tabu list length of Improvement between Different Path parts was 30. Additionally, the condition set for stopping calculations for the programming of both paths was that the solution had not been updated or improved in the most recent 50 generations. The condition set for stopping the path improvement part was that the optimal solution had not been updated nor improved in the second iteration. After all of these settings were in place, the paths were assigned to robot cars for them to run on the designated site with the paths programmed.

5.3.1. Maximum Distance Limit = 170 m

The test started by randomly setting up 15 target points on the e-map of the Remote Monitoring Station. We assumed that the maximum travelable distance limit was 170 m. If we were sending a single robot car to travel all target points, the optimal (shortest) path was 270.6 m (as shown in Figure 10). But this path obviously had exceeded the maximum distance limit that a single car can handle for completing a task; hence three cars were sent for this test. The system soon provided new paths (as shown in Figure 11), with the shortest distance by using the algorithm developed in this research. In this test, the CPU time of the Remote Monitoring Station computer was only 0.79 s. With this solution, the A-path was 81.3 m; the B-path was 164.7 m; and the C-path was 139.9 m. None of the paths exceeded the maximum distance limit. The shortest total distance was 385.9 m. Then the system automatically dispatched those three routes to robot cars to visit their assigned target points. We recorded the path trajectory of each robot car, as shown in Figure 12. The actual total distance was 401.1 m. The actual distance of the A path was 86.9 m; for the B path it was 167.9 m; and for the C path it was 146.3 m. The maximum error of all robot cars was 6.8% of car A. The error of total distance was 3.9% (Table 2). This error was mainly caused by GPS drifting, a bumpy surface and steering center offset factors.

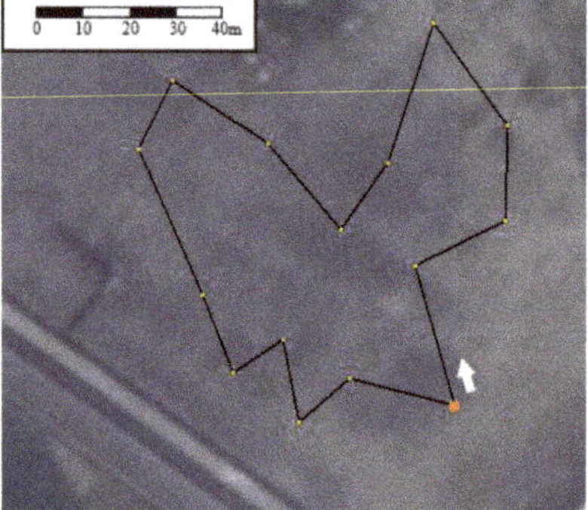

Figure 10. Shortest Single Distance Path.

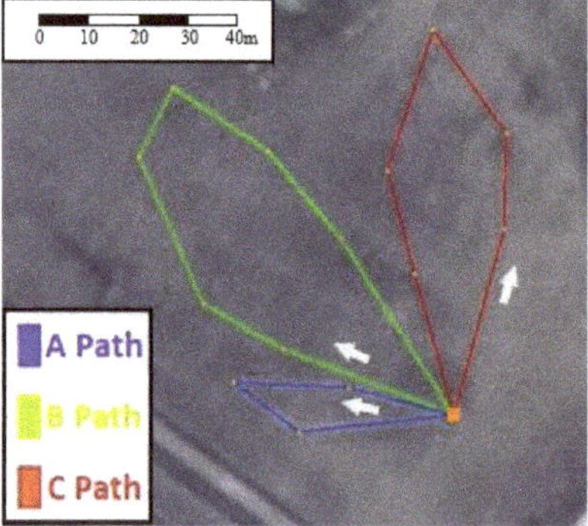

Figure 11. Shortest Total Distance Path When the Maximum Distance Limit Is 170 m.

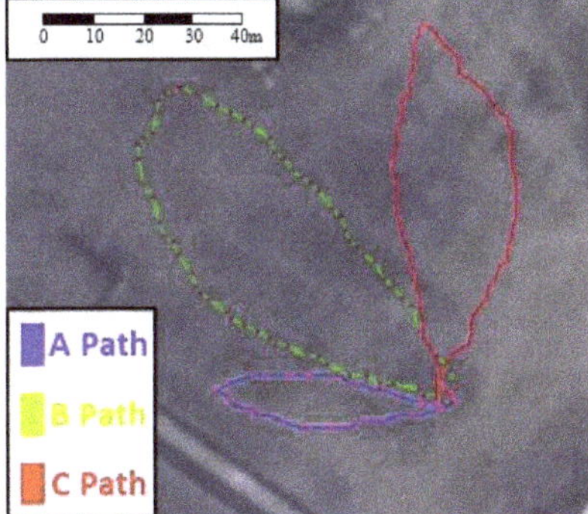

Figure 12. Experimental Trajectory When the Maximum Distance Limit Is 170 m.

Table 2. The Comparison of Shortest Total Distance (Theoretical) and Total Experimental Trajectory (Actual) under the Constraint of the Maximum Distance Limit of 170 m for Each Car.

	A	B	C	Total
Theoretical	81.3	164.7	139.9	385.9
Actual	86.9	167.9	146.3	401.1
Error	+5.6	+3.2	+6.4	+15.2
Error%	+6.8%	+1.9%	+4.5%	+3.9%
Path programming CPU time = 0.79 s				

5.3.2. Maximum Distance Limit = 164 m

With the same location target points set up, we tuned down the maximum distance limit as 164 m. The algorithm converged a new solution set (as shown in Figure 13) in a very short (0.75 s) CPU time. For this solution, the A path was 105.2 m; the B path was 159.9 m; and the C path was 139.9 m. None of paths exceeded the maximum distance limit. The total distance, however, increased to 405 m, because the maximum travelable distance of each car was compressed, which made the solution more "load-balanced". The completion time was relatively less. After the solution was converted, the system immediately dispatched those three routes to robot cars to visit their assigned target points. We recorded the path trajectory of each robot car, as shown in Figure 14. The actual total distance was 412.5 m. The actual distance of the A path was 106.8 m; for the B path it was 160.9 m; and for the C path it was 144.8 m. The maximum error of all robot cars was 3.5% of car C. The error of total distance was 1.8% (see Table 3). It is worth mentioning that none of the robots exceeded the range limitation.

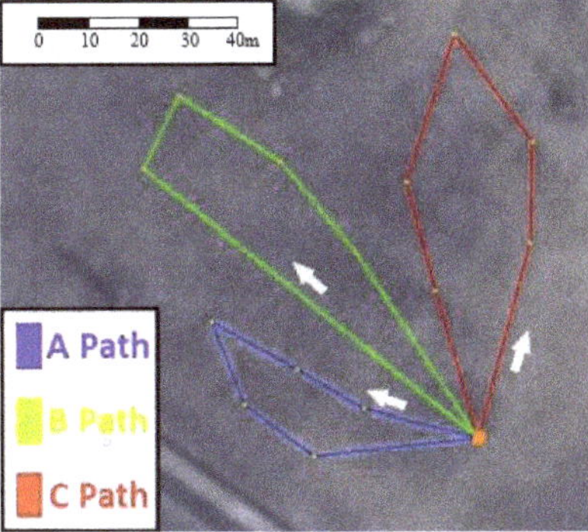

Figure 13. Shortest Total Distance Path When the Maximum Distance Limit Is 164 m.

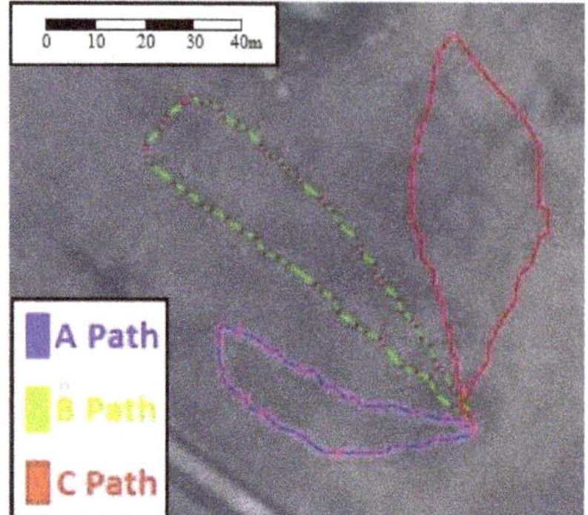

Figure 14. Experimental Trajectory When the Maximum Distance Limit Is 164 m.

Table 3. The Comparison of Shortest Total Distance (Theoretical) and Total Experimental Trajectory (Actual) under the Constraint of Maximum Distance Limit 164 m for Each Car.

	A	B	C	Total
Theoretical	105.2	159.9	139.9	405.0
Actual	106.8	160.9	144.8	412.5
Error	+1.6	+1.0	+4.9	+7.5
Error %	+1.5%	+0.6%	+3.5%	1.8%
Path programming CPU time = 0.75 s				

5.4. Discussion

From the bench test, compared to other algorithms, the 2TS+2OPT hybrid algorithm proposed in this research has very high advantages in both path optimization and CPU time, which are crucial for the practical applications of unmanned system. Besides, when examining the results generated from this hybrid algorithm of path programming, it was found that the total distance was inversely proportional to the maximum distance limit value. Thus, the lower the maximum distance limit value was, the longer the total distance would be. Relatively speaking, the lower the maximum distance limit value, the shorter the mission time. In general, a tighter margin in onboard energy capacity yielded a "load-balanced" situation for each robot in the group, which means every robot had equal loading.

From the field experiments, we obtained the error of maximum total distance at 3.9%, and the maximum error of each robot car at 6.8%. These are both minor and representative of a rather satisfying result, as we expected. In addition, we also found a few minor errors in Algorithm 1; Table 1, which were due to the following three factors:

1. GPS drifting

The GPS device adopted in this research was designed for general commercial purposes, with couple meters measuring error. This GPS drifting may be easily fixed with higher-level equipment, such as that of a real-time, kinematic GPS.

2. Pavement condition

The paths programmed by the system were generated based on a smooth ground surface for car operation. But in fact, there some unexpected surface conditions, like tiny pebbles on the pavement that may have caused an offset to the route.

3. Offset of steering center

The steering mounted on the robot car consisted of a servo motor and a steering mechanism; this unit may have been offset while it was traveling among target points during a long run. This offset might have somehow led to the car moving off the path, and thus resulted in a few minor errors at the end. In this research, we tried to minimize the effect of this factor by regularly correcting the steering.

In terms of the case that set the maximum distance limit at 170 m, since the longest path was 164.7 m (which was very close to the maximum distance limit), the robot car could run and exceed the limit as long as any error occurred. For example, the car with a maximum error of 6.8% in this experiment was obviously not able to complete the task. Thus, reserving 10% as an allowance margin when setting the maximum distance limit is recommended (maximum distance limit: 10%).

6. Conclusions

In this research, a hybrid 2TS+2OPT algorithm with limited range constraints for the path programming of multiple robot vehicles was successfully developed. This innovative algorithm is superior in both better path optimization and shorter CPU time, compared to other mTSP algorithm solutions. In the last stage of this research, a general scenario was presented to show the whole process of the multi-robot mission planning, in which three robots were deployed in a field to complete a series of wide area multi-waypoint tasks which were far beyond a single car's endurance capabilities. Those tests validated that the algorithm can successfully optimize robots' routes to visit assigned target points within their range limitations.

In real-world, unmanned vehicles practices, optimized paths based on onboard energy capacity (fuel or battery) constraints are critical to multiple-agent system applications of this type, including large area multi-points surveillance exercises, robot swarm deliveries, multi-drone attacks, and so on. This research could contribute to those types of instances.

Author Contributions: Conceptualization, M.-T.L.; Formal analysis, B.-Y.C., M.-T.L. and Y.-C.L.; Methodology, M.-T.L.; Project administration, M.-T.L.; Software, B.-Y.C. and M.-T.L.; Writing—original draft, B.-Y.C. and M.-T.L.; Writing—review & editing, M.-T.L. and Y.-C.L. All authors have read and agreed to the published version of the manuscript.

Funding: This research was funded by the Ministry of Science and Technology, Taiwan (ROC), under grant number MOST-107-EPA-F-010-001.

Conflicts of Interest: The authors declare no conflict of interest.

References

1. Luan, Y.; Xue, H.; Song, B. The Simulation of the Human-Machine Partnership in UCAV Operation. In Proceedings of the 26th International Congress of the Aeronautical Sciences, Anchorage, AK, USA, 14–19 September 2008.
2. Cai, B. Path Programming for Autonomous Unmanned Vehicle System. Bachelor's Thesis, National Formosa University (NFU), Huwei, Taiwan, 2015.
3. Atanasov, N.; Ny, J.L.; Daniilidis, K.; Pappas, G.J. Decentralized Active Information Acquisition: Theory and Application to Multi-robot SLAM. In Proceedings of the 2015 IEEE International Conference on Robotics and Automation (ICRA), Seattle, WA, USA, 26–30 May 2015; pp. 4775–4782.
4. Ravankar, A.; Ravankar, A.A.; Kobayashi, Y.; Emaru, T. Hitchhiking Robots: A Collaborative Approach for Efficient Multi-Robot Navigation in Indoor Environments. *Sensors* **2017**, *17*, 1878. [CrossRef] [PubMed]
5. Maity, D.S.; Goswami, S. Multipath Data Transmission with Minimization of Congestion Using Ant Colony Optimization for mTSP and Total Queue Length. *IJLRST* **2013**, *2*, 109–114.
6. Sariel-Talay, S.; Balch, T.R.; Erdogan, N. Multiple Traveling Robot Problem: A Solution Based on Dynamic Task Selection and Robust Execution. *IEEE/ASME Trans. Mechatron.* **2009**, *14*, 198–206. [CrossRef]
7. Ahmadvand, M.; Yousefikhoshbakht, M.; Mahmoodi Darani, N. Solving the Traveling Salesman Problem by an Efficient Hybrid Metaheuristic Algorithm. *JACR* **2012**, *3*, 75–84.
8. Atashpaz-Gargari, E.; Lucas, C. Imperialist Competitive Algorithm: An Algorithm for Optimization Inspired by Imperialistic Competition. In Proceedings of the IEEE CEC 2007, Singapore, 25–28 September 2007.
9. Larki, H.; Yousefikhoshbakht, M. Solving the Multiple Traveling Salesman Problem by a Novel Metaheuristic Algorithm. *JOIE* **2014**, *16*, 55–63.
10. Dorigo, M.; Maniezzo, V.; Colorni, A. Ant System: Optimization by a Colony of Cooperating Agents. *IEEE Trans. Syst. Man Cybern. Part B (Cybern.)* **1996**, *26*, 29–41. [CrossRef] [PubMed]

11. Yousefikhoshbakht, M.; Didehvar, F.; Rahmati, F. Modification of the Ant Colony Optimization for Solving the Multiple Traveling Salesman Problem. *ROMJIST* **2013**, *16*, 65–80.

12. Necula, R.; Breaban, M.; Raschip, M. Performance Evaluation of Ant Colony System for the Single-Depot Multiple Traveling Salesman Problem. In Proceedings of the 10th International Conference on Hybrid Artificial Intelligence Systems, Bilbao, Spain, 22–24 June 2015; pp. 257–268.

13. Xu, M.; Li, S.; Guo, J. Optimization of Multiple Traveling Salesman Problem Based on Simulated Annealing Genetic Algorithm. *MATEC Web Conf.* 2017. [CrossRef]

14. Côté, J.F.; Potvin, J.Y. A Tabu Search Heuristic for the Vehicle Routing Problem with Private Fleet and Common Carrier. *EJOR* **2009**, *198*, 464–469. [CrossRef]

15. Liao, T.; Huang, W. A Study of Dynamic Logistics Based on Two-Phased Method. *IJAIT* **2008**, *2*, 76–94.

16. Lin, W.; Cheng, F.; Tsay, M. An Improved Tabu Search for Economic Dispatch with Multiple Minima. *IEEE Trans. Power Syst.* **2002**, *17*, 108–112. [CrossRef]

17. Bektas, T. The Multiple Traveling Salesman Problem: An Overview of Formulations and Solution Procedures. *Omega* **2006**, *34*, 209–219. [CrossRef]

18. Hung, F. Vehicle Routing Problem of Integrated Supply Medical Materials in Strategic Alliance Hospitals—A Study of One Medical Center. Master's Thesis, National Yunlin University of Science and Technology, Yunlin, Taiwan, 2004; p. 15, Unpublished.

19. Lin, S. Computer Solutions of the Traveling Salesman Problem. *BSTJ* **1965**, *44*, 2245–2269. [CrossRef]

20. Khamis, A.; Hussein, A.; Elmogy, A. *Multi-robot Task Allocation: A Review of the State-of-the-Art, Cooperative Robots and Sensor Networks 2015*; Springer: Berlin/Heidelberg, Germany, 2015; pp. 31–51.

21. Balch, T.R. Social Entropy: A New Metric for Learning Multi-Robot Teams. In Proceedings of the 10th International FLAIRS Conference (FLAIRS-97), Daytona, FL, USA, 11–14 May 1997.

22. Lai, L.; Hsieh, Y. The Research in Real-time Dynamic Path Programming for Multiple Autonomous Ground Vehicle Mission. Bachelor's Thesis, National Formosa University (NFU), Huwei, Taiwan, 2016.

23. Tang, T.; Liu, J. Multiple Traveling Salesman Problem Model for Hot Rolling Scheduling in Shanghai Baoshan Iron & Steel Complex. *EJOR* **2000**, *24*, 267–282.

24. Junjie, P.; Dingwei, W. An Ant Colony Optimization Algorithm for Multiple Traveling Salesman Problem. In Proceedings of the First International Conference on Innovative Computing, Information and Control (ICICIC '06), Beijing, China, 30 August–1 September 2006; pp. 210–213.

25. Yousefikhoshbakht, M.; Sedighpour, M. A Combination of Sweep Algorithm and Elite Ant Colony Optimization for Solving the Multiple Traveling Salesman Problem. *P. Pomanian. Acad. A* **2012**, *13*, 295–302.

 *electronics*

Article

Research on Anti-Radiation Noise Interference of High Definition Multimedia Interface Circuit Layout of a Laptop

Wei Chien [1,2,*], Yu-Ting Cheng [3], Chiuan-Fu Hsiao [3], Kai-Xu Han [1] and Chien-Ching Chiu [3,*]

[1] Department of School of Electric and Information Engineering, Beibu Gulf University, Qinzhou 535000, Guangxi, China; frog0696@163.com
[2] Ningde QianWei Industry Technology Co., Ningde 352000, Fujian , China
[3] Department of Electrical and Computer and Engineering Department, Tamkang University, New Taipei 25137, Taiwan; rainsstop@gmail.com (Y.-T.C.); vawrp@yahoo.com.tw (C.-F.H.)
* Correspondence: air180@seed.net.tw (W.C.); chiu@ee.tku.edu.tw (C.-C.C.)

Received: 28 December 2019; Accepted: 21 February 2020; Published: 3 March 2020

Abstract: In this paper, several aspects were studied, including the effect of an electromagnetic interference (EMI) noise interference strategy with High Definition Multimedia Interface (HDMI) 1.4, the analysis of a test on a printed circuit board (PCB) layout, and a comparison of the near field intensity radiation distribution between an EMI with a modified HDMI layout and an original layout. In this study, the near field detection instrument of APREL EM-ISight was employed to analyze the distribution of the strength of an electromagnetic noise field. After the practical validation, we found that the PCB layout complies with the standards after the modifications. Meanwhile, the PCB layout satisfies the requirements of most laptop HDMI-related products for EMI.

Keywords: high-definition multimedia interface; PCB layout; electromagnetic interference; radiation resistance

---

1. Introduction

At present, electromagnetic compatibility has attracted much attention from the science and technology industry. How to eliminate the electromagnetic interference while reaching a balance in signal quality is a difficult challenge facing every electronic researcher. Nowadays, technology has entered the era of Artificial Intelligence (AI) and the Internet of Things (IoT). As the functions performed by electronic products and facilities function are made increasingly complex and diversified, the mitigation of noise interference plays an important role. If the manufacturing process it is not consideration during the design of electronic products unforeseeable conditions can arise. It is possible for the electromagnetic interference to cause dysfunction, and even result in the crash or burnout of the product.

Currently, there is only limited literature that has focused on the problems caused by electromagnetic interference [1–5], which is a challenge faced by every single electronics developer. The design of electronic devices is subject to various limitations in the initial stage, which increases the difficulty that arises from reaching a balance between the quality of signal and electromagnetic interference. It is difficult to ensure that the signal is free from distortion and that the noise meets the existing requirements set out by the international electro technical association.

With respect to the test project of high definition multimedia interface (HDMI) electromagnetic interference (EMI), the product can meet the standard in the design stage, but not satisfy the requirements specified for mass production, which forces the production line to be suspended. Even worse, in some cases, it is necessary to redesign the product, which is not desired by the company. Usually this problem arises from the design stage. The engineers fail to give full consideration to the potential impact of electromagnetic interference, and the choice of countermeasure is unavailable. In addition, these problems result from the design flaws and also from the countermeasure selection. In terms of a general EMI verification project, test results are usually restricted to below the regulatory limit of −0 dB, but the error value can exist in the practical measurement. Therefore, it is essential to identify a proper countermeasure for these factors to be avoided.

For the general EMI test error, the error value can range between −1 dB and −3 dB. As shown in Figure 1 [6], in order to avoid the circumstance where the shipment is affected by a lag in the test schedule as a result of test error, some demanding customers set the verification limitation value below −6 dB at the time of shipment. However, combined with the test errors, as mentioned above, there is a difference of up to 9 dB. Therefore, this is a significant topic for engineers [7–9].

F.1   General

This annex shows the basis for the acceptability criterion of ±4 dB for the normalized site attenuation measurements required in 5.4.

F.2   Error analysis

The error analysis in Table F.1 applies to the normalized site attenuation measurement methods given in 5.4. The total estimated errors are the basis for the ±4 dB site acceptability criterion consisting of approximately 3 dB measurement uncertainty and an additional allowable 1 dB for site imperfections.

The error budget in Table F.1 does not include uncertainties in the amplitude stability of the signal generator, tracking generator, or any amplifiers that may be used, nor does it include the potential errors in measurement technique. The output level of most signal and tracking generators will drift with time and temperature, and the gain of many amplifiers will drift as temperature changes. It is imperative that these sources of error be held to an insignificant amount or corrected in making the measurements, otherwise the site may fail to meet the acceptability criterion due to instrumentation problems alone.

Table F.1 – Error budget

Error item	Measurement method	
	Discrete frequency method dB	Swept frequency method dB
Antenna factor (Tx)[a]	±1	±1
Antenna factor (Rx)[a]	±1	±1
Voltmeter	0	±1,6[b]
Attenuator	±1	0
Site imperfections	±1	±1
Totals	±4	±4,6

[a]   At frequencies above 800 MHz, F_a errors may approach ±1,5 dB.

[b]   From the operating instructions.

Figure 1. CISPR 16 1–4 site error description.

With regards to the specifications of HDMI 1.4, its resolution is supposed to support 4K2K with 30 Hz refresh rate. However, according to the verification requirements specified by the customer, HMDI noise at 891 MHz frequency is frequently incapable of meeting the customers' requirements for EMI of −6 dB, as shown in Figure 2. * means that the customers' requirements. The 891 MHz frequency point is merely −2.578 dB. Although the measurement results comply with the EN55022 regulations, they do not conform to the customers' requirements. Therefore, this paper presents a practical study of a printed circuit board (PCB) layout based on satisfying customers' requirements for HDMI noise.

Engineer :	
Site : Site6	Time : 2015/09/07 - 14:11
Limit : CISPR_B_10M_QP	Margin : 6
EUT : Notebook PC	Probe : Site6_MD_10M - HORIZONTAL
Power : AC 230V/50Hz	Note : M/N:E15S,HDMI Extend Mode,1920*1080/60Hz+3840*2160/30Hz,單機+monitor only

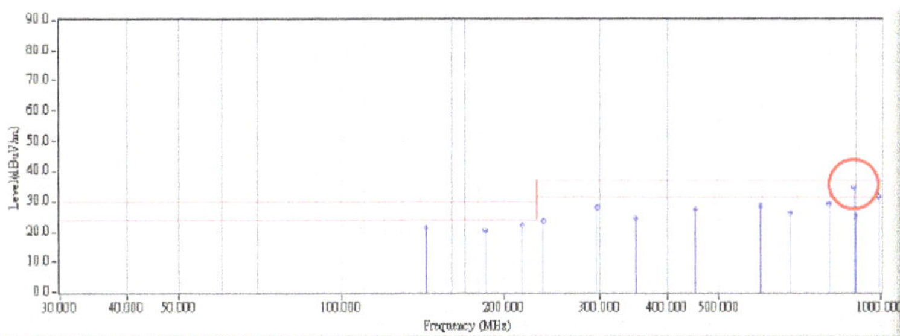

		Frequency (MHz)	Correct Factor (dB)	Reading Level (dBuV)	Measure Level (dBuV/m)	Margin (dB)	Limit (dBuV/m)	Detector Type
1		144.000	-20.604	42.100	21.496	-8.504	30.000	QUASIPEAK
2		184.870	-22.381	42.800	20.420	-9.580	30.000	QUASIPEAK
3		216.000	-20.989	43.100	22.110	-7.890	30.000	QUASIPEAK
4		236.100	-19.268	42.900	23.633	-13.367	37.000	QUASIPEAK
5		297.000	-16.067	44.300	28.233	-8.767	37.000	QUASIPEAK
6		348.500	-15.279	39.800	24.521	-12.479	37.000	QUASIPEAK
7		449.800	-11.926	39.200	27.274	-9.726	37.000	QUASIPEAK
8		594.000	-8.342	36.900	28.559	-8.441	37.000	QUASIPEAK
9		674.300	-6.504	32.800	26.297	-10.703	37.000	QUASIPEAK
10		799.800	-2.604	31.900	29.296	-7.704	37.000	QUASIPEAK
11	·	890.970	-3.678	38.100	34.422	-2.578	37.000	QUASIPEAK
12		899.800	-3.489	28.500	25.011	-11.989	37.000	QUASIPEAK
13		989.990	-2.020	33.600	31.580	-5.420	37.000	QUASIPEAK

Figure 2. Radiation 30 MHz to 1GHz measurement result data.

2. Theoretical Formulation

During the development of an electronic product, when the electrical engineer or the mechanical engineer lacks the relevant knowledge of electromagnetic compatibility (EMC), a flawed design is unavoidable in the initial research and development stage. The occurrence of electromagnetic radiation during the process of production is not allowed. The most common fatal problem results from the method of wiring.

Generically, when the electronic circuits are designed, electronic engineers take into consideration only signal integrity (SI), power integrity (PI), and impedance matching accuracy (IMA), which means the poor design of conducting wires in PCB is ignored, thus causing an EMI problem. However, most mechanical engineers only give consideration to the stack structure and the minimum line-to-line distance (line length affects price), and therefore overlook whether the wiring and wiring mode crosses in the vicinity of the radio frequency emitting chip, as shown in Figures 3 and 4. Invariably, a bad design results in phenomena such as radiation divergence and coupling, either directly or indirectly. In addition, the grounding property of the basic mechanical structure can also be ignored. All these problems contribute to electromagnetic noise. Finally, the EMC measurement results of products are affected. In consideration of this, the relevant issues are worthy of discussion before the design stage in order to prevent serious problems from arising.

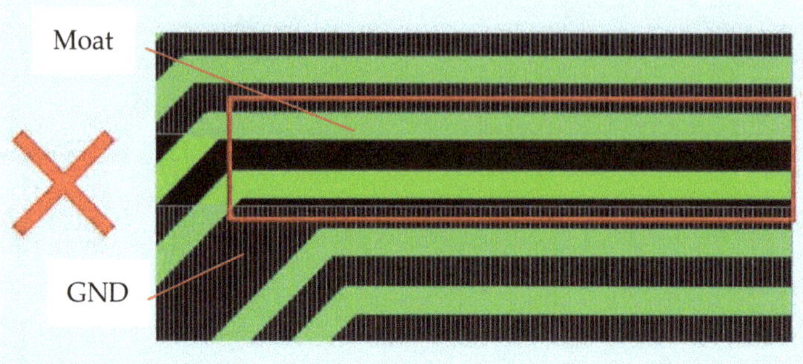

Figure 3. Poor design of signal line across cutting groove.

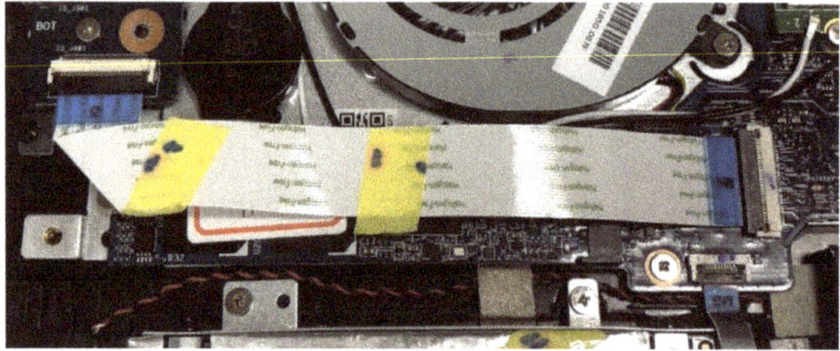

Figure 4. Poor design of overlapping of soft wiring and Solid-state drive.

Nevertheless, in order to eliminate electromagnetic interference, no more than three elements of electromagnetic interference should be taken into consideration. Although it is unnecessary to address the problem from any aspect at the start, it is recommended to deal with the source of the disturbance at the beginning. An effective design for the wiring can resolve approximately 70% to 80% of the radiation problems. Not only does it reduce the strategy cost incurred by this problem, it also saves a substantial amount of staffing costs related to research and development.

In the general layout of a circuit board (hereinafter referred to as the PCB layout), not only is consideration given to the signal quality and whether the signal function is normal, attention is also given to line width, line spacing, return path, and signal line reference plane. In addition, if the return path is excessively large, the magnetic field effect is produced that creates the EMI radiation area.

In addition, it is necessary to prevent the circumstance where irrelevant through holes pass the high speed or sensitive signal lines. This is because the overly strong noise source makes it easy for signal interference and coupling mechanism to occur, as shown in Figure 5. For example, regardless of whether a PCB is designed with either four or six layers, if the pulse signal line is shifted from the top layer to the bottom layer, and the pulse signal line is adjacent to the through hole of the high-speed signal line, it is highly likely to cause signal interference. This type of wiring occurs on a frequent basis, which requires developers to pay special attention to this issue.

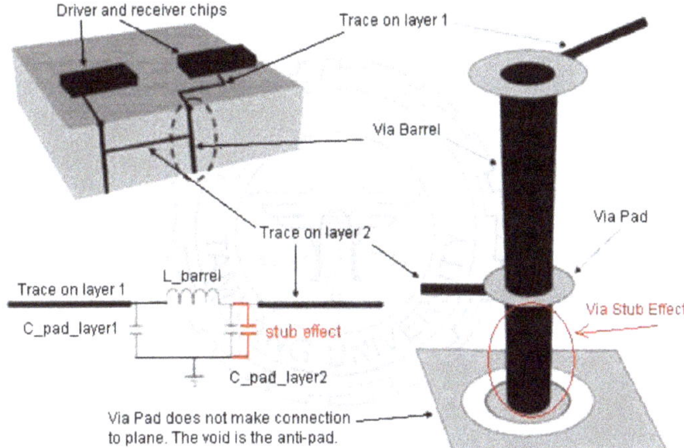

Figure 5. Electromagnetic interference through hole.

In addition to eliminating the interference of signal line through holes and other signal lines, it is also necessary to take into account the conditions of the return path when the signal lines cannot refer to the same reference plane (GND). In general, a decoupled capacitor is employed, or a reference plane is added to achieve reinforcement by the nearby signal line through hole, as shown in Figure 6.

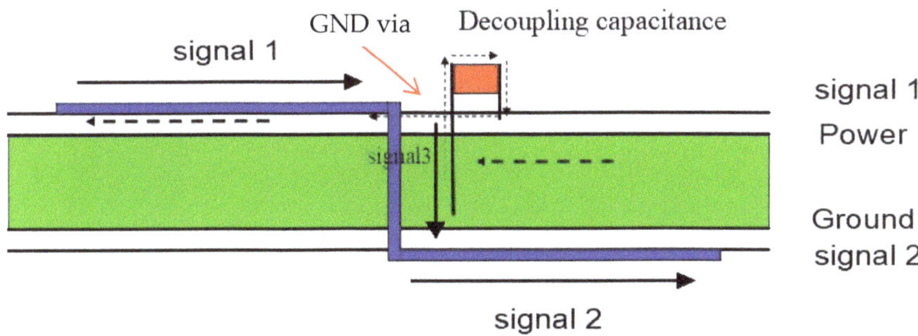

Figure 6. Schematic diagram of decoupling capacitance.

Therefore, it is a necessity that the 3W rule is adhered to as much as possible as part of the wiring requirements [7]. The clearance between lines should be three times the width of lines, as shown in Figure 7. It should be measured by midline, or the clearance between lines should be twice as wide as a single line, and snake-like wiring needs to be reduced, as marked in yellow in Figure 8. In addition, for the signals with high speed and high sensitivity, it is necessary to avoid the clock or the power of the through hole as much as possible. It is also necessary to avoid the front end of the HDMI port and the HDMI signal online. Meanwhile, the PCB factory production line and the line testing point must be detected. If the PCB itself with HDMI signal noise has the opportunity to couple radiation to an HDMI port, what is borrowed from HDMI cable will noise out the screen and join in the high-speed signal transfer layer path GND via design. Moreover, the GND via advice from a signal line shows a penetration hole distance of 25 mil, which makes it necessary not only to shorten the return path and area but also to reduce the formation of the radiation field. These issues affect EMI test results on a frequent basis.

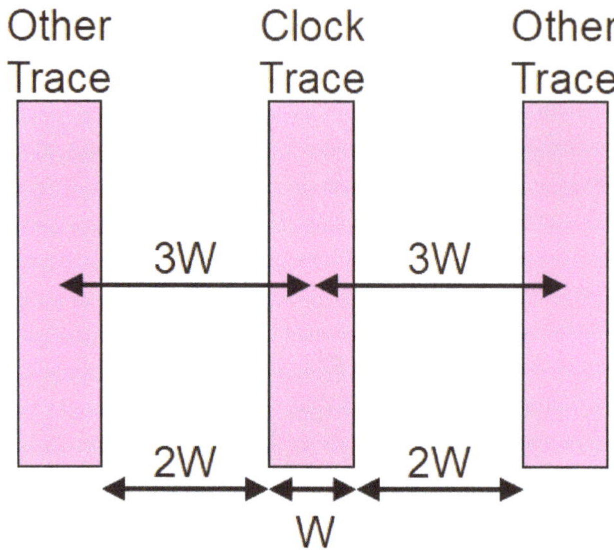

Figure 7. Principle of 3 times line width.

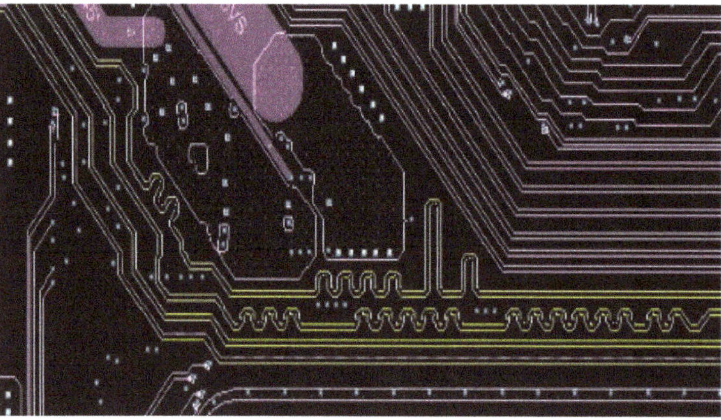

Figure 8. Schematic diagram of snake route.

3. EMI Test Environment and Attention

3.1. EMI Validation Environment

According to the formal EMI EN55022 regulations (30 m~1 GHZ) for validation field, there are two kinds of spaces. One is a relatively open space, commonly known as the open site, as shown in Figure 9. In some cases, open site testing leads to the results of validation such as the AM/FM radio band in Taiwan from 88 MHZ to 108 MHZ due to the environmental noise effects. If the noise falls into the radio frequency band, the system noise signal is broadcast band and the phenomenon of cover occurs. As a result, the validation gives rise to a grey area, and it is impractical to ensure the authenticity of the products. Due of this, additionally, another indoor dark room space was created, and it is known as the 10 m chamber, as shown in Figure 10. Since the enclosed space can be isolated from the external noise interference, it ensures that full spectrum noise occurs in electronic products.

Although it is not easy to misjudge the noise signal, the venue is costly. Therefore, the chamber has a 10 m fewer third-party certification unit. In addition, to make an appointment at the site, the cost is approximately three to four times higher as compared with the open site.

Figure 9. Schematic diagram of the open site.

Figure 10. Schematic diagram of the 10 m chamber.

3.2. EMI Site Configuration Map

In accordance with the regulations set out for an EMI measured erection diagram, as shown in Figure 11, it is comprised of a pair of conical log periodic antennae. The antenna rises between 1 m and 4 m, while the antenna signal is distinguishable between two vertical and horizontal polarities. It receives the material under test (equipment under test) within 30 MHZ–1 GHZ fan questions around noise. In addition, it locates the EUT ten meters clear of the table and performs the EUT functions. The tunntable is rotated 360 degree, once it receives signals back and forth, to make sure the EUT signals have no dead angles from the antenna. Meanwhile, an examination is carried out on whether the EUT is in compliance with the requirements.

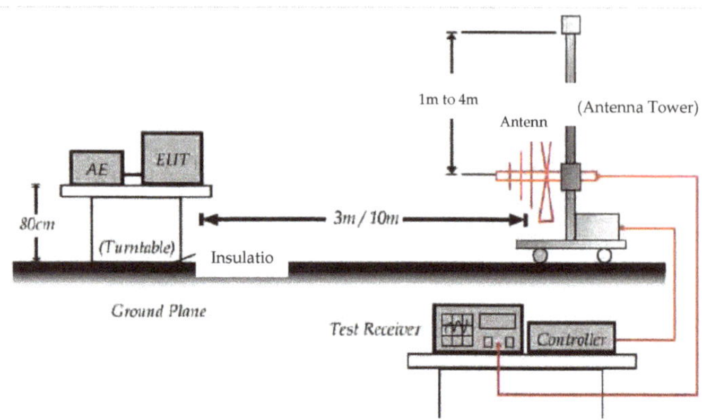

Figure 11. EN55022 electromagnetic interference (EMI) official site configuration map.

3.3. EMI Verification Attention

During the developmental phase, in order to eliminate errors and the impact of environmental factors, artificial factors should be taken into consideration. When validation issues are considered for the differences in the EUT itself, it is necessary to pay more attention to laboratory space between the measurement error values, and each set up and validation. In general, an engineering personnel certification experimental unit is required to comply with the regulations set out in the standards set.

Every erection is different. Nevertheless, although they are seemingly identical, they are not identical. For example, every time the validation is set u, the first and the second are consistent, and it verifies whether the wire is surrounding the suitability problem. The most common validation that occurs it is the wire surrounding the overlap, as shown in Figure 12. It is red box space. In some cases, it produces a signal or noise coupling antenna effect, which makes the measurement result far from accurate. If the EMC engineer in research and development validation fails to notice these details, the validation results are inconsistent. A common criticism of the industry is that research and development (R & D) staff need to pay more attention to the erection of validation status to prevent unnecessary problems. In doing so, it ensures that the validation phase results are consistent with mass production.

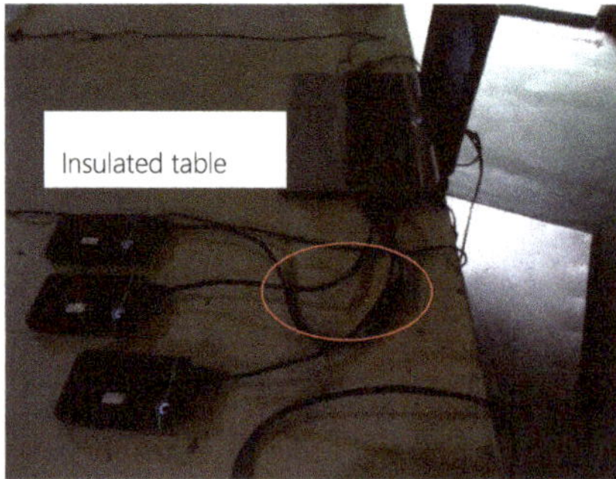

Figure 12. Schematic diagram of staggered wire around the object to be measured.

4. Measurement Data and Experimental Results

In this paper, practicality is introduced to analyze the differences of EMI noise interference countermeasures, as listed in HDMI 1.4 specifications, as follows:

For the discussion and experiment on PCB wiring, this paper mainly focused on examining the distribution of intensity radiation in the near field and EMI results before and after the improvement of HDMI wiring.

Experiment 1 *To examine the near field radiation distribution and EMI measurement results of PCB HDMI region;*

Experiment 2 *To address PCBHDMI noise problem, PCB wiring improvement and measurement of the near field intensity radiation distribution and EMI measurement results are studied.*

4.1. HDMI Circuit Diagram and PCB Layout in This Paper

The HDMI circuit diagram, for this paper, is shown in Figure 13, and the HDMI circuit layout is marked as yellow in Figure 14.

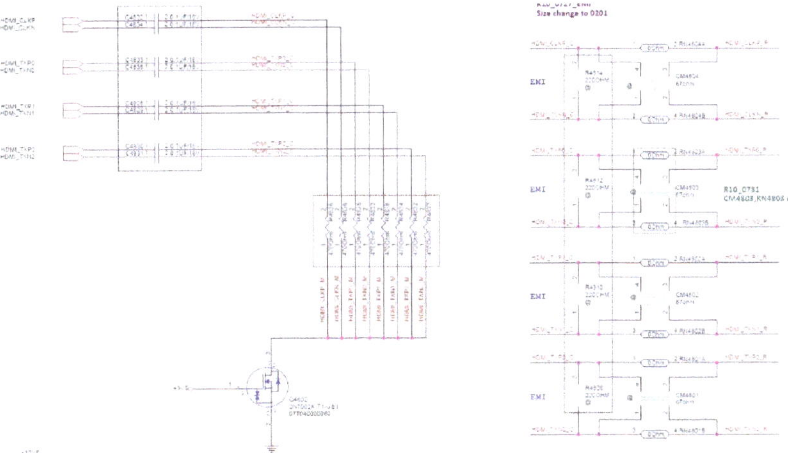

Figure 13. High Definition Multimedia Interface (HDMI) circuit diagram.

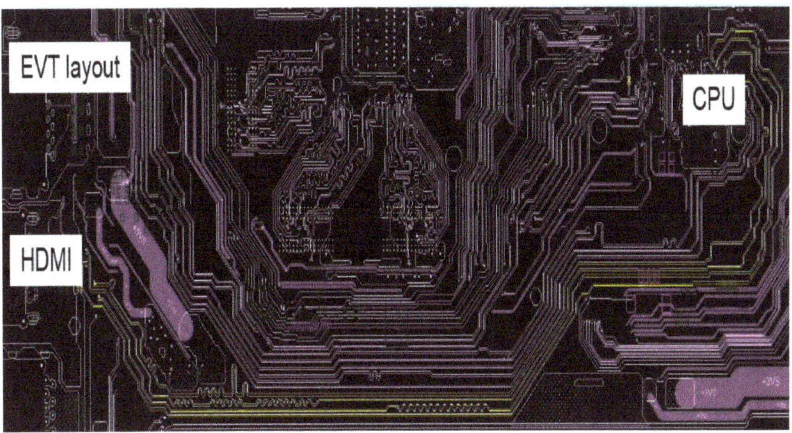

Figure 14. Printed circuit board (PCB) HDMI wiring diagram.

In order to resolve EMI problems, it is common for EMC engineers to employ the near-field probe to identify the source of PCB noise, in order to reduce cost due to the use of a third-party laboratory for a long time. After identification, the tested objects are transferred to the formal site for verification purposes.

In this paper, APREL EM—ISight was applied to perform the near field measurement, as shown in Figure 15 [10]. The instrument relies on mechanical arm instead of hand to hold the receiving antenna (commonly known as the probe). During the process of measurement, the probe will have two directional scannings with a range from 0 degrees to 90 degrees, to simulate the far-field antenna vertical level of polarity, and the area of mechanical arm measurement is set according to the size of the PCB, the displacement of the Probe every time distance, and the minimum moving distance of 1 mm.

4.2. Experiment 1

As revealed by the PCB EMI measurement results, the measured spectrum of only HDMI at 297 MHZ and 891 MHZ EMI energy is strong, as shown in Table 1. While the original design meets the requirements of the laws and regulations, it cannot satisfy the requirements of customers' demand for a limit of –6 dB. Therefore, the EM—ISight for the near field measurement is applied, and the measurement area is shown in Figure 15. The red box marks the HDMI, and almost every intensity distribution is shown in Figure 16.

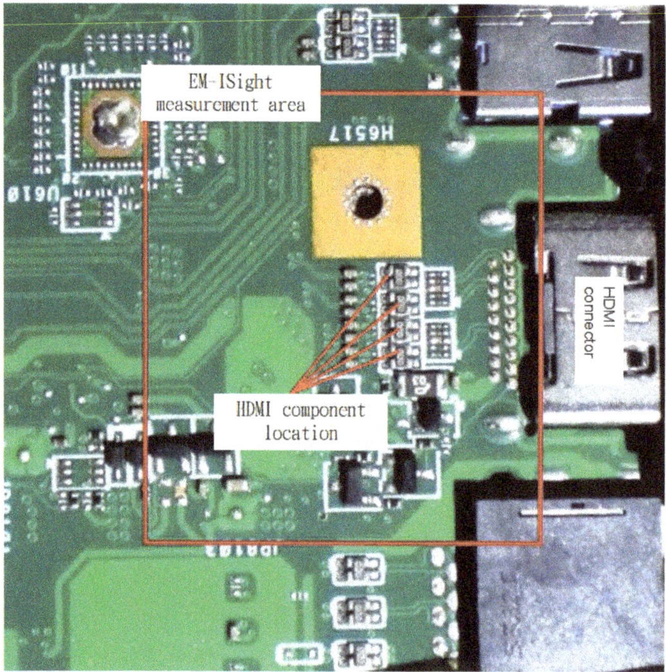

Figure 15. HDMI near field measurement area diagram of PCB initial version.

Layer 1 Peak Magnitude Plots: 890.125000 MHz

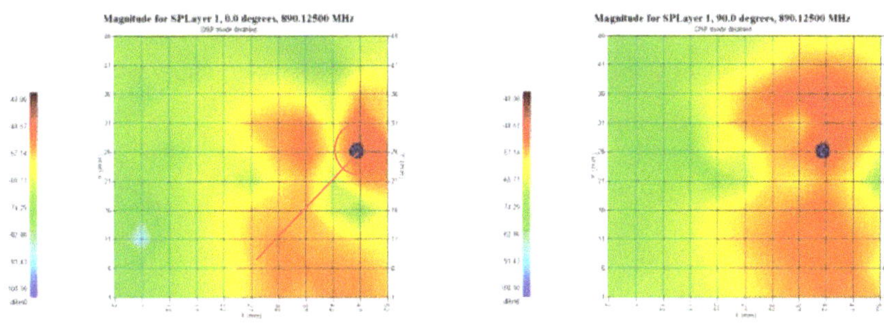

Peak Magnitude: -46.21 dBm0
Loc (X,Y,Z0): 46.49, 25.80, 5.00 (mm)

Peak Magnitude: -48.64 dBm0
Loc (X,Y,Z0): 41.49, 25.81, 5.00 (mm)

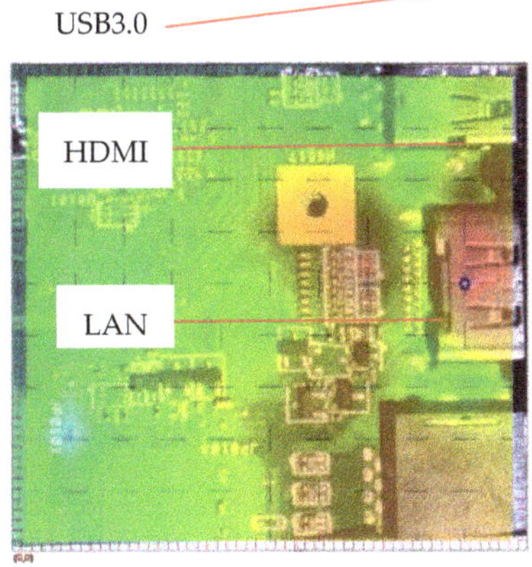

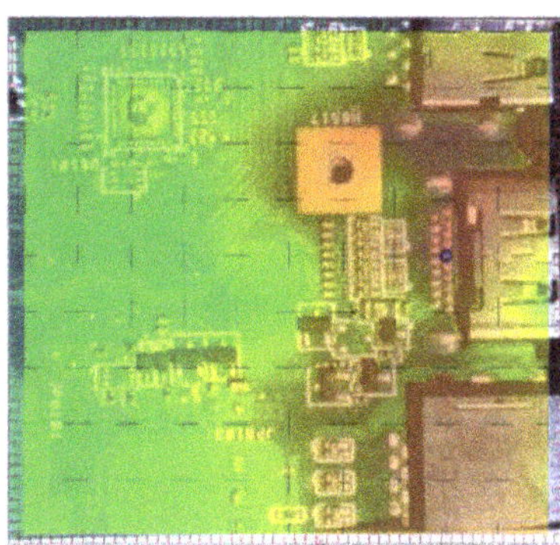

Figure 16. HDMI field intensity distribution map of PCB initial version.

Before its improvement, the PCB layout for EM—ISight instruments is applied to examine the distribution of recent games intensity to find the HDMI port, and there is also a substantial amount of EMI radiation energy. The scattering region has covered the HDMI and the adjacent IO port (LAN and start), and therefore the noise through the wire around the test is easily out of the noise, as shown in Figure 16. As for its color, the red represents more energy, and the blue dot in the graph indicates the strongest power in the area.

4.3. Experiment 2

According to the original PCB layout design, we discovered that there is a large number of snaking wires in the HDMI wiring, power supply, and clock via on the side, as well as extra winding, fine tuning, and straightening. In doing so, the bus length can be reduced, and GND can be added at the differential signal wiring change layer. After the improvement of the PCB HDMI layout, the near field measurement area of the HDMI is indicated by the red box, as shown in Figure 17.

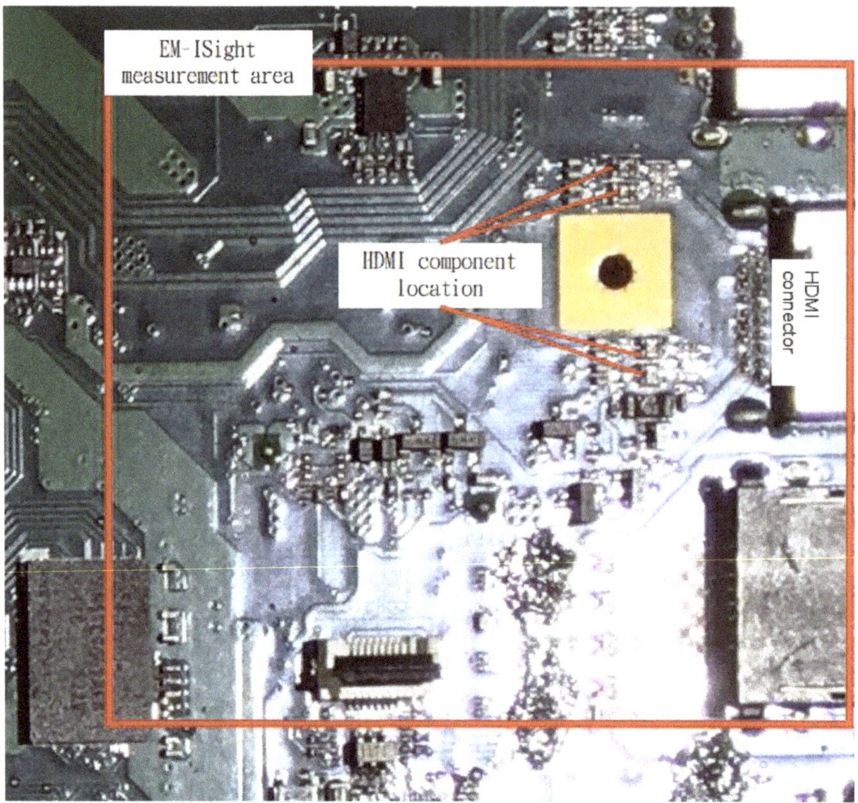

Figure 17. The improved near field measurement area diagram of HDMI layout.

As shown in Figure 18, in addition to the obvious narrowing of the EMI radiation region, the scattering region also concentrates in the HDMI port, and its energy is not coupled to other ports. In doing so, most EMI problems are resolved. After the improvement of the layout, the EMI measurements conform to the customers' requirements of −6 db HDMI frequency band points, as shown in Table 2.

Table 1. EMI measurement data of PCB first edition.

Noise Frequency	297 MHz		594 MHz		891 MHz	
Antenna polarization	H	V	H	V	H	V
Measurement results	−8.767	−1.667	−8.441	−11.815	−2.578	−4.728
						Unit: dBuV/m

Table 2. EMI measurement data after HDMI layout improvement.

Noise Frequency	297 MHz		594 MHz		891 MHz	
Antenna polarization	H	V	H	V	H	V
Measurement results	−6.567	−7.767	−9.741	−10.515	−10.777	−7.629
						Unit: dBuV/m

Layer 1 Peak Magnitude Plots: 890.125000 MHz

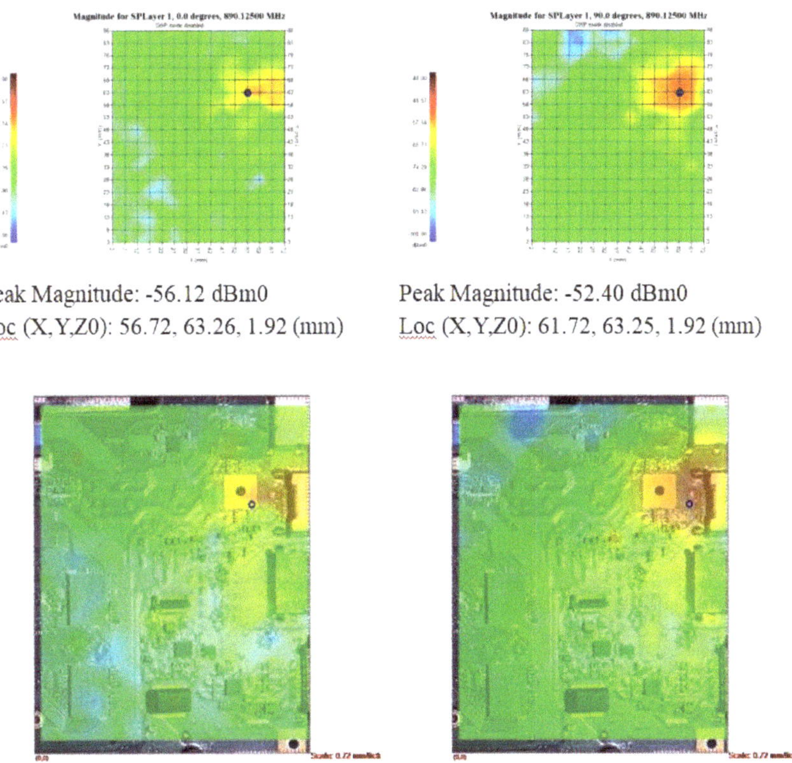

Peak Magnitude: -56.12 dBm0
Loc (X,Y,Z0): 56.72, 63.26, 1.92 (mm)

Peak Magnitude: -52.40 dBm0
Loc (X,Y,Z0): 61.72, 63.25, 1.92 (mm)

Figure 18. Intensity distribution of the near field after HDMI layout improvement.

5. Conclusions

In the laptop industry, low levels of profitability and high-quality design are the norm, but it is challenging to ensure design precision and quality while avoiding the waste of cost. We believe this is a difficulty facing each electronic engineer with common HDMI noise as a starting point. In this paper, the following conclusions are drawn:

1. To resolve the EMC noise problem, first, engineers must clarify and confirm the core of the problem and these problems can be analyzed from three aspects, which are institutional structure, hardware design, and software design;
2. PCB wiring plays a significant role in hardware design. A good PCB layout is effective in reducing the interference of EMI noise radiation or EMS.

In this paper, an APREL EM-Isight near field detection instrument is applied to assist in analyzing the distribution of electromagnetic noise intensity in the near-field. After verification, the PCB wiring proves to be compliant with the specification after adjustments made to the design, and with the EMI requirements for most HD multimedia interface related products.

Author Contributions: W.C., Y.C. and C.-F.H. contributed to the conception of the study. W.C., Y.-T.C. and C.-F.H. contributed significantly to analysis and manuscript preparation; K.-X.H. and C.-F.H. performed the data analyses and wrote the manuscript; C.-C.C. and Y.-T.C. helped perform the analysis with constructive discussions. All authors have read and agreed to the published version of the manuscript.

Funding: This research was funded by Research Initiation Project of Introducing High-level Talents from Beibu Gulf University of China, with grant number as 2017KYQD209.

Conflicts of Interest: The authors declare no conflict of interest

References

1. Hao, X.; Xie, S.; Chen, Z. A Parametric Conducted Emission Modeling Method of a Switching Model Power Supply (SMPS) Chip by a Developed Vector Fitting Algorithm. *Electronics* **2019**, *8*, 725. [CrossRef]
2. Kim, M. Analytical Modeling of Metamaterial Differential Transmission Line Using Corrugated Ground Planes in High-Speed Printed Circuit Boards. *Electronics* **2019**, *8*, 299. [CrossRef]
3. Capriglione, D.; Chiariello, A.G.; Maffucci, A. Accurate Models for Evaluating the Direct Conducted and Radiated Emissions from Integrated Circuits. *Appl. Sci.* **2018**, *8*, 477. [CrossRef]
4. Ali, I.; Rikhan, B.S.; Kim, D.G.; Lee, D.S.; Rehman, M.R.; Abbasizadeh, H.; Asif, M.; Lee, M.; Hwang, K.C.; Yang, Y.; et al. Design of a Low-Power, Small-Area AEC-Q100-Compliant SENT Transmitter in Signal Conditioning IC for Automotive Pressure and Temperature Complex Sensors in 180 Nm CMOS Technology. *Sensors* **2018**, *18*, 1555. [CrossRef] [PubMed]
5. Huang, C.-Y.; Chen, J.-H. Development of Dual-Axis MEMS Accelerometers for Machine Tools Vibration Monitoring. *Appl. Sci.* **2016**, *6*, 201. [CrossRef]
6. IEC Website. Available online: https://webstore.iec.ch (accessed on 1 July 2019).
7. Mike, EMC Practical Design. Available online: http://in.ncu.edu.tw/ncume_ee/emc9620142design.pdf (accessed on 1 July 2019).
8. Kwon, S.Y.; Yoo, M. Evaluation of Dynamic Soil-Pile-Structure Interactive Behavior in Dry Sand by 3D Numerical Simulation. *Appl. Sci.* **2019**, *9*, 2612. [CrossRef]
9. Kuo, M.-T.; Tsou, M.-C. Novel Frequency Swapping Technique for Conducted Electromagnetic Interference Suppression in Power Converter Applications. *Energies* **2017**, *10*, 24. [CrossRef]
10. APREL Website. Available online: https://www.aprel.com/em-isight (accessed on 1 July 2019).

 *electronics*

Article

Stereo Vision-Based Object Recognition and Manipulation by Regions with Convolutional Neural Network

Yi-Chun Du, Muslikhin Muslikhin, Tsung-Han Hsieh and Ming-Shyan Wang *

Department of Electrical Engineering, Southern Taiwan University of Science and Technology, 1 Nan-Tai St., Yung Kang District, Tainan City 710, Taiwan; terrydu@stust.edu.tw (Y.-C.D.); muslikhin@uny.ac.id (M.M.); henry129123@gmail.com (T.-H.H.)
* Correspondence: mswang@stust.edu.tw; Tel.: +886-6-2533131 (ext. 3328)

Received: 10 December 2019; Accepted: 20 January 2020; Published: 24 January 2020

Abstract: This paper develops a hybrid algorithm of adaptive network-based fuzzy inference system (ANFIS) and regions with convolutional neural network (R-CNN) for stereo vision-based object recognition and manipulation. The stereo camera at an eye-to-hand configuration firstly captures the image of the target object. Then, the shape, features, and centroid of the object are estimated. Similar pixels are segmented by the image segmentation method, and similar regions are merged through selective search. The eye-to-hand calibration is based on ANFIS to reduce computing burden. A six-degree-of-freedom (6-DOF) robot arm with a gripper will conduct experiments to demonstrate the effectiveness of the proposed system.

Keywords: regions with convolutional neural network (R-CNN); adaptive network-based fuzzy inference system (ANFIS); 6-DOF robot arm

1. Introduction

Various types of vision technology, such as image measurement, stereo vision, structured light, time of flight, and laser triangulation are widely used in the field of robotics [1,2]. Due to its superior features in safety, scope, and accuracy, stereo vision is more commonly used. Stereoscopic vision is an imaging technique that compares two images of the same scene and takes object depth from the camera image [3–5]. It has been used in industrial automation and applications, for example, box picking and placing, three-dimension (3D) object positioning and recognition, as well as volume measurement [6,7].

Applying stereo vision to a robotic manipulation system typically requires camera calibration and coordinate frame transformation between the stereo camera and the robotic arm. Through MATLAB, the intrinsic and extrinsic parameters required for camera calibration are obtained [1]. The eye-to-hand calibration is used to calculate the relative 3D position and orientation between the camera and the robot arm [8,9]. On identifying objects, there are many techniques for object detection proposed in the literature, for example, sliding window classifiers, pictorial structures, constellation models, and implicit shape models [10]. Sliding window classifiers have been widely used in the fields of detection of faces, pedestrians, and cars because they are especially well suited for rigid objects. Subsequently, the convolutional neural network (CNN) [11–13] is one of the most common algorithms. It extracts the image features through the convolutional layer and marks them. The CNN is a special class of neural networks that is best suited for the intelligent processing of visual data. It is a variation of the architecture of a multilayer neural network and generally includes the convolutional layer, pooling layer, flatten layer, fully connected layer, and output layer.

Since a fixed-size frame is used to sweep the entire image one by one, and the size of the target object is unpredictable, it is necessary to use a lot of convolutional layers to perform operations, which result in a longer operation time. As a consequence, the methodology of regions with CNN (R-CNN) [14] is proposed. Regions have not been popular as features due to their sensitivity to segmentation errors. However, region features are appealing because they encode the shape and scale information of objects naturally and are only mildly affected by background clutter [15]. Similar pixels are segmented by the image segmentation method [16], and then the similar regions are merged by selective search [17]. These regions are finally merged into one. The approximate number of frames generated during the merge process is 2000. This method can reduce the amount of input data to speed up the training time. An effective region-based solution for saliency detection is first introduced. Then, the achieved saliency map is applied to better encode the image features for solving object recognition tasks [18]. Superpixels based on an adaptive mean shift algorithm as the basic elements for saliency detection are extracted to find the perceptually and semantically meaningful salient regions. In addition, the Gaussian mixture model (GMM) clustering is used to calculate spatial compactness to measure the saliency of each superpixel. A region-based object recognition (RBOR) method is proposed to identify objects from complex real-world scenes via performing color image segmentation by a simplified pulse-coupled neural network (SPCNN) for the object model image and test image. Then, a region-based matching between them is conducted [19]. Cai et al. [20] proposed a mitosis detection method for breast cancer histopathology images of TUPAC (Tumor Proliferation Assessment Challenge) 2016 and ICPR (International Conference on Pattern Recognition) 2014 datasets by applying the modified R-CNN whose backbone feature extractor is the Resnet-101 network pre-trained on the ImageNet dataset. For traffic surveillance systems, Murugan et al. [21] employed techniques of box filter-based background subtraction to identify the moving objects by smoothing the pixel variations due to the movement of vehicles and R-CNN for the classification of variant moving vehicles. Moreover, region proposals and support vector machine (SVM) classifier are used to reduce the computational complexity and the recognition of vehicles. In order to improve the efficiency of the service robot's target capture task, Shi et al. [22] used Light-Head R-CNN to replace the mask branch into the Mask R-CNN network, increased R-CNN subnet and regions of interest (RoI) warping, and adjusted the proportion of the anchor in the region proposals network (RPN). They claimed that the detection time is reduced by more than two times. For recognizing values of pointer meters, He et al. [23] proposed the Mask R-CNN with a principal component analysis (PCA) algorithm to fit the pointer binary mask and PrRoIPooling to improve the instance segmentation accuracy.

Consequently, in this paper, we decided to use a stereo camera for R-CNN. The main reason is that the verification process is at least to recognize one target by pair cameras. On the other hand, with a stereo camera, the target detection process will be more accurate because of the triangulation among the right camera, left camera, and dataset. In practice, we use the hybrid object recognition algorithm, firstly using R-CNN to determine the triangle or square target. Secondly, the object recognition algorithm is also used to ascertain the coordinates of triangles or squares in an image frame, including the bounding box area (height and width). The stereo camera at an eye-to-hand configuration and a six-degree-of-freedom (6-DOF) robot arm with the gripper are firstly applied to capture the images of the target object in this paper. The eye-to-hand calibration is based on adaptive network-based fuzzy inference system (ANFIS). An algorithm of regions with convolutional neural network (R-CNN) is developed for image processing to extract the specific features of the target object, such as the shape, features, and centroid of the object. Therefore, we confer on a high accuracy to estimate the XYZ position using ANFIS and the ability of the system to distinguish triangles, squares, or other objects according to the environment settings using R-CNN. Finally, the experiments demonstrate the effectiveness of the proposed system.

In this paper, the stereo vision-based object manipulation is introduced in Section 2. In Section 3, the method of regions with convolutional neural network is described. Experimental results are shown in Section 4. Finally, conclusions are given in Section 5.

2. Stereo Vision-Based Object Manipulation

The stereo vision-based object manipulation system includes four tasks, stereo camera calibration, object feature extraction, pose estimation, and eye-to-hand calibration using adaptive network-based fuzzy inference system (ANFIS). The configuration scheme of the stereo vision is shown in Figure 1 [24]. This consists of two cameras with the same parameters to be obtained by stereo camera calibration in MATLAB [1]. Given a reference point $P(X_p, Y_p, Z_p)$, the projections in image plan 1 is $p_1(x_1, y_1)$ and in image plan 2 is $p_2(x_2, y_2)$, where f is the focal length; $d = x_1 + x_2$ is the parallax; and b is the distance of two camera's optical centers [25]. Referring to Figure 2 and the principle of similar triangles, we can get the depth Z and the X and Y coordinates of point P from Equations (1) to (3), respectively:

$$Z = b * f / d \tag{1}$$

$$X = Z * x_1 / f \tag{2}$$

$$Y = Z * y_1 / f \tag{3}$$

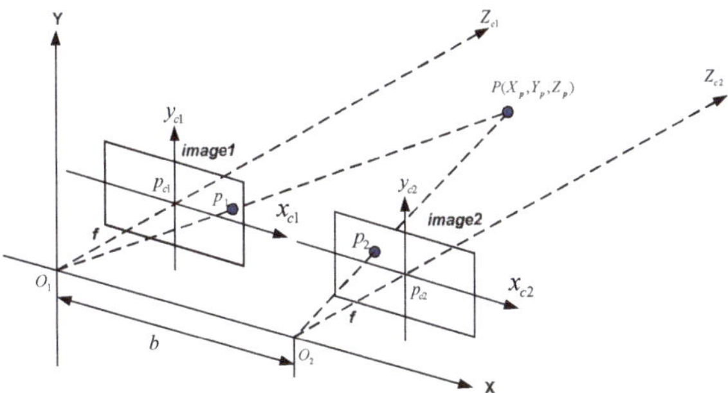

Figure 1. Configuration scheme of stereo vision.

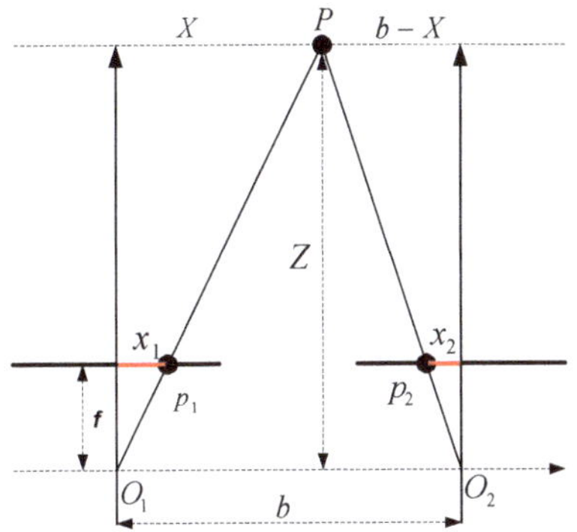

Figure 2. Triangulation scheme of stereo vision.

Before estimating the actual object distances, camera calibration is essential for determining the intrinsic and extrinsic camera parameters in computer vision tasks. (α, β, γ, u_0, v_0) stand for the intrinsic parameters, where an image plane includes u and v axes; (u_0, v_0) are the coordinates of the principal point; α and β are the axial scale factors; and γ is the parameter describing the skewness. (R, t) represent the extrinsic parameters, meaning the rotation and translation of the right camera with respect to (w.r.t.) the left camera, respectively [26].

The coordinate frame systems of the stereo vision-based object manipulation system and their relationships ($^{B}\xi_{E}$: end-effector coordinate frame w.r.t. robot base frame, $^{E}\xi_{G}$: gripper to end-effector, $^{C}\xi_{T}$: targeted object to camera, and $^{B}\xi_{T}$: targeted object to robot base) are depicted in Figure 3 [24]. $^{B}\xi_{C}$ is the camera coordinate w.r.t. robot base that will be obtained using ANFIS so that the targeted object to robot base $^{B}\xi_{T}$ can be found based on the information of $^{C}\xi_{T}$, as shown in Figure 4.

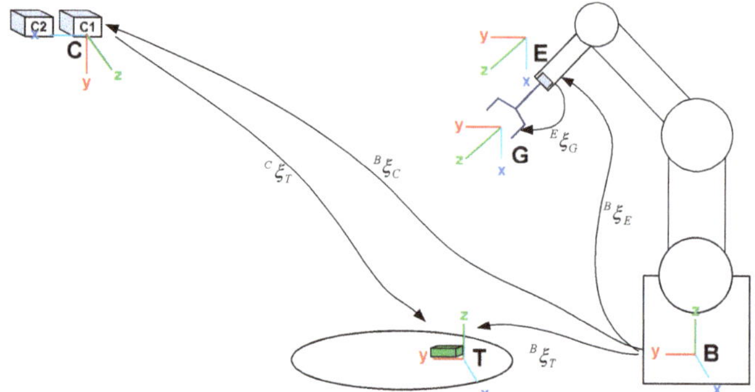

Figure 3. Coordinate transformation relationship.

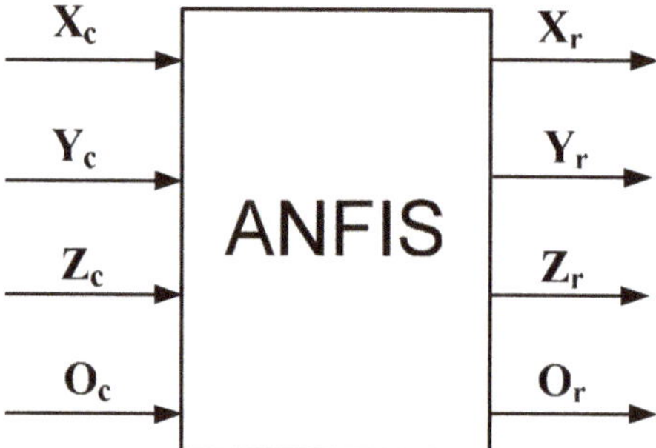

Figure 4. Block of the proposed adaptive network-based fuzzy inference system (ANFIS).

The ANFIS architecture consists of a fuzzy layer, product layer, normalized layer, de-fuzzy layer, and summation layer. Figure 5 [24] shows the structure of a two-input type-3 ANFIS with Takagi-Sugeno if-then rule [3] as follows, in which the circle and square respectively indicate a fixed node and an adjustable node,

$$\text{IF } x \text{ is } A_i \text{ and } y \text{ is } B_i \text{ and } z \text{ is } C_i \text{ THEN } Z_i = Z_i + p_i x + q_i y + r_i z + s_i \tag{4}$$

where x and y stand for input variables; A_i and B_i ($I = 1,2$) are linguistic variables that cover the input variable universe of discourse; $z_i(x, y)$ ($i = 1 : 4$) mean output variables; and p_i, q_i and r_i ($I = 1{:}4$) are linear consequent parameters. The layers and their functions can be described as follows:
Layer 1: Fuzzification Layer

The fuzzification is realized by the corresponding membership function, denoted by the node. The membership functions generally include adjustable parameters to provide adaptation. The Gaussian membership functions (MFs) of fuzzy sets A_i and B_i, ($i = 1, 2$), $\mu_{Ai(x)}$, $\mu_{Bi(y)}$, and $\mu_{Ci(z)}$, are considered here and shown in Equation (5),

$$gaussmf(x, c_i, s_i) = e^{-\frac{(x-c_i)^2}{2s_i^2}} \tag{5}$$

where x is the input, and c_i and s_i are the center and standard deviation that change the shape of the MF.

Layer 2: Product Layer

The T-norm operation is used to calculate the firing strength of a rule via multiplication:

$$\omega_i = \mu_{Ai(x)}\mu_{Bi(y)}\mu_{Ci(z)}. \tag{6}$$

Layer 3: Normalization Layer

The ratio of a rule's firing strength to the total of all firing strengths is calculated via:

$$\overline{\omega_i} = \frac{\omega_i}{\sum_{i=1}^{6}\omega_i} = \frac{\omega_i}{\omega_1 + \omega_2 + \omega_3 + \omega_4 + \omega_5 + \omega_6}. \tag{7}$$

Layer 4: Defuzzification Layer

The linear compound is obtained from the inputs of the system as THEN part of fuzzy rules as:

$$\overline{\omega_i}Z_i(x, y, z) = \overline{\omega_i}(p_i x + q_i y + r_i z + s_i) \tag{8}$$

where $\overline{\omega_i}$ is the output of layer 3 and $\{p_i x + q_i y + r_i z + s_i\}$ is the consequent parameter set.

Layer 5: Summation Layer

A fixed node calculates the overall output as the summation of all incoming inputs:

$$Z = \sum_{i=1}^{6} \overline{\omega_i}Z_i(x, y, z) = \frac{\omega_1 Z_1 + \omega_2 Z_2 + \omega_3 Z_3 + \omega_4 Z_4 + \omega_5\omega Z_5 + \omega_6 Z_6}{\omega_1 + \omega_2 + \omega_3 + \omega_4 + \omega_5 + \omega_6}. \tag{9}$$

The ANFIS block shown in Figure 4 consists of the inputs of $^C\xi_T$ or (X_c, Y_c, Z_c, O_c) which are the orientation and coordinates of the targeted object base w.r.t. the camera coordinate frame, which is found by the stereo vision system for computing the solution of camera to robot arm calibration, and the outputs of $^B\xi_T$ or (X_r, Y_r, Z_r, O_r), which are the orientation and coordinates of the targeted object w.r.t. the robot base, which is acquired by positioning the end-effector to the desired object position using the teaching box of the robot arm controller. In addition, $^B\xi_C$ is the camera coordinate w.r.t. the robot base that will be obtained by training the ANFIS.

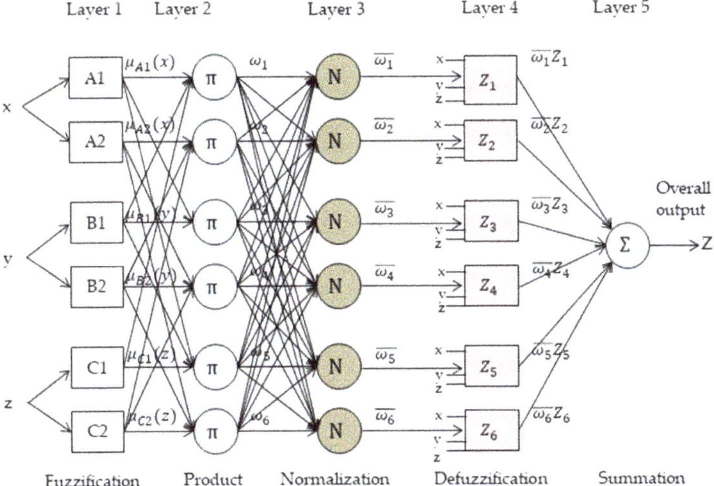

Figure 5. ANFIS structure.

3. Regions with Convolutional Neural Network (R-CNN)

The CNN is a special class of neural networks that is best suited for the intelligent processing of visual data. It is a variation of the architecture of a multilayer neural network and is generally composed of many neural layers, including a convolutional layer, pooling layer, fully connected layer, and output layer, as shown in Figure 6 [27]. On identifying the object from a picture and marking the location, the easiest way is to use the concept of sliding a window, which is a fixed-size frame, sweeping the entire picture one by one. The output is dropped into the CNN each time to determine the classes. However, the number and size of objects to be identified are unpredictable. In order to maintain high spatial resolution, the CNN usually has two or more convolutional layers and pooling layers, which possess huge data at each layer input and result in computational complexity during processing.

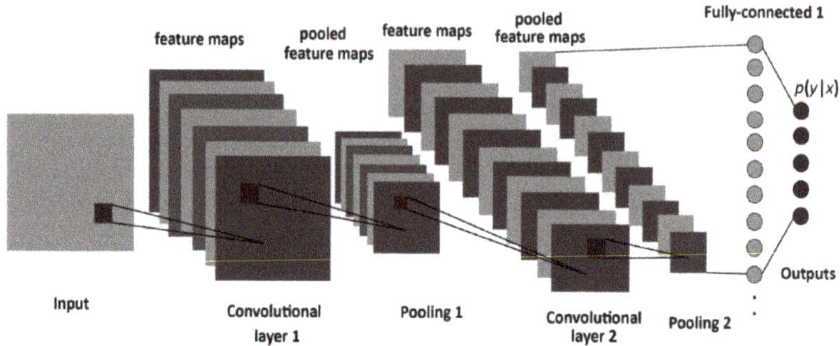

Figure 6. General configuration of a convolutional neural network (CNN).

A convolutional layer has a set of matrix filters that are applied to images and isolate a feature. A combination of several of these layers will build up new signs for the previous ones with signs of a lower order. In practice, this means that the network is trained to see complex features, which is a composition of simpler ones. In the process, the rectified linear unit (ReLU) is used to remove negative values for a sharper object shape. The sub-sample layer represents a layer without training, where the images are filtered with the highest value of the pixel in the window and the others ignore it. Thus, the image decreases in size and only the most significant features are left, regardless of the location. The last layer is a fully connected network, where each neuron takes in the inputs from all the outputs of the neurons of the previous layer. The obtained feature map is reduced by pooling to reduce the size of the data. The most commonly used method is max pooling. During the pooling process, there is no impact on the image, and it has a good anti-aliasing function. Before entering the fully connected layer, it is necessary to flatten it and turn the data into a straight line.

The method of region identification of regions with CNN (R-CNN) [15] is used to solve the above-mentioned problem of CNN. The image segmentation method [16] is used for selective search [17] on the input image. Then, about 2000 region proposals are selected and act as the inputs to convolutional neural networks to extract features and distinguish the regions. In this paper, in principle, we do something similar to [16] and [17], which conduct segmentation to find the centroid (XY coordinate) of an object. However, as an additional proposal in this paper, R-CNN is used to distinguish the triangle and square blocks captured from the stereo camera. Finally, the regression is used to correct the position of the frame.

An image is formed by interconnecting pixels. The pixels are also called vertices (V). The lines connecting pixels are called edges (E). Let G = (V, E) be an undirected graph with vertices $v_i \in V$, and edges $\left(v_i, v_j\right) \in E$ corresponding to adjacent vertices, each having a weight $w\left(v_i, v_j\right)$. There are paths at any two vertices in the graph, but those without loops are called trees. The tree with the smallest sum of the edges' weights is called the minimum spanning tree (MST). The image segmentation method initializes each pixel as an independent vertex at the initialization time, and uses Equation (10) to calculate the similarity between each pixel,

$$pixel\ distance = \sqrt{(r_1 - r_2)^2 + (g_1 - g_2)^2 + (b_1 - b_2)^2} \tag{10}$$

where r_i, g_i, and b_i are the three color values of the pixel, respectively. To identify the similarity between two regions or a region and a pixel, a threshold is set to consider the similarity between two parts. Below the threshold, the two regions are merged into one region; thus, the threshold needs to be changed in accordance to different areas. The intra-class variation of Equation (11) is used to find the largest dissimilarity in the MST, which is also the largest luminance difference in a region,

$$Int(V) = \max w\big(v_i, v_j\big). \tag{11}$$

The inter-class difference method of Equation (12) will obtain the dissimilarity of the edges with the least dissimilarity between the two regions, that is, the most similar in the two regions,

$$Diff(V_1, V_2) = \min w\big(v_i, v_j\big). \tag{12}$$

$Int(V_1)$ and $Int(V_2)$ are the maximum differences that can be accepted by the regions V_1 and V_2, respectively, and they are larger than or equal to $Diff(V_1, V_2)$. When both regions can meet the requirements, they are merged into one region. Otherwise, they cannot be merged. Finally, using the above method, the original image can be divided into different color regions for segmentation.

The selective search first uses image segmentation to get the color regions $R = \{r_1, \ldots, r_n\}$ in the image; then, it calculates the similarity $s\big(r_i, r_j\big)$ of each adjacent region and merges the two regions with the highest degree each time. The entire image is finally merged into a few regions. The algorithm for the similarity of each region may be based on color, texture, size, and fit.

On the training data, we mark the target regions and use the labeled regions as positive samples. A selective search is used to generate the hypothetical region of the target. The region with the overlap degree between 20% and 50% of the target label region is marked as a negative sample. Then, the extracted feature is input for training. The false positive is added to the training samples to increase the number of difficult samples after each training finishes. Then, training is conducted again until convergence happens.

As for verifying, Precision (Equation (13)), Recall (Equation (14)), and Accuracy (Equation (15)) are used to describe the performance [26], where TP denotes the number of true positives, FP denotes the number of false positives, FN denotes the number of false negatives, and TN indicates the correct rejection of results (triangle or square), respectively [28],

$$Precision = \frac{TP}{TP + FP} \tag{13}$$

$$Recall = \frac{TP}{TP + FN} \tag{14}$$

$$Accuracy = \frac{TP + TN}{TP + TN + FP + FN}. \tag{15}$$

Hence, the disparity between our previous research [24] and the current one is that we include R-CNN to recognize objects. If the object is identified successfully, the gripper will grasp; see Table 1.

Table 1. Comparison of previous with the proposed method. 6-DOF: six degrees of freedom, ANFIS: adaptive network-based fuzzy inference system, R-CNN: regions with convolutional neural network.

	Previous Method [24]	**Proposed Method**
Arm robotic control	inverse kinematic-6 DOF	inverse kinematic-6 DOF
Vision structure	eye to hand (stereo camera)	eye to hand (stereo camera)
Pose estimation	ANFIS	ANFIS
Centroid detection	HSV masking	image boundary
Object recognition	image segmenting	R-CNN
Object characteristic	cylindrical with some colors	triangle and square (same color)

4. Experimental Results

We conducted experiments to validate the proposed method. The experimental setup is shown in Figure 7, which includes a set of stereo cameras consisting of two identical Logitech C310 cameras and the targets placed anywhere in the work area. On the other hand, the robotic arm controller uses the built-in software development commands of MATLAB to drive the robotic arm through a serial communication interface and implement the proposed method through a graphical user interface (GUI). We estimated the pose of the object by using a calibrated stereo vision system and its coordinates relative to the base frame of the robot. Then, a target grabbing task is performed using a three-finger gripper and a 6-degree-of-freedom (6-DOF) robot to confirm the performance of the 3D target pose estimation in the robotic coordinate system.

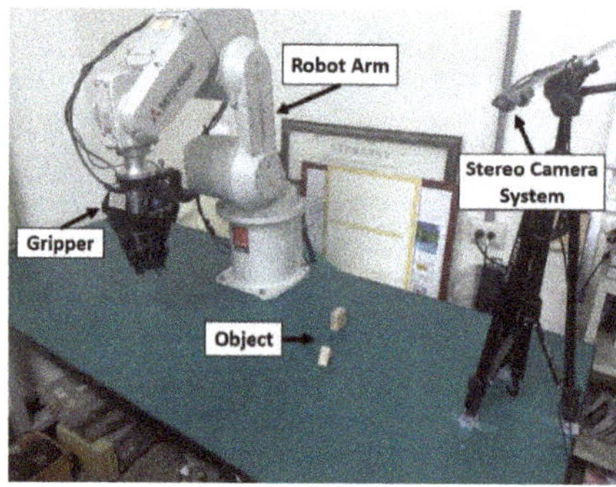

Figure 7. Stereo camera, robot arm, gripper, and their working space.

According to the experimental results, the calibration of the stereo camera is successful and the internal and external parameters can be used for the triangulation process. The intrinsic and extrinsic camera parameters are first computed by stereo camera calibration, and then eye-to-eye calibration is performed. In this paper, the method proposed in [2] and the classic black-and-white checkerboard is used to calibrate a stereo camera system, which was built with two cameras with a baseline of 92 mm. The checkerboard has 63 square blocks (9×7 patterns) whose dimensions are 40 mm $\times$ 40 mm. For the calibration process, each camera captures 16 different positions and orientations of the 640×480 pixels image of the board and loads them into MATLAB. The corners of the checkerboard are detected by sub-pixel precision as input to the calibration method. The outputs include the internal matrix and the outer matrix of the two cameras and perspective transformation matrix. All of them are required to re-project the depth data to the real-world coordinates. Camera calibration is an essential part of robotic vision, but it is only a portion of this study. On the other hand, calibration is necessary to reset the camera back into its standard conditions (α, β, γ, u_0, v_0, R, t). Thus, the right and left camera are valid in estimating the position of $X_t Y_t Z_t$ (target world). In the end, the pose estimation of the left camera, the pose estimation of the right camera, and the target world dataset will be compared and triangulated as

$X_r Y_r Z_r$(robot world). To make sure the stereo camera is working validly, we include the stereo camera parameters in each, taking a picture. In other words, we do not use autofocus when snapping targets.

Figure 8 illustrates the difference in orientation (angle), which in this paper is known by calculating a number of the major ellipse axis to the x-axis. After calibrating the stereo camera, we take pictures from different angles to identify the target object as shown in Figure 9 and use the built-in Image Labeler of MATLAB to capture the region of interest (ROI), in which the object will be identified. As shown in Figure 10, the R-CNN is used to distinguish objects to be grasped, which are a triangle or square. After the training by R-CNN is completed, it can be tested to recognize at any position or at different angles of the object and the possibility of the targeted object (confidence). In terms of the level of confidence in our study, we set a minimum limit of 80%. If the detection results are less than that value, then the target will not be held by the gripper.

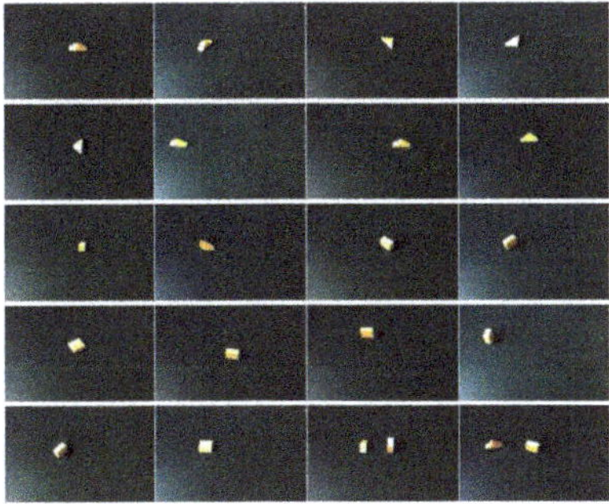

Figure 8. Views from various angles of the object.

After calibrating the internal and external camera parameters of the stereo camera, the image processing system will perform the tasks of feature extraction and pose estimation. Figure 11 shows estimations of postures at each step of the target feature extraction in the two cameras. First, two cameras capture the image pair at the same time, and, based on the color, the HSV (Hue, Saturation, Value) space threshold is used to extract the target from the image and locate the boundary. Next, the boundary target and the centroid of the positioned target in the image pair are searched. Finally, the position of the target object will be determined based on the estimated centroid.

The ANFIS structure of the first-order Sugeno fuzzy system is used to perform eye-to-hand calibration training, and three, five, and seven Gaussian membership functions are respectively used to calibrate the position of the stereo camera relative to the robot arm. The centroid point of the three-dimensional object is calculated by the stereo vision system and input into ANFIS for training.

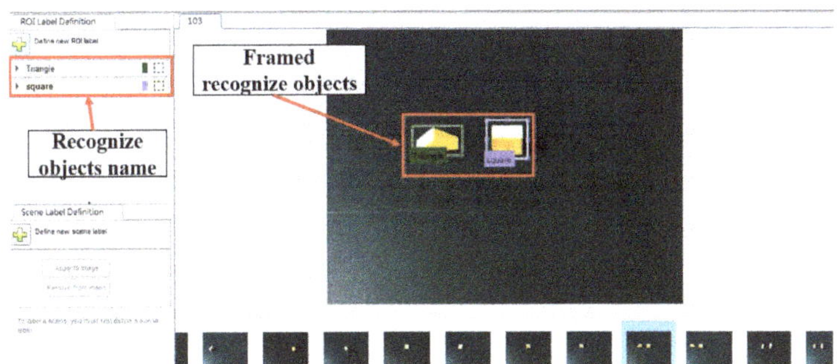

Figure 9. Marking objects.

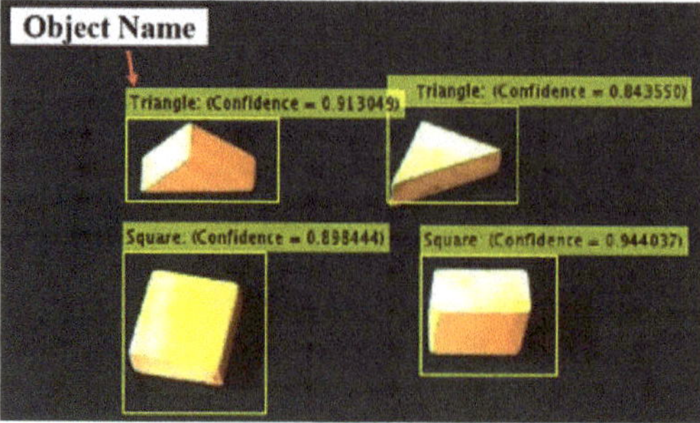

Figure 10. Training results.

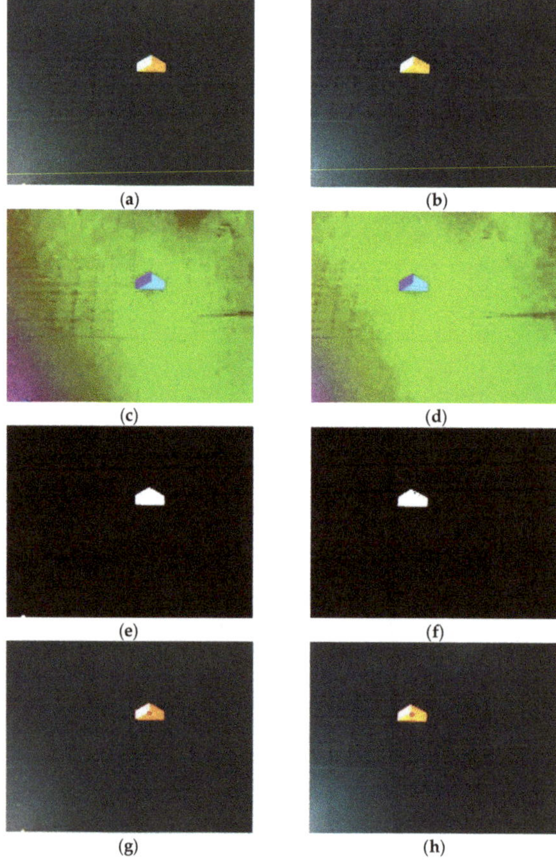

(a)

(b)

(c)

(d)

(e)

(f)

(g)

(h)

Figure 11. Target feature extraction and pose estimation process: (**a**) Color image taken by the left camera, (**b**) Color image taken by the right camera, (**c**) Image of the left camera on the HSV space, (**d**) Image of the right camera on the HSV space, (**e**) Filtered image of the left camera, (**f**) Filtered image of the right camera, (**g**) Object pose estimation on the left camera, and (**h**) Object pose estimation on the right camera.

After the training process is finished, the ANFIS will learn the input and output mapping and test it with different test data. Table 2 shows the comparison of details and errors between different MF training results. It is found that the training error results obtained using the five membership functions were the smallest compared to the other cases. Figure 12 shows that the training error of the orientation data is 0.28923 and is reached in approximately 7000 epochs during ANFIS training. This value indicates that the target direction can be estimated with the ANFIS structure.

Table 2. ANFIS training error.

No. of Input Data	No. of MFs	Training Error (mm)			Average Error (mm)
		X	Y	Z	
	3	0.086907	0.08204	0.029948	0.066298
81	5	0.002016	0.002831	0.002224	0.002357
	7	0.014919	0.015392	0.013941	0.014751

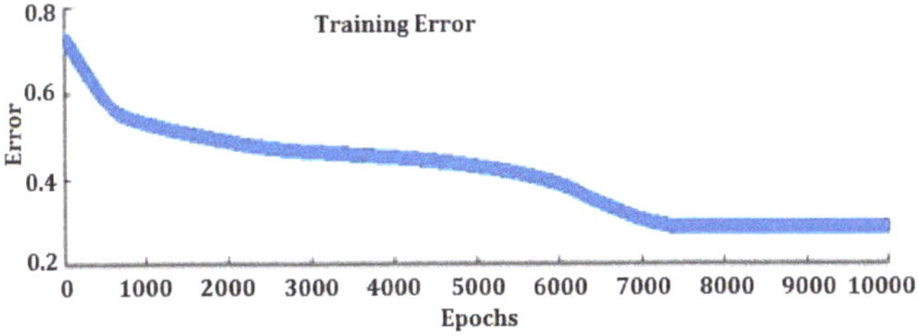

Figure 12. Minimum training error of ANFIS in the target orientation.

The target object identification and pose estimation experiments are conducted to validate the system performance and its orientation in the camera coordinate system, as shown in Figures 13 and 14. Figure 13 shows the object name, location, and the direction estimates, that is, triangle (object name), 38.9 mm (x axis), 37.1 mm (y axis), 686.5 mm (z axis), −2.8° (orientation). Since the x, y, and z coordinates are known, then with inverse kinematics, these three points are enough to be transformed into six movements at each joint of the 6-DOF manipulators. Two different objects at any position and orientation within the workspace of the camera coordinate system have been successfully identified and detected in Figure 14. A triangular object was detected at −69.9 mm (x axis), −16.4 mm (y axis), 717.6 mm (z axis), and 30.8° (orientation), and a square object (blocks) was detected at 33.3 mm (x axis), 43.9 mm (y axis), 642.2 mm (z axis), and 7.3° (orientation). Then, the tasks of target grabbing, picking, and placing are shown in Figure 15.

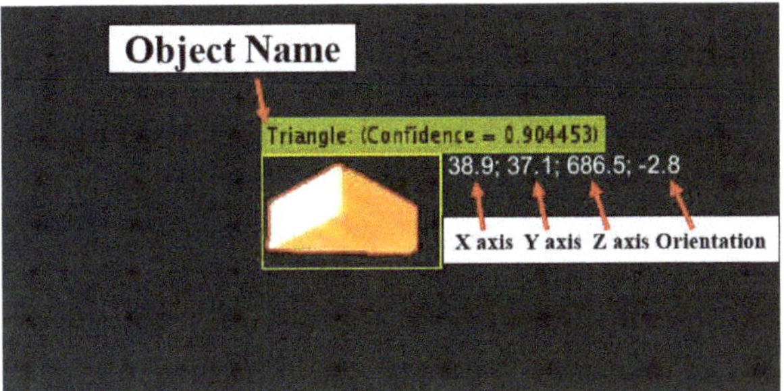

Figure 13. Object name and location and their orientation estimation results.

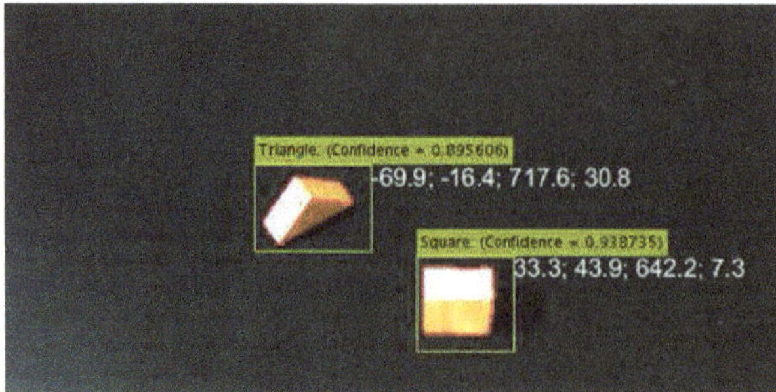

Figure 14. Estimation results of two object names and positions and their orientations.

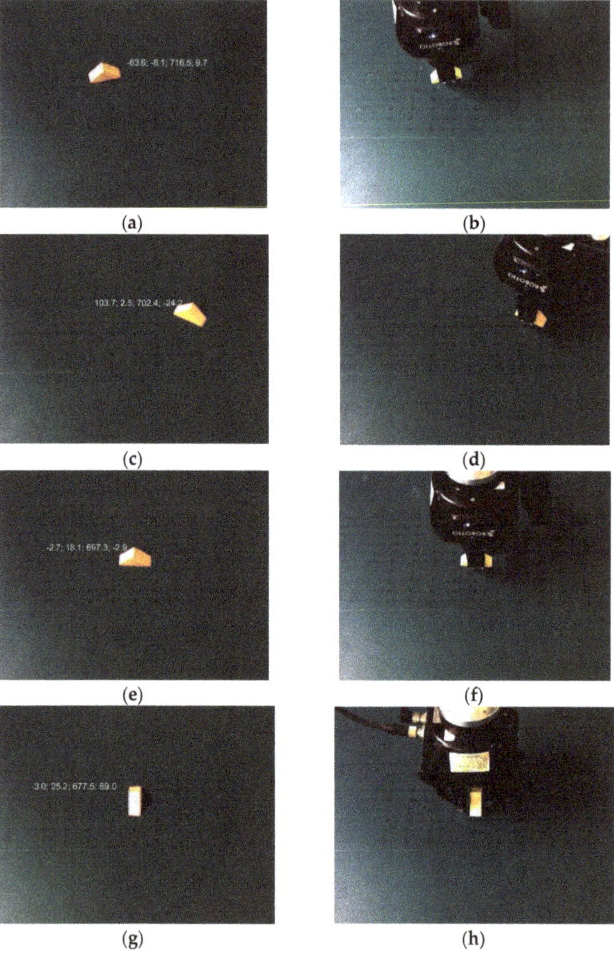

Figure 15. Estimations of the object and the gripper (**a**) The measured orientation of the object is 85.85°, (**b**) The orientation of the gripper is 83.71°, (**c**) The measured orientation of the object is 117°, (**d**) The orientation of the gripper is 119.95°, (**e**) The measured orientation of the object is 93.78°, (**f**) The orientation of the gripper is 93°, (**g**) The measured orientation of the object is 6.23°, (**h**) The orientation of the gripper is 6°.

Table 3 shows the actual values and measured values for four different cases. The absolute error of the orientation and averaged absolute position error are also included. The results demonstrated that the gripper can successfully reach the target object according to the measurements of the position and orientation of the object. The manipulation system shows good performance of the 3D object pose estimation and grabbing in applications. The corresponding GUI user interface is shown in Figure 16.

Table 3. Test results of object position and orientation.

No.	Actual Coordinates and Orientation				Measured Coordinates and Orientation				Absolute Orientation Error (°)	Absolute Averaged Position Error (mm)
	X	Y	Z	Or	X	Y	Z	Or		
1	267.91	327.52	289.49	−83.71	264.83	325.45	289.49	−85.85	2.14	2.575
2	106.12	345.96	294.68	−119.95	108.32	347.74	294.68	−117	2.95	1.99
3	210	360.59	287.75	−93	209.33	360.59	287.75	−93.78	0.78	0.335
4	213.52	367.92	280.85	−6	210.04	368.09	280.85	−6.23	0.23	1.655

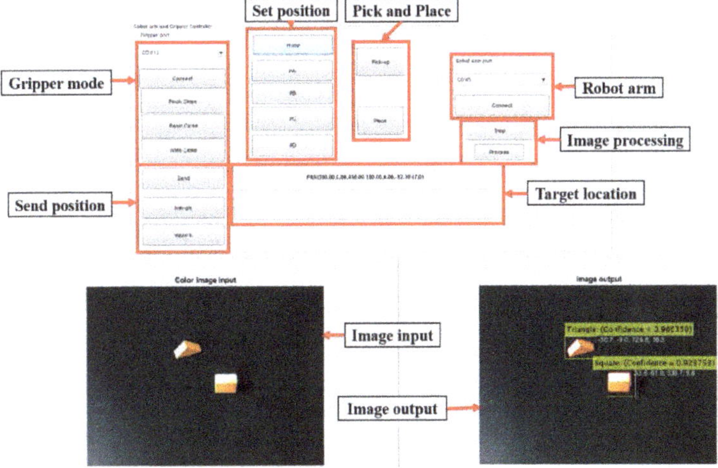

Figure 16. User interface.

After determining the scope of the work area, image processing techniques will be used to distinguish all the objects in the range and the background. The coordinates of the camera relative to the object are obtained by triangulation. The names of all the objects in the range can be known through the use of R-CNN. ANFIS will convert the camera coordinates to the coordinates of the robotic arm. Figure 17 shows the sequence in estimating the position of XYZ + O. In the beginning, we called the stereo camera parameters from the camera calibration results. It is followed by the stereo camera taking pictures for both the right and left cameras. The second result of the image is processed to determine the object area using HSV and color thresholding. Since some color thresholding results sometimes omit the noise especially when lighting is feeble, noise removal is necessary by a median filter and a morphological filter. When the two images are completely clear from noise, using the centroid feature in MATLAB, the center of the object can be seen from two perspectives (right and left cameras). As a result, the two centroid points are triangulated with the dataset to estimate the actual position, see Figure 11g–h. Finally, the arm is driven to pick up the object and place it in a preset style and position, as shown in Figures 18–20.

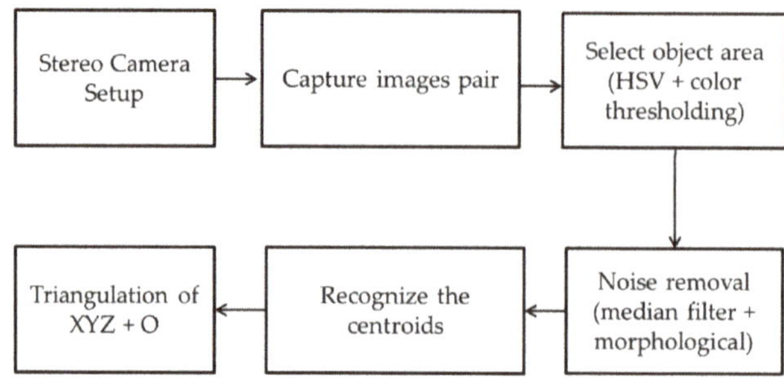

Figure 17. The sequence of processes in estimating the position of XYZ + O.

The number of datasets in the R-CNN training was 120 images and 64 images for testing. The performance of our method is very reliable that it is capable of recognizing triangles at 100% for precision, recall, and accuracy, as listed in Table 4. The tendency for our method to recognize objects is only with one bounding box result. If there are many bounding boxes, then the decision to make blocks will be based on the most significant coordinate position. Examples such as Figure 18b were a square block (15.1, 61.6, 634.9, 3.0) and triangle (−99.8, −15.7, 724.2, 21.7); then, the rectangle will be grasped first by the gripper. Meanwhile, to prove the absolute errors of the estimated position and orientation, our system is tested with a scenario of setting up buildings from blocks. As shown in Figure 19, the robot arm can execute commands based on ANFIS estimation results captured from a stereo camera. In a piling position, if the error is high, it is impossible to complete the final layout such as a house.

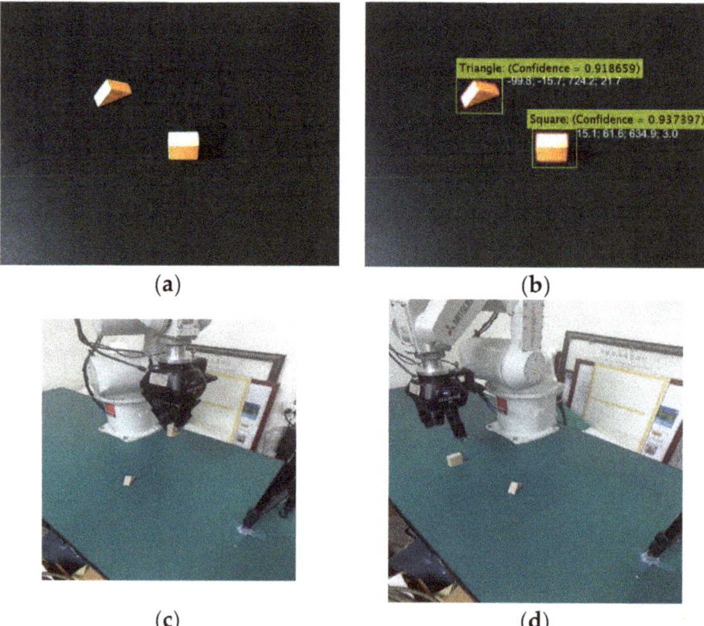

(a)   (b)

(c)   (d)

Figure 18. *Cont.*

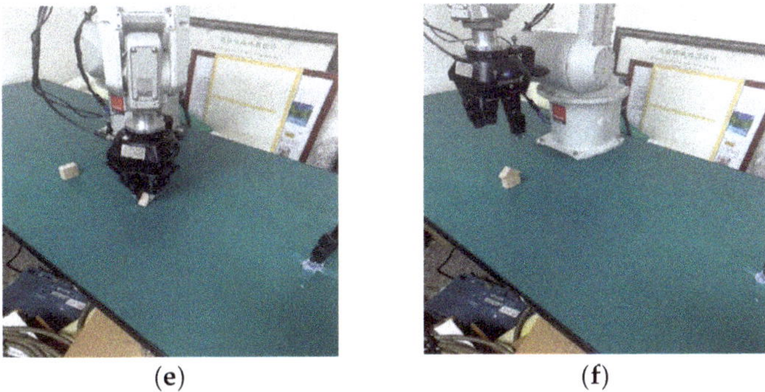

Figure 18. Operation process: (**a**) Get the image in the working range, (**b**) Image processing, (**c**)–(**f**) Pick and place objects intentionally for the first specified layout style.

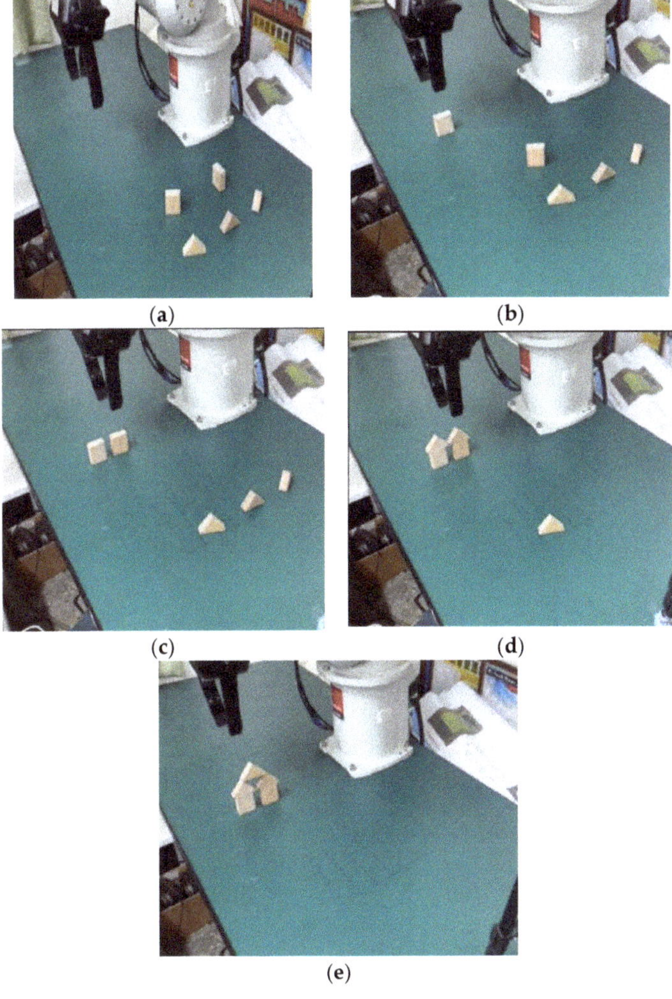

Figure 19. Operation process: (**a**)–(**e**) The second layout style.

Table 4. Performance of R-CNN to recognize triangle and square block.

	Triangle	Square
Precision (%)	100	96.68
Recall (%)	100	100
Accuracy (%)	100	98.44

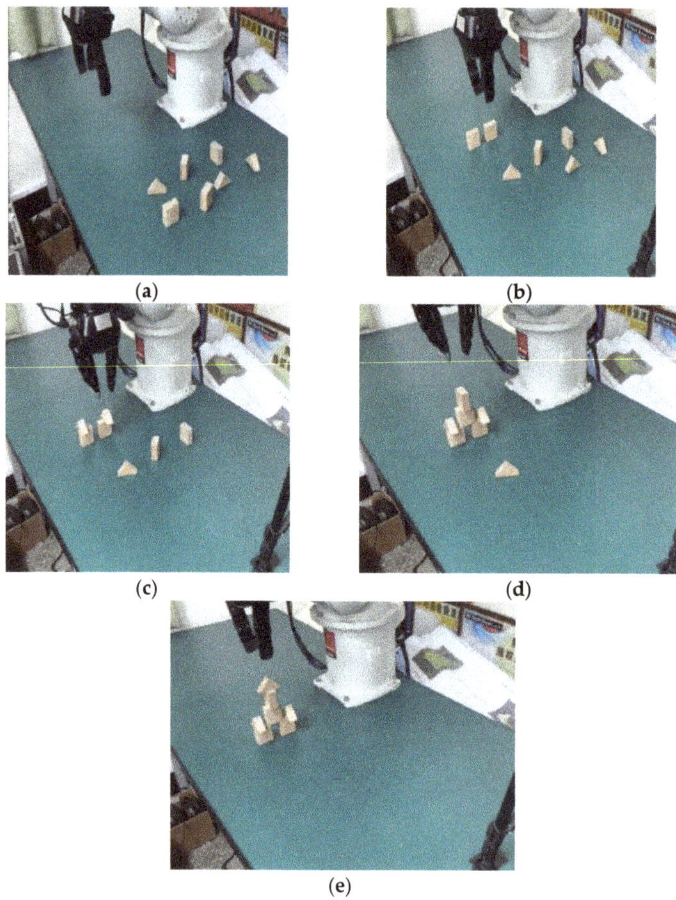

Figure 20. Operation process: (**a**)–(**e**) The third layout style.

5. Conclusions

In this paper, the coordinate frame systems of the stereo vision-based object manipulation system and their relationships are first introduced. The camera coordinate with respect to the robot base is obtained using ANFIS so that the targeted object to robot base can be easily found based on the information of the targeted object to the camera. The ANFIS architecture consists of a fuzzy layer, product layer, normalized layer, de-fuzzy layer, and summation layer, wherein the two-input type-3 first-order Sugeno fuzzy system is used to perform eye-to-hand calibration training, and three, five, and seven Gaussian membership functions are respectively used to calibrate the position of the stereo camera relative to the robot arm. From the training data, it can find that the errors are small, as shown in Table 2 and Figure 12. Based from the above results and the operation of R-CNN in the three experiments of picking and placing various numbers of blocks for specified styles and positions shown in Figures 18–20, the ability of XYZ coordinate estimation with the highest error at 2575 mm can be seen. Subsequently, for orientation of 2.14 degrees, this condition is still at an acceptable level because the system is able to form a construction, as proven in Figures 19 and 20. The application of R-CNN to recognize triangle blocks has precision, recall, and accuracy of 100% each. Meanwhile, percentages to identify square blocks are slightly lower, the precision is 96.88%, the recall is 100%, and the accuracy is 98.44%. Based on the testing results, we conclude the effectiveness of the proposed system.

Author Contributions: Y.-C.D. and M.-S.W. conceived and designed the experiments; M.M. and T.-H.H. performed the experiments; M.-S.W. and T.-H.H. analyzed the data; M.-S.W. and Y.-C.D. contributed materials and analytical tools; M.-S.W. wrote the paper. All authors have read and agreed to the published version of the manuscript.

Funding: This research was funded by Higher Education Sprout and Ministry of Science and Technology, the Ministry of Education, Taiwan and contract No. of MOST 108-2622-E-218-006-CC2, Ministry of Science and Technology.

Conflicts of Interest: The authors declare no conflict of interest. The funders had no role in the design of the study; in the collection, analyses, or interpretation of data; in the writing of the manuscript, or in the decision to publish the results.

References

1. Bouguet, J.-Y. Matlab Camera Calibration Toolbox. Available online: http://www.vision.caltech.edu/bouguetj/calib_doc (accessed on 10 July 2019).
2. Borangiu, T.; Dumitrache, A. *Robot Arms with 3D Vision Capabilities*; Intech Open Access Publisher: Rijeka, Croatia, 2010.
3. Jang, J.-S.R. ANFIS: Adaptive-network-based fuzzy inference system. *IEEE Trans. Syst. Man Cybern.* **1993**, *23*, 665–685. [CrossRef]
4. Tsai, R.Y.; Lenz, R.K. A new technique for fully autonomous and efficient 3D robotics hand/eye calibration. *IEEE Trans. Robot. Autom.* **1989**, *5*, 345–358. [CrossRef]
5. Abe, S. *Neural Networks and Fuzzy Systems: Theory and Applications*; Springer Science & Business Media: New York, NY, USA, 2012.
6. Kucuk, S.; Bingul, Z. *Robot kinematics: Forward and inverse kinematics*; Intech Open Access Publisher: Rijeka, Croatia, 2006.
7. Mitsubishi, I.R. *CRnQ/CRnD Controller Instruction Manual Detailed Explanations of Functions and Operations*; The University of Tokyo: Tokyo, Japan, 2010.
8. Buragohain, M.; Mahanta, C. A novel approach for ANFIS modeling based on full factorial design. *Appl. Soft Comput.* **2008**, *8*, 609–625. [CrossRef]
9. Zhang, Z. A flexible new technique for camera calibration. *IEEE Trans. Pattern Anal. Mach. Intell.* **2000**, *22*, 1330–1334. [CrossRef]
10. Maji, S.; Malik, J. Object detection using a max-margin hough transform. In Proceedings of the 2009 IEEE Conference on Computer Vision and Pattern Recognition (CVPR), Miami, FL, USA, 20–25 June 2009.
11. Lecun, Y.; Bottou, L.; Bengio, Y.; Haffner, P. Gradient-based learning applied to document recognition. *Proc. IEEE* **1998**, *86*, 2278–2324. [CrossRef]
12. LeCun, Y.; Kavukcuoglu, K.; Farabet, C. Convolutional networks and applications in vision. In Proceedings of the 2010 IEEE International Symposium on Circuits and Systems (ISCAS), Paris, France, 30 May–2 June 2010; pp. 253–256.
13. Krizhevsky, A.; Sutskever, I.; Hinton, G.E. Imagenet classification with deep convolutional neural networks. *Adv. Neural Inf. Process. Syst.* **2012**, *2012*, 1097–1105. [CrossRef]
14. Girshick, R.; Donahue, J.; Darrell, T.; Malik, J. Region-Based Convolutional Networks for Accurate Object Detection and Segmentation. *IEEE Trans. Pattern Anal. Mach. Intell.* **2016**, *38*, 142–158. [CrossRef] [PubMed]
15. Gu, C.; Lim, J.J.; Arbel´aez, P.; Malik, J. Recognition using Regions. In Proceedings of the 2009 IEEE Conference on Computer Vision and Pattern Recognition, Miami, FL, USA, 20–25 June 2009; pp. 1030–1037.
16. Felzenszwalb, P.F.; Huttenlocher, D.P. Efficient Graph-Based Image Segmentation. *Int. J. Comput. Vis.* **2004**, *59*, 167. [CrossRef]

17. Sande, K.E.A.; van de Uijlings, J.R.R.; Gevers, T.; Smeulders, A.W.M. Segmentation as selective search for object recognition. In Proceedings of the 2011 International Conference on Computer Vision, Barcelona, Spain, 6–13 November 2011; pp. 1879–1886.
18. Chen, Y.; Ma, Y.; Kim, D.H.; Park, S.-K. Region-Based Object Recognition by Color Segmentation Using a Simplified PCNN. *IEEE Trans. Neural Netw. Learn. Syst.* **2015**, *26*, 1682–1697. [CrossRef] [PubMed]
19. Ren, Z.; Gao, S.; Chia, L.-T.; Tsang, I.W.-H. Region-Based Saliency Detection and Its Application in Object Recognition. *IEEE Trans. Circuits Syst. Video Technol.* **2014**, *24*, 769–779. [CrossRef]
20. Cai, D.; Sun, X.; Zhou, N.; Han, X.; Yao, J. Efficient Mitosis Detection in Breast Cancer Histology Images by RCNN. In Proceedings of the 2019 IEEE 16th International Symposium on Biomedical Imaging (ISBI 2019), Venice, Italy, 8–11 April 2019; pp. 919–922.
21. Murugan, V.; Vijaykumar, V.R.; Nidhila, A. A Deep Learning RCNN Approach for Vehicle Recognition in Traffic Surveillance System. In Proceedings of the 2019 International Conference on Communication and Signal Processing (ICCSP), India, 4–6 April 2019; pp. 157–160.
22. Shi, J.; Zhou, Y.; Xia, W.; Zhang, Q. Target Detection Based on Improved Mask Rcnn in Service Robot. In Proceedings of the 2019 Chinese Control Conference (CCC), Guangzhou, China, 27–30 July 2019; pp. 8519–8524.
23. He, P.; Zuo, L.; Zhang, C.; Zhang, Z. A Value Recognition Algorithm for Pointer Meter Based on Improved Mask-RCNN. In Proceedings of the 2019 9th International Conference on Information Science and Technology (ICIST), Hulunbuir, China, 2–5 August 2019; pp. 108–113.
24. Taryudi; Wang, M.-S. Eye to Hand Calibration Using ANFIS for Stereo Vision-Based Object Manipulation System. *Microsyst. Technol.* **2018**, *24*, 305–317. [CrossRef]
25. Liu, Z.; Chen, T. Distance Measurement System Based on Binocular Stereo Vision. In Proceedings of the 2009 International Joint Conference Artificial Intelligence (JCAI), Hainan Island, China, 25–26 April 2009; pp. 456–459.
26. Zhang, Z.; Matsushita, Y.; Ma, Y. Camera calibration with lens distortion from low-rank textures. In Proceedings of the 2011 IEEE Conference on Computer Vision and Pattern Recognition (CVPR), Providence, RI, USA, 20–25 June 2011; pp. 2321–2328.
27. Szegedy, C.; Liu, W.; Jia, Y. Going deeper with convolutions. In Proceedings of the IEEE Conference on Computer Vision and Pattern Recognition (CVPR '15), Boston, Mass, USA, 7–12 June 2015; pp. 1–9.
28. Zhao, W.; Ma, W.; Jiao, L.; Chen, P.; Yang, S.; Hou, B. Multi-Scale Image Block-Level F-CNN for Remote Sensing Images Object Detection. *IEEE Access* **2019**, *7*, 43607–43621. [CrossRef]

Article

An Electromagnetic Lock Actuated by a Mobile Phone Equipped with a Self-Made Laser Pointer

Jau-Woei Perng [1] and Tung-Li Hsieh [1,2,*]

[1] Department of Mechanical and Electromechanical Engineering, National Sun Yat-sen University, Kaohsiung 80424, Taiwan; jwperng@faculty.nsysu.edu.tw
[2] General Education Center of Wenzao Ursuline University of Languages, Wenzao Ursuline University, Kaohsiung 80793, Taiwan
* Correspondence: tunglihsieh@gmail.com; Tel.: +886-7-342-6031

Received: 26 October 2019; Accepted: 10 December 2019; Published: 11 December 2019

Abstract: The main purpose of this study was to create an acousto-optic control lock device to convert electrical signals with a specific sound command using an acousto-optic conversion module, thereby improving the reliability and safety of opening or closing remote controlled door locks, such as car central locks or rolling doors. We used music playing through a smart phone speaker to create a special laser pointer to connect with the smart phone's auxiliary input. The laser pointer (wavelength of 630–650 nm and maximum output of 5 mw) lights up when the smart phone's music starts playing at a music frequency matching the light frequency. When the solar panel receives light, it converts the frequency of the light signal into an electrical frequency signal. The current is amplified using the power amplifier and then the amplified current flows to the sound recognition module. The sound recognition module performs audio comparison on the set sound signal, and once the comparison is correct, the output voltage activates the electromagnetic switch on the door to open or close it.

Keywords: laser pointer; electromagnetic lock; sound recognition module

1. Introduction

Traditional electronic remote controls [1,2] have been widely used in iron rolling doors, car central locks, scooter electronic locks, and various types of door locks. The remote control is used to control the lock. A signal transmitter is configured in the remote control to send the control signal to open or close the door to a signal receiver. The signal receiver is usually connected to an actuation circuit, which then turns the connection mechanism on or off to open or close the door. Most remote controls use an infrared signal or a radio frequency (RF) signal [3–5] as the control signal to open or close the door. However, this requires the user to carry an additional device, which can be lost, stolen, or damaged easily. Losing or forgetting the remote control would result in being locked out and the electronic remote controls fail when their battery power is depleted. With technological advancement, electronic remote controls are susceptible to replication and theft, and their security is gradually being challenged. Such incidences cause inconvenience to the users of traditional remote controls.

Traditional electronic remote controls need to be improved to eliminate the inconvenience and lack of security. To overcome these issues, in this study we formulated a method so that users do not need to carry an electronic remote control with them at all.

2. Materials and Methods

The proposed model is based on a control lock device: an audio-coupled, laser-actuated, electromagnetic lock device mounted on a door plank used to open or close the door using audio. The audio-coupled, laser-actuated, electromagnetic lock device consists of a transmitting unit, a receiving unit, a recognition unit, and an actuation unit. The transmitting unit of the proposed device is composed of an acoustic receiver and an acousto-optic conversion module, and the receiving unit is configured with an optical receiver. The acousto-optic conversion module converts a specific audio command signal into an electrical signal, thereby improving the reliability and safety of door locking or unlocking. The transmitting unit can be embedded in a mobile phone to open or close the door lock. Thus, users would not need to carry the remote control, increasing convenience of use.

2.1. Apparatus

The energy conversion module contains a power amplifying circuit, signal input and output terminals, and a power terminal. When the energy conversion module is operating, its temperature increases. The positive and negative poles of a solar panel are connected to the input pins of the energy conversion module using wires. Because ambient light in the environment can affect a solar panel and generate noise, the panel was modified to include a masking structure. The energy conversion module adopts a low-voltage LM386 chip (U.S. National Semiconductor, city, state abbrv. if USA, country) [6,7], which is equipped with a pin function (Figure 1). These chips are commonly used in low-voltage consumer products to minimize peripheral chip components. The LM386 chip's voltage gain is typically set to 20. However, by adding an external resistor and a capacitor between pins 1 and 8, the voltage gain can be arbitrarily adjusted up to a maximum value of 200 (Figure 2).

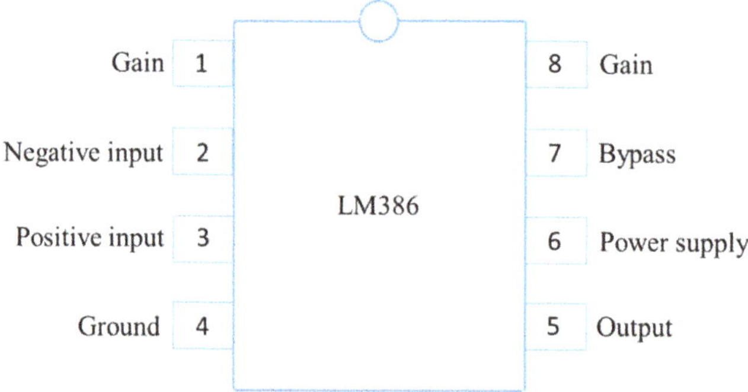

Figure 1. LM386 chip function pin, which features low static power consumption (approximately 4 mA) that is suitable for battery power supply, a wide operating voltage range (4–12 V or 5–18 V), fewer peripheral components, an adjustable voltage gain of 20–200 V, and a low distortion rate.

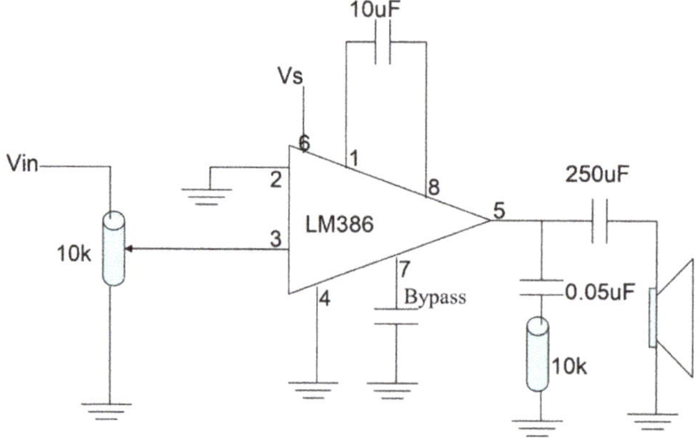

Figure 2. Circuit diagram with a gain of 200 V.

2.2. Electromagnetic Lock Mechanism Modeling

Figure 3a presents the mathematical model [8] with physical parameters, with the equation of motion as follows:

$$m\ddot{x} + c\dot{x} + kx = f(t) \tag{1}$$

where m, c, and k represent the mass, damping coefficient, and spring constant, respectively, which are the physical quantities of the system's mass, damping, and spring components, respectively. Figure 3b presents a mathematical model with modal parameters. By dividing Equation (1) by m and substituting the variables, the equation of motion for the physical parameters can be rewritten in modal parameter form as follows:

$$\ddot{q} + 2\xi\omega_n\dot{q} + \omega_n^2 q = N(t) \tag{2}$$

where

$$\omega_n = \sqrt{\frac{k}{m}}, \xi = \frac{c}{c_c}, c_c = 2m\omega_n = 2\sqrt{mk} \; q(t) = x(t), \text{ and } N(t) = \frac{f(t)}{m},$$

where c_c is the critical damping coefficient, which denotes the damping ratio, ω_n denotes the natural frequency, $x(t)$ is the physical coordinate, $f(t)$ defines the physical force, and $q(t)$ is the modal coordinate.

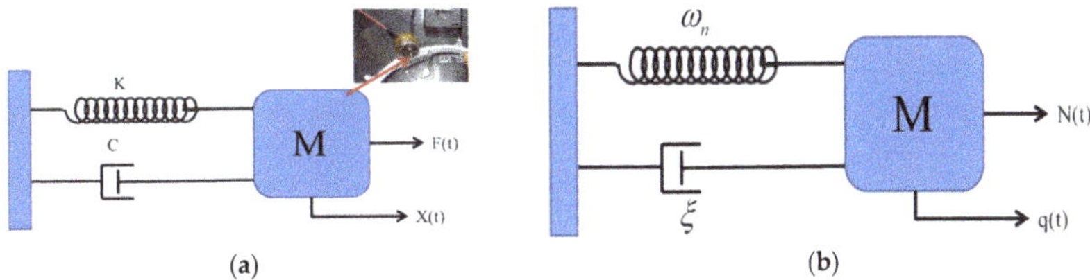

Figure 3. Mathematical model with (**a**) physical parameters and (**b**) with modal parameters.

Each conversion between domains, such as the conversion from a mathematical model with physical parameters to one with modal parameters, or the conversion of a mathematical model with modal parameters into one with frequency parameters, has a corresponding equation (Figure 4).

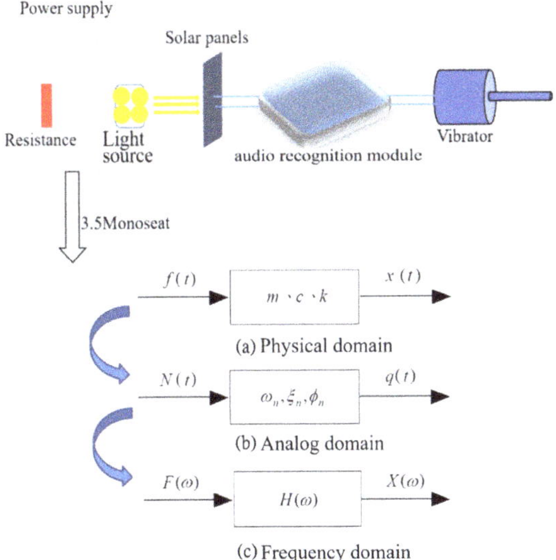

Figure 4. Circuit diagram of an audio-coupled, laser-actuated, electromagnetic lock device.

2.3. Response Analysis

Harmonic excitation force was set as the external force, $f(t) = Fe^{i\omega t}$. By substituting $x(t) = Xe^{i\omega t}$ into the equation of motion, the frequency response function is obtained:

$$H(\omega) = \frac{X}{F} = \frac{1}{(k - m\omega^2) + i(\omega c)} = \frac{1/m}{(\omega_n^2 - \omega^2) + i(2\xi\omega_n\omega)} \tag{3}$$

Using Equation (3), the frequency response function can be converted into a function of physical or modal parameters, which is related to the harmonic excitation frequency ω.

For the transient response analysis, if the input condition $f(t)$, initial conditions x_0 and v_0, and system content (e.g., the physical and modal parameters) are known, the time-domain output response of the system can be obtained using the following equation:

$$
\begin{aligned}
&x(t) = e^{-\xi\omega_n t}(A\cos\omega_d t + B\sin\omega_d t) + \int_0^t f(t)h(t - \tau)d\tau \\
&\omega_d = \omega_n\sqrt{1 - \xi^2}; h(t) = \frac{1}{m\omega_d}e^{-\xi\omega_n t}\sin\omega_d t
\end{aligned}
\tag{4}
$$

where ω_d denotes the damped natural frequency; A and B are random constants defined by the initial condition; and $h(t)$ represents the unit impulse response function of the system.

For the spectrum response analysis, if the frequency domain can be presented as the power spectral density (PSD) [9] function $x(t)$ and the frequency response function is known, the frequency domain response can be obtained through spectrum response analysis:

$$G_{xx}(\omega) = |H(\omega)|^2 G_{ff}(\omega) \tag{5}$$

where $G_{xx}(\omega)$ is the physical coordinate and $x(t)$ denotes the PSD function.

Figure 4 depicts the circuit of an audio-coupled, laser-actuated, electromagnetic lock device. In the creation of the proposed method, the audio-coupled, laser-actuated, electromagnetic lock included a light-emitting diode (LED) and a modulation circuit. The LED was used to generate an optical signal, and the modulation circuit was electrically connected to the audio receiver and LED. Next, the receiving unit, which included an amplification circuit, was electrically connected to an optical receiver to amplify its electrical output signal. In the third implementation, the optical receiver, namely a solar panel, was included. In the fourth implementation, the solar panel and door lock were mounted on a door plank. Then, the recognition unit was configured with a comparator to compare the electrical signal with a preset signal, which generated an actuation signal if the comparison result was the same. Next, the recognition unit, which includes storage, stores the preset signal. In the seventh implementation, the actuation unit was created, which included an electromagnetic actuator for opening or closing the door lock. In the eighth implementation, the transmitting unit was built into a portable electronic product, either a smartphone or a tablet. In the ninth implementation, the audio receiver, a monophonic plug, was inserted in the sound port of a portable electronic product.

As described earlier, the acousto-optic control lock device of the proposed model can open or close the door lock using a specific audio command signal, thereby improving the reliability and safety of the door lock, as well as preventing a lock key from being replicated or stolen. The transmitting unit embedded in the mobile phone enables the lock to be opened or closed without the use of a remote control, increasing convenience of use for the user. The proposed model is suitable for long-distance transmission because an audio command signal is replaced with an optical signal and a specific audio signal is transmitted, thereby increasing convenience.

The objective of the modal system analysis and testing was to develop specific analysis procedures and testing and measurement methods to obtain a mathematical model of the sound vibration mechanism. This model can be presented in the form of a mathematical model with modal or physical parameters. The analysis process flowchart is presented in Figure 5.

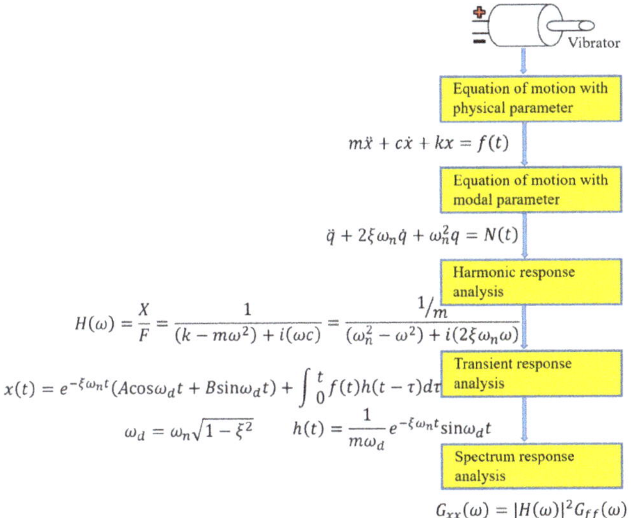

$$m\ddot{x} + c\dot{x} + kx = f(t)$$

$$\ddot{q} + 2\xi\omega_n\dot{q} + \omega_n^2 q = N(t)$$

$$H(\omega) = \frac{X}{F} = \frac{1}{(k - m\omega^2) + i(\omega c)} = \frac{1/m}{(\omega_n^2 - \omega^2) + i(2\xi\omega_n\omega)}$$

$$x(t) = e^{-\xi\omega_n t}(A\cos\omega_d t + B\sin\omega_d t) + \int_0^t f(t)h(t-\tau)d\tau$$

$$\omega_d = \omega_n\sqrt{1 - \xi^2} \qquad h(t) = \frac{1}{m\omega_d}e^{-\xi\omega_n t}\sin\omega_d t$$

$$G_{xx}(\omega) = |H(\omega)|^2 G_{ff}(\omega)$$

Figure 5. Flowchart of analysis of the sound vibration mechanism.

3. Results

3.1. Speech Endpoint Detection

We adopted a single channel in this experiment. Because different volumes generate different amplitudes, audio signals are normalized in order to unify amplitudes and obtain a valid frame range for subsequent signal processing. Human speech often contains aspiration or friction, which can complicate the detection of sound energy due to its low energy. The zero crossing rate in the endpoint detection method [10] can be employed to correct the endpoint range and extract an entire syllable (Figure 6).

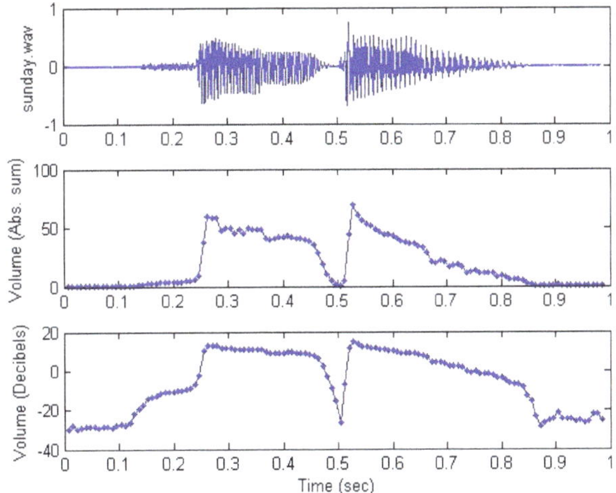

Figure 6. Sampled voice signals in the time domain after ZCR calculation.

3.2. Feature Extraction

Signal characteristics are difficult to recognize by observing changes in the amplitude of an audio signal over time (Figure 7). Converting an audio signal into a spectrogram (Figure 8) allows sonic characteristics to be identified [11]. The amplitude value of an audio signal in the time domain is often converted into an energy distribution in the frequency domain for observation. Various energy distributions in the frequency domain represent different speech characteristics. A signal changes rapidly and constantly over time, leading to inaccurate observation. Therefore, the most commonly used technique is conversion of the audio signal from the time domain to the frequency domain, enabling the identification of the spectral characteristics of various sounds through their energy distribution. The spectrum is a representation of a time domain signal in the frequency domain and can be obtained by performing FFT on the signal [12–17]. The result is presented as a spectrogram, with the amplitude or phase as the vertical axis and the frequency as the horizontal axis (Figure 9).

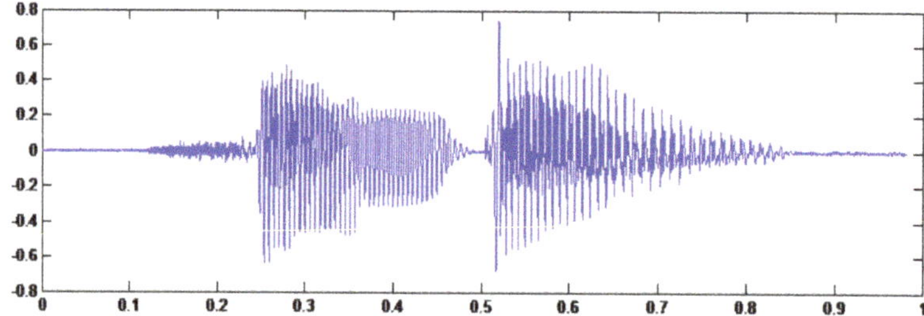

Figure 7. Amplitude value of audio signal, where *fs* (44,100) is the sampled frequency.

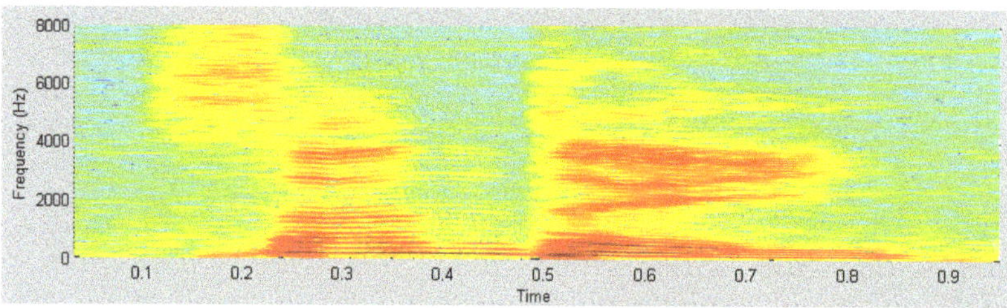

Figure 8. Audio signal spectrogram.

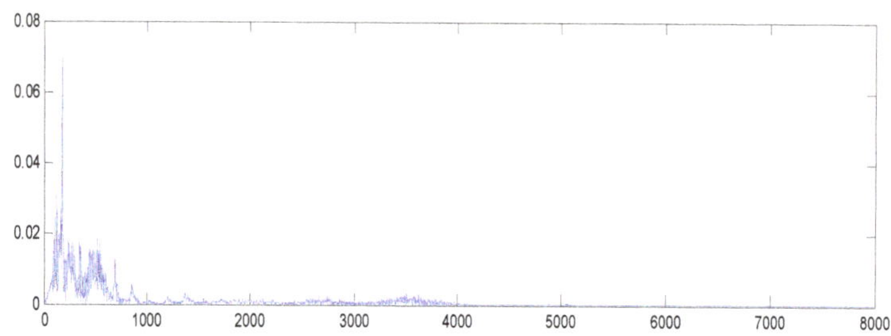

Figure 9. Audio signal spectrum.

3.3. Filter Design

Figure 10 shows the design of a Butterworth filter [18–22] with a cutoff frequency of 1000 Hz. The plot depicts the magnitude frequency response of the filter.

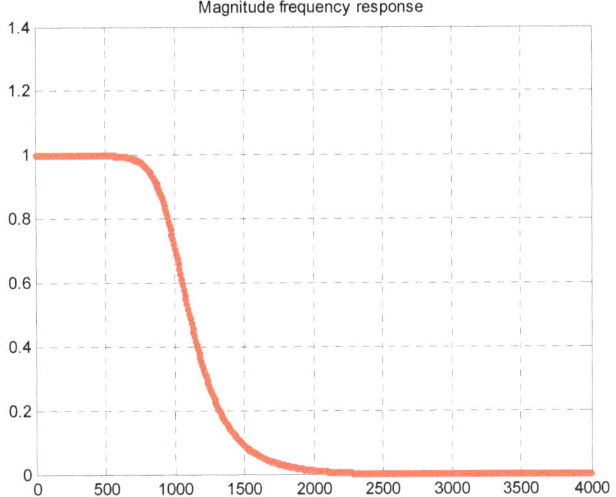

Figure 10. Magnitude frequency response of the filter.

When the order of the filter is larger, the filtering effect is better at the cost of longer computation time. A lower order leads to a shorter computation time and a less desirable filtering effect. Figure 11 demonstrates the magnitude frequency response as a function of the order of the Butterworth filter. Figure 11 shows that when the order increases from one to eight, the magnitude frequency response sharpens at the cutoff frequency of 1000 Hz.

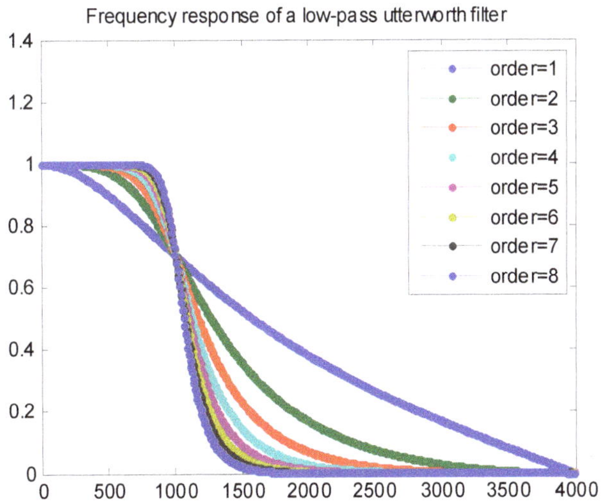

Figure 11. Frequency response of a low-pass Butterworth filter.

The original audio was passed through a low-pass filter with a cutoff frequency of 1000 Hz to test whether it can filter the treble (Figure 12). Figure 13 depicts the difference between the original signal and the filter output signal, indicating that the treble is almost removed. So, we can apply the audio to see if the high-frequency components can be removed.

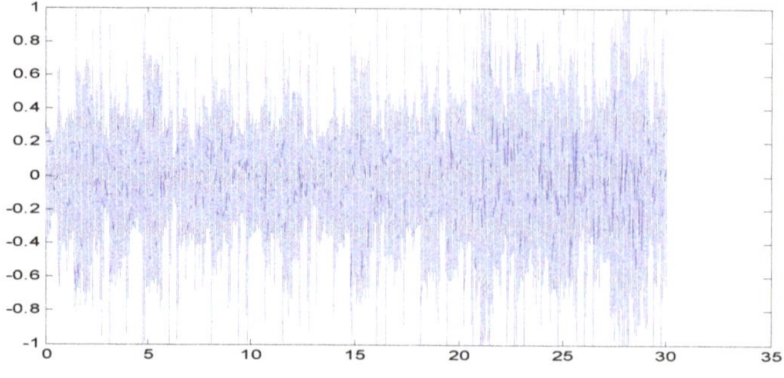

Figure 12. Original audio.

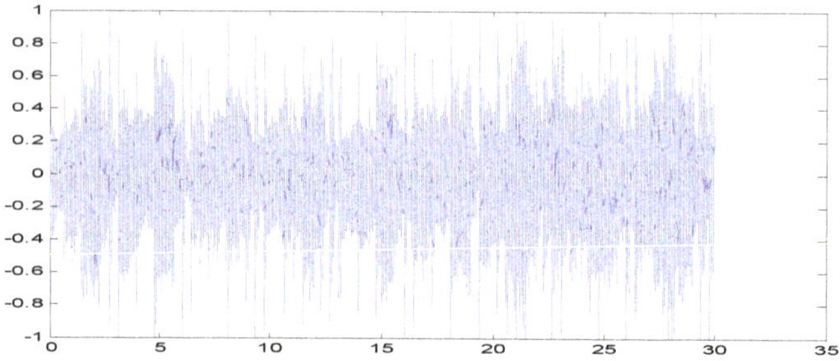

Figure 13. A low-pass filter with a cutoff frequency of 1000 Hz.

If we set the cutoff frequency to 100 Hz, then the output signal is almost inaudible unless we use a speaker with a subwoofer, as shown in Figure 14. After applying the low-pass filter at a cutoff frequency of 100 Hz, most of the sounds were removed, except the bass.

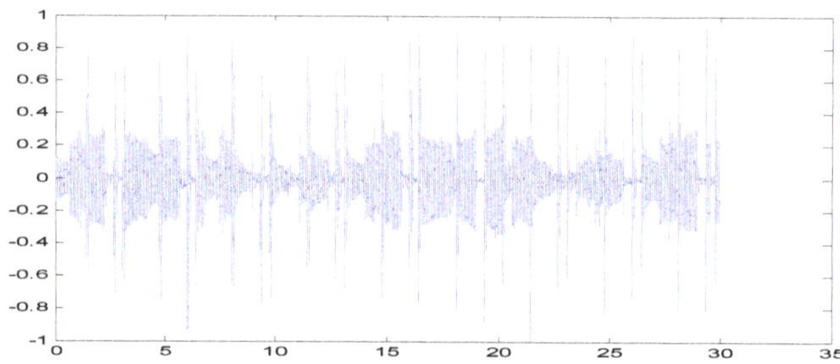

Figure 14. A low-pass filter with a cutoff frequency of 100 Hz.

The low-pass chopper functions to pass the low-frequency signal and attenuate the high-frequency signal, which is suitable for high-frequency noise. For example, if the temperature or flow sensor has a low frequency, the low-pass chopper can be used to remove electrical noise generated by a motor [23–26].

A digital filter can be represented by two parameter vectors, a and b, where the lengths of a and b are p and q, respectively, and the first element of a is always 1, as follows:

$$a = \left[1, a_2, \ldots a_p\right]$$

$$b = \left[b_1, b_2, \ldots b_q\right]$$

If we apply a digital filter with parameters a and b to a stream of discrete-time signal x[n], the output y[n] should satisfy the following equation:

$$y[n] + a_2 y[n-1] + a_3 y[n-2] + \ldots + a_p x[n-p+1] = b_1 x[n] + b_2 x[n-1] + \ldots + b_q x[n-q+1]$$

Or equivalently, we can express y[n] explicitly as:

$$y[n] = b_1 x[n] + b_2 x[n-1] + \ldots + b_q x[n-q+1] - a_2 y[n-1] - a_3 y[n-2] - \ldots - a_p x[n-p+1]$$

The preceding equation is somewhat complicated. We provide some more specific examples to facilitate understanding.

First, if we have a filter with the parameters a = [1] and b = [1/5, 1/5, 1/5, 1/5, 1/5], then the output of the filter is y[n] = (x[n] + x[n − 1] + x[n − 2] + x[n − 3] + x[n − 4])/5.

Figure 15 shows the different levels of filtering effects at different cutoff frequencies. The treble part is almost deleted after filtering. Only the sound of the bass survives, and then the bass sound that appears through these regular rules allows us to track the beat.

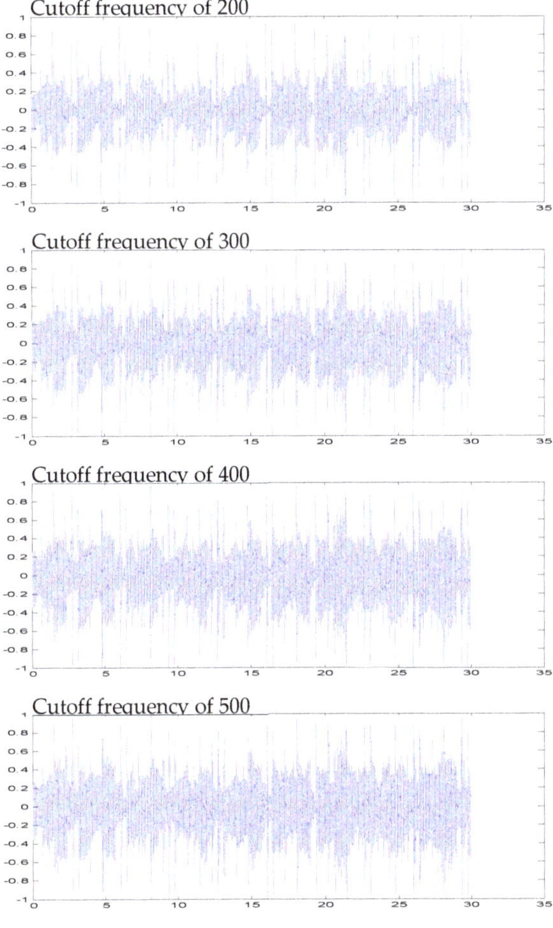

Figure 15. The different levels of filtering effects at different cutoff frequencies.

The result is a simple filter set y[n] as the average of the preceding five points of the input signals. In fact, this is a low-pass filter [27–31], since after the averaging operator, the high-frequency component is averaged out, while the low-frequency component is retained. The effect of a low-pass filter is like putting a paper cup over one's mouth while speaking, generating a murmuring-like, indistinct sound.

4. Discussion

The acousto-optic control lock (Figure 16) in the proposed device can open or close a door lock using a specific audio command signal, thereby improving the reliability and safety of the door lock and preventing a key from being replicated or stolen. The transmitting unit, embedded in a mobile phone, enables a door lock to be opened or closed without a remote control, which is convenient for the user. This model is suitable for long-distance transmission because the audio command signal is replaced with an optical signal and a specific audio signal is transmitted, thereby further enhancing convenience. The benefits of the proposed model are as follows: (1) elimination of inconvenience caused when keys are forgotten; (2) elimination of the need to carry multiple keys due to the laser-actuated electromagnetic lock device being coupled with mobile audio; (3) avoidance of the need to open a door with a key, as with a mobile phone a user can open multiple doors and set different opening passwords for each door; and (4) a significant reduction in the need for metallic materials for creating keys and elimination of the requirement for electroplating, which damages the environment, thus mitigating environmental pollution. The performance factors of an audio-coupled, laser-actuated, electromagnetic lock device are as follows: (1) An audio file is equivalent to a key. Users can easily set up or change different audio files on their own as the key. It is even more convenient than changing keys for traditional door locks. (2) The device allows for high confidentiality. As the audio file in the mobile phone is outputted through the laser carrier waves emitted by the laser pointer at the same audio frequency, only laser light is emitted when the phone plays the audio file, whereas no sound is generated. Hence, external remote sensors cannot detect the audio information. (3) Compared to traditional door locks, this device enables users to carry fewer bulky keys with them. (4) For traditional door locks, once the key is lost, both the lock and key have to be replaced. However, the photoelectric lock and laser pointer of this device are independent of one another and need not be replaced simultaneously. (5) The effective distance of the laser pointer (wavelength of 630–650 nm and maximum output of 5 mw) is <100 m and the time latency is about 30 ms. Table 1 shows the performance comparison of various locks

Table 1. The performance comparison of various locks.

	Traditional Door Locks	Radio Frequency (RF) Electronic Remote Lock	Optical Electromagnetic Lock
Unlock mode	key	RF signal	laser pointer
Remote open	X	V (10 m)	V (100 m)
safety	Easy to destroy and copy	RF is susceptible to strong wave disturbances Easy to copy and steal	Laser is not easily interfered with or copied
Convenience	Door locks or keys are damaged and must be replaced together	It is not easy to copy the program after losing it	Audio can be easily copied
Security	Easy copying from mobile audio	Easy to destroy from the outside	Because the electromagnetic lock is in the door, it is difficult to break
Interactivity	X	V	V
cost	NT $100–$200	NT $3000–$4000	NT $1000 or less

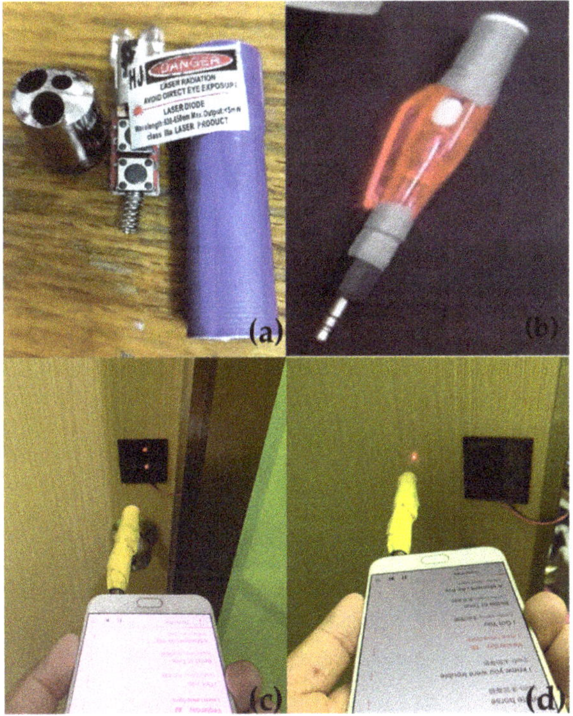

Figure 16. The composition and application of acousto-optic control locks. (**a**) Disassembly of a laser diode. (**b**) The circuit combination of a 3.5 monoseat and laser diode. (**c**) Laser pointer combined with mobile phone audio for unlocking. (**d**) The door lock is opened.

Author Contributions: T.-L.H. and J.-W.P. conceived of the presented idea. T.-L.H. developed the theory and performed the computations. T.-L.H. verified the analytical methods. J.-W.P. encouraged investigation and supervised the work. All authors discussed the results and contributed to the final manuscript.

Funding: This research received no external funding.

Acknowledgments: This research has been reviewed by the IEEE ECICE 2019 committee, where it was recommended to be submitted to *Electronics*.

Conflicts of Interest: The authors declare no conflict of interest.

References

1. Josephine, M.M. Design and Construction of a remote control switching device for household appliances application. *ASTESJ* **2017**, *2*, 154–164. [CrossRef]
2. Delgado, A.R.; Picking, R.; Grout, V. *Remote-Controlled Home Automation Systems with Different Network Technologies*; University of Wales: Wrexham, UK, 2013.
3. Hossain, E.; Mamun, N.; Faisal, M.F. Vehicle to Vehicle Communication Using RF and IR Technology. In Proceedings of the 2nd International Conference on Electrical & Electronic Engineering (ICEEE), Rajshahi, Bangladesh, 27–29 December 2017. [CrossRef]
4. Bakare, B.I.; Nwakpang, F.M.; Desire, A.E. Propagation Analysis of Radio Frequency (RF) Signal of Love FM Transmitter in Port Harcourt, Nigeria. *IOSR J. Electron. Commun. Eng. (IOSR-JECE)* **2019**, *14*, 5–12. [CrossRef]
5. Akar, F.; AŞKIN, O. Design and implementation of IR and laser-based electronic ciphering systems. *Turk. J. Electr. Eng. Comput. Sci.* **2015**, *23*, 17–27. [CrossRef]

6. Vimal Raj, V.; Aravindan, S.; Agnishwar Jayaprakash Harith, M.; Manivannan, R.; Bhuvaneshwaran, G. Comparative Study of Audio Amplifiers. *J. Electron. Commun. Eng. (JECE)* **2017**, *12*, 32–36. [CrossRef]

7. Jiang, M. Using Three Op Amp Simple AM/FM Radio Circuit. *Open Access Libr. J.* **2018**, *5*, 1. [CrossRef]

8. Munyazikwiye, B.B.; Robbersmyr, K.G.; Karimi, H.R. A state-space approach to mathematical modeling and parameters identification of vehicle frontal crash. *Syst. Sci. Control Eng.* **2014**, *2*, 351–361. [CrossRef]

9. Sun, Y.; Wang, X. Power spectral density of pulse train over random time scaling. *IET Signal Process* **2014**, *8*, 601–605. [CrossRef]

10. Jiea, L.; Pingb, Z.; Xinxingc, J.; Zhiran, D. Speech End point Detection Method Based on TEO in Noisy Environment. *Procedia Eng.* **2012**, *29*, 2655–2660. [CrossRef]

11. Müller, M.; Ellis, D.P.W.; Klapuri, A.; Richard, G. Signal Processing for Music Analysis. *IEEE J. Sel. Top. Signal Process.* **2011**, *5*, 1088–1110. [CrossRef]

12. Jiang, D.N.; Cai, L.H. Speech emotion classification with the combination of statistic features and temporal features. *IEEE ICME* **2004**, *3*, 1967–1970.

13. Schuller, B.; Rigoll, G. Timing levels in segment-based speech emotion recognition. In Proceedings of the Interspeech 2006—ICSLP, Pittsburgh, PA, USA, 17–21 September 2006.

14. Reynolds, D.A. Large Population Speaker Identification Using Clean and Telephone Speech. *IEEE Signal Process. Lett.* **1995**, *2*, 46–48. [CrossRef]

15. Mashao, D.J.; Baloyi, N.T. Improvements in The Speaker Identification Rate Using Feature-Sets on A Large Population Database. In Proceedings of the Seventh European Conference on Speech Communication and Technology, Aalborg, Denmark, 3–7 September 2001.

16. Chetouani, M.; Mahdhaoui, A.; Ringeval, F. Time-scale feature extractions for emotional speech characterization. *Cognit. Comput.* **2009**, *1*, 194–201. [CrossRef]

17. Vondra, M.; Vích, R. Recognition of emotions in german speech using gaussian mixture models. In *Multimodal Signals: Cognitive and Algorithmic Issues*; Springer: New York, NY, USA, 2009.

18. Junbo, L.; Xiangning, F.; Kuan, B. A 4th-order active-Gm-RC low-pass filter with RC time constant auto-tuning for reconfigurable wireless receivers. In Proceedings of the 2012 IEEE Solid-State and Integrated Circuit Technology (ICSICT), Xi'an, China, 29 October–21 November 2012.

19. Jiang, C.; Xie, R.; Li, W.; Huang, Y.; Hong, Z. Reconfigurable low pass filter with Automatic Frequency Tuning for WCDMA and GSM application. In Proceedings of the 2011 IEEE ASIC (ASICON), Xiamen, China, 25–28 October 2011.

20. de Matteis, M.; Pezzotta, A.; D'Amico, S.; Baschirotto, A. A 33MHz 70dB-SNR super-source-follower-based low-pass analog filter. *IEEE J. Solid State Circuits* **2015**, *50*, 1516–1524. [CrossRef]

21. Xu, Y.; Muhlestein, J.; Moon, U.K. A 0.65 mW 20 MHz 5th-order low-pass filter with +28.8dBm IIP3 using source follower coupling. In Proceedings of the IEEE Custom Integrated Circuits Conference (CICC), Austin, TX, USA, 30 April–3 May 2017.

22. Ding, Y.; Liu, L.; Li, R.; Zhang, X.; Liu, L. A 5–80 MHz CMOS Gm-C Low-Pass Filter with On-Chip Automatic Tuning. In Proceedings of the 2011 International Conference on Digital Manufacturing and Automation (ICDMA), Hunan, China, 5–7 August 2011.

23. Bao, K.; Fan, X.; Wang, Z. A 0.18-μm-CMOS Low-Power Reconfigurable Low Pass Filter for Multi-Standard Receivers. In Proceedings of the 2013 IEEE Advanced Technologies for Communications (ATC), Ho Chi Minh City, Vietnam, 16–18 October 2013.

24. Oliveira, M.S.; de Aguirre, P.C.; Severo, L.C.; Girardi, A.G.; Susin, A.A. A Digitally Tunable 4th-order Gm-C Low-Pass Filter for Multi-Standards Receivers. In Proceedings of the 2016 IEEE Integrated Circuits and Systems Design (SBCCI), Belo Horizonte, Brazil, 29 August–3 September 2016.

25. Gao, J.; Jiang, H.; Zhang, L.; Dong, J.; Wang, Z. A programmable low-pass filter with adaptive miller compensation for zero-IF transceiver. In Proceedings of the 2012 IEEE Circuits and Systems (MWSCAS), Boise, ID, USA, 5–8 August 2012.

26. Jin, X.; Dai, F.F. A 6th order zero capacitor spread 1 MHz–10 MHz tunable CMOS active-RC low pass filter with fast tuning scheme. In Proceedings of the 2012 IEEE International Symposium on Circuits and Systems (ISCAS), Seoul, Korea, 20–23 May 2012.

27. Lee, I.Y.; Im, D.; Ko, J.; Lee, S.G. A 50–450 MHz Tunable RF Biquad Filter Based on a Wideband Source Follower With >26 dBm IIP$_3$, +12 dBm P$_{1dB}$, and 15 dB Noise Figure. *IEEE J. Solid State Circuits* **2015**, *50*, 2294–2305. [CrossRef]

28. D'Amico, S.; De Blasi, M.; De Matteis, M.; Baschirotto, A. A 255 MHz Programmable Gain Amplifier and Low-Pass Filter for Ultra Low Power Impulse-Radio UWB Receivers. *IEEE Trans. Circuits Syst. I Regul. Pap.* **2012**, *59*, 337–345. [CrossRef]

29. Houfaf, F.; Egot, M.; Kaiser, A.; Cathelin, A.; Nauta, B. A 65 nm CMOS 1-to-10 GHz tunable continuous-time low-pass filter for highdata-rate communications. In Proceedings of the 2012 IEEE International Solid-State Circuits Conference, San Francisco, CA, USA, 19–23 February 2012; pp. 362–364.

30. Cheng, Y.W.; Wu, J.W.; Lin, T.H.; Wang, C.J. A Low-Pass Filter by Using a Meandered Slot. In Proceedings of the 2014 IEEE International Workshop on Electromagnetics (iWEM), Sapporo, Japan, 4–6 August 2014.

31. Wenxiang, X.; Xinfeng, L.; Yongzhong, H. Design of Compact Ultra Wide-Stopband Lowpass Filter. In Proceedings of the International Conference Computational Problem-Solving (ICCP), Chengdu, China, 21–23 October 2011; pp. 265–268.

 *electronics*

Article

Electrostatic-Discharge-Immunity Impacts in 300 V nLDMOS by Comprehensive Drift-Region Engineering

Po-Lin Lin, Shen-Li Chen * and Sheng-Kai Fan

Department of Electronic Engineering, National United University, Miaoli City 36003, Taiwan;
linda1518b@yahoo.com.tw (P.-L.L.); ke21vin21@gmail.com (S.-K.F.)
* Correspondence: jackchen@nuu.edu.tw; Tel.: +886-37-382525

Received: 1 November 2019; Accepted: 30 November 2019; Published: 3 December 2019

Abstract: Electrostatic discharge (ESD) events are the main factors impacting the reliability of Integrated circuits (ICs); therefore, the ESD immunity level of these ICs is an important index. This paper focuses on comprehensive drift-region engineering for ultra-high-voltage (UHV) circular n-channel lateral diffusion metal-oxide-semiconductor transistor (nLDMOS) devices used to investigate impacts on ESD ability. Under the condition of fixed layout area, there are four kinds of modulation in the drift region. First, by floating a polysilicon stripe above the drift region, the breakdown voltage and secondary breakdown current of this modulation can be increased. Second, adjusting the width of the field-oxide layer in the drift region when the width of the field-oxide layer is 5.8 μm will result in the minimum breakdown voltage (105 V) but the best secondary breakdown current (6.84 A). Third, by adjusting the discrete unit cell and its spacing, the corresponding improved trigger voltage, holding voltage, and secondary breakdown current can be obtained. According to the experimental results, the holding voltage of all devices under test (DUTs) is greater than that of the reference group, so the discrete HV N-Well (HVNW) layer can effectively improve its latch-up immunity. Finally, by embedding different P-Well lengths, the findings suggest that when the embedded P-Well length is 9 μm, it will have the highest ESD ability and latch-up immunity.

Keywords: drift region; electrostatic discharge (ESD); holding voltage (V_h); lateral diffusion MOS (LDMOS); transmission-line pulse system (TLP system)

1. Introduction

In recent years, the UHV LDMOS has been implemented in power electronics, Microelectromechanical systems (MEMS) domains, power management circuits, and internet of things (IoT) applications [1–16]. The power management circuit is also an indispensable project of the internet of things. The internet of things is facing the tricky problem of battery endurance, but it can be improved through the power management circuit [14]. However, high-voltage ICs pose serious risks to electrostatic discharge (ESD), and according to the statistics, the ratio of component failure is nearly half due to ESD damage, so ESD protection for the silicon chips is needed to reduce the number of ESD failures. To achieve effective ESD protection, according to the ESD design window shown in Figure 1, there are three important parameters: trigger voltage (V_{t1}), holding voltage (V_h), and secondary breakdown current (I_{t2}). The trigger voltage must be lower than the core circuit breakdown voltage. However, if a high-voltage transient is injected into a circuit, the protection device should be turned on to bypass the heavy current in order to avoid core circuit destruction. Additionally, these protection devices need to be turned on quickly and can sustain a heavy current. The Float Cum Boost Charger (FCBC) architecture [17] itself is susceptible to large current damage, so a contactor is commonly used to discharge its large voltage/current. The ESD protection component in our paper is also the same as the contactor role, so the UHV nLDMOS component response time (V_{t1} related) is a key factor. Therefore, we can find out the V_{t1} of the component by using the transmission-line pulse (TLP) system to determine whether the protection component can be turned on quickly under a large voltage/current bombardment to prevent the circuit from being damaged by the instantaneous large voltage and current transient. The holding voltage must be higher than positive supply voltage (V_{DD}), otherwise there is a latch-up risk. The secondary breakdown current is as high as possible because it is defined as a device of ESD ability. Additionally, the whole-chip ESD protection design has been proposed to suggest where the chip should be protected to reduce the ESD risk [18]. The protect method is divided into two types—the first is to design ESD protection circuits, which protect chips by using the gate-couple technique [19,20] and the substrate-trigger technique [20,21]; the second is to design ESD protection devices, which use silicon controlled rectifier (SCR) [22,23], Grounded-gate nMOS (GGnMOS) [24], stacked field-oxide device (FOD) [25], and diodes [26] to protect a chip.

The current crowding effect can often reduce the reliability level of UHV LDMOS [27–29]. For example, when the n-type heavily doped (N^+) junction edge of the drain side is adjacent to the field oxide, as shown in Figure 2a, a large ESD spike will inject into the N^+ junction and then crowd at the edge of the field oxide when an ESD event occurs. If the ESD current is too large because the power is equal to the voltage multiplied by the current, the power dissipation will increase when the current is increased. This causes heat generation at the current gathering location and a current crowding effect can occur, which causes device damage.

In this paper, four kinds of novel modulations are proposed for drift-region engineering in order to strengthen the ESD ability of UHV nLDMOSs. (1) By using a floating polysilicon stripe above the field-oxide layer; (2) shortening the width of the field-oxide layer; (3) discrete HV N-Well layer; and (4) the embedded P-Well, the ESD protection ability can be effectively improved.

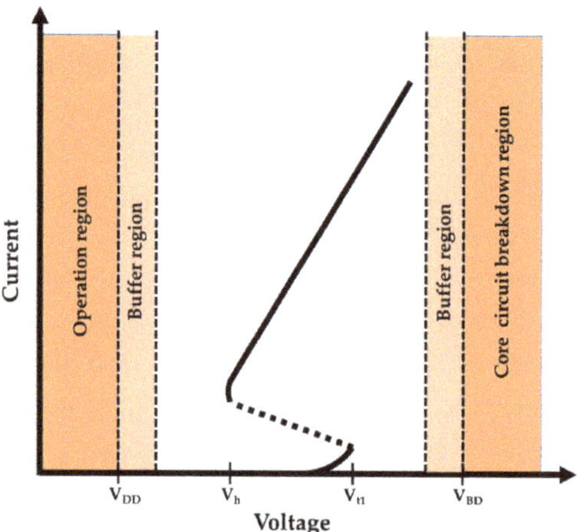

Figure 1. Electrostatic discharge (ESD) protection window of a LDMOS.

2. Layout of UHV Circular nLDMOS Devices under Test (DUTs)

2.1. UHV Circular nLDMOS Reference Group

The cross-sectional view and layout top view of an UHV circular nLDMOS are shown in Figure 2a,b, respectively. Due to the process specifications, a field-oxide layer (FOX) is fabricated above the drift region to enhance the breakdown of the electric field. Due to the operational voltage of UHV applications, the n-type lightly doped HVNW layer is used in the drift region. The PBody and the deep P-Well (DPW) form a RESURF structure, which causes the drift region to be completely depleted and increases the breakdown voltage of the device without increasing the length of the drift region [30]. The polysilicon-stripe (poly2) above the drift region is used to reduce the peak value of the electric field. The traditional UHV ESD protection device adopts the elliptical layout type, but the layout area is huge. In this paper, a circular layout type is adopted, which reduces the layout area and makes the voltage distribution more uniform [31,32]. In order to ensure the normal operation of the device characteristics, a semiconductor curve tracer is used to measure the current-voltage (I-V) curve and the breakdown voltage to assure that the DUTs have the correct output characteristics and the correct breakdown voltage value. When the UHV nLDMOS transistor acts as an ESD protection device, its device configuration forms a GGnMOS structure by grounding the gate electrode, which can discharge the ESD current beneath the parasitic Bipolar junction transistor (BJT). In this paper, all the DUTs are fabricated via a TSMC 0.5 μm BCD process. The channel length (L) is 4 μm, the channel width (W) is 394.4 μm, and the drift region length is 29 μm.

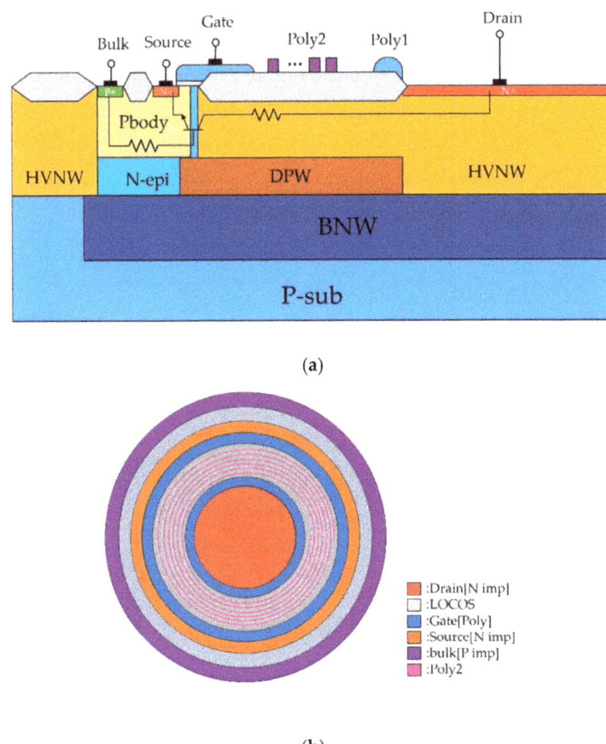

(a)

(b)

Figure 2. (**a**) Cross-sectional view and (**b**) layout top view of a circular lateral diffusion MOS (nLDMOS).

2.2. UHV Circular nLDMOS—Polysilicon-Stripe Modulation above the Drift Region

In this structure, the layout of the polysilicon-stripe varies from spiral type to concentric circle type, as shown in Figure 3. The spiral poly2 starts from the drain-side edge, then passes above the field-oxide layer in the drift region, and finally connects to the source electrode. Initially, when an ESD event occurs, the poly2 contact is damaged due to the excessive current density, which reduces the ESD capability. The concentric poly2 type has multiple concentric circles that float above the drift region. The concentric poly2 type reduces the peak value of the electric field below the FOX and increases the breakdown voltage. Additionally, the floating concentric circle poly2 has no contacts, so it avoids the risk of contact damage.

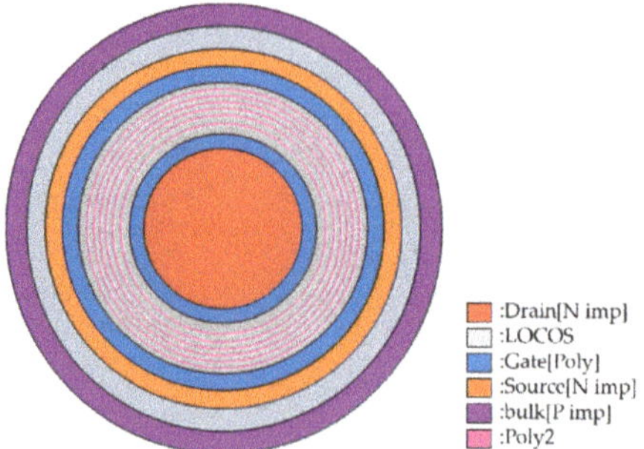

Figure 3. Layout top view of a circular nLDMOS with concentric polysilicon-stripe (poly2) circles.

2.3. UHV Circular nLDMOS—Field-Oxide Width Modulation in the Drift Region

A cross-sectional view and layout top view of the field-oxide width modulation in the drift region are shown in Figure 4a,b, respectively. By shortening the width of the field-oxide layer in the drift region, the equivalent series resistance of this device decreases. The purpose of this is to reduce the device impedance, so the breakdown voltage is also reduced. Due to a strong correlation between the breakdown voltage and the trigger voltage, these devices can be applied for the desired operating voltage applications and are fabricated by the same process. The cell names of the modulation parameter are shown in Table 1.

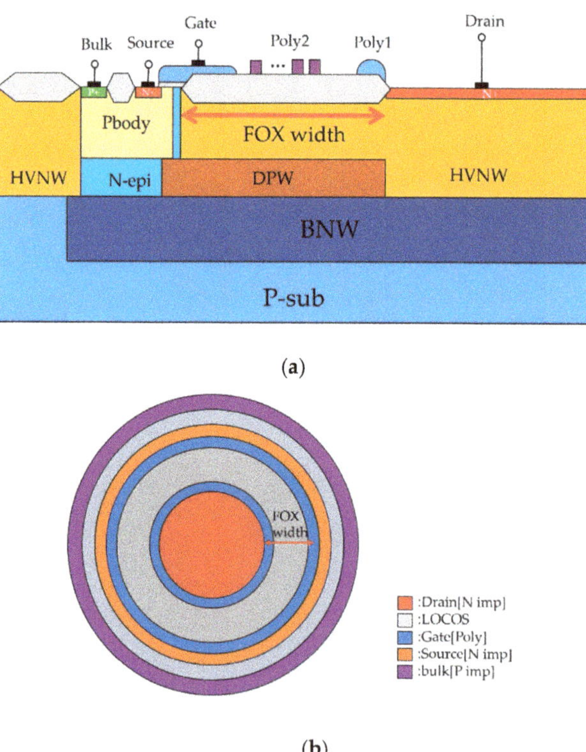

(**a**)

(**b**)

Figure 4. (**a**) Cross-sectional view and (**b**) layout top view of a circular nLDMOS with a field-oxide width modulation in the drift region.

Table 1. Cell names of the field-oxide width modulation.

Samples Name	Field-Oxide Width (µm)
Ref.	29
FOX_1	23.2
FOX_2	17.4
FOX_3	11.6
FOX_4	5.8

2.4. UHV Circular nLDMOS—Discrete HV N-Well (HVNW) Layer Modulation in the Drift Region

The cross-sectional view and the layout top view of the discrete HVNW layer modulation in the drift region are shown in Figure 5a,b, respectively. In this architecture, the drift region is designed to be discrete and independent by using layout skills. Furthermore, the poly2 layer is changed to a concentric circle form to evaluate the influence of poly2 concentric circles on the discrete HVNW layer. Due to the fact that the parasitic resistance of n-epi is larger than the HVNW layer, the discrete HVNW layer can upgrade the equivalent resistance of the drift region. The DUTs are divided into two modulation types: a unit cell-size modulation (three cell sizes: 1, 2, and 3 µm) and a unit-cell spacing modulation (cell spacings: 1.34, 2.68, and 4.02 µm). Cell names of the discrete HVNW layer modulation are shown in Table 2.

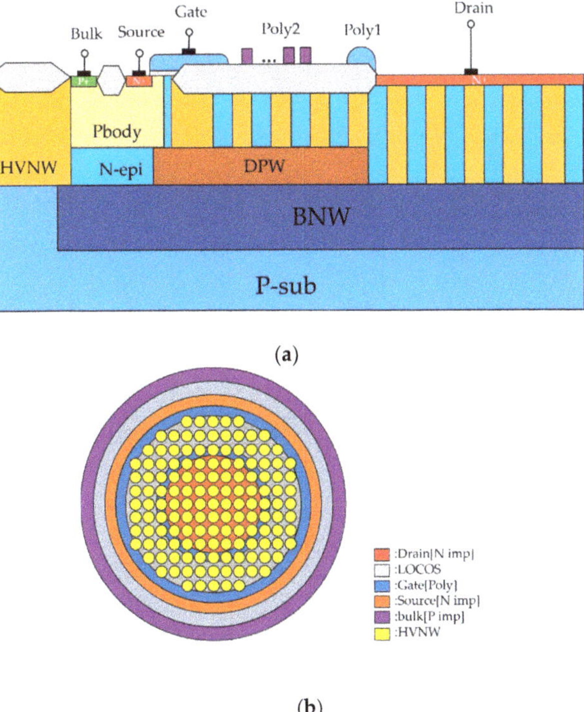

(a)

(b)

Figure 5. (a) Cross-sectional view and (b) layout top view of a circular nLDMOS with discrete HV N-Well (HVNW) layer modulation.

Table 2. Cell names of the discrete HVNW layer modulation.

Space \ Size	1 μm	2 μm	3 μm
1.34 μm	dis10	dis20	dis30
2.68 μm	dis11	dis21	dis31
4.02 μm	dis12	dis22	dis32

2.5. UHV Circular nLDMOS—Embedded P-Well Length Modulation in the Drift Region

The cross-sectional and layout top views of the embedded P-Well length modulation in the drift region are shown in Figure 6a,b, respectively. Starting from the drain side, the N^+ junction edge extends into the local oxidation of silicon (LOCOS) region with an embedded P-Well layer, and the lengths of the extended P-Well (K) are 5, 7, 9, and 11 μm. Since the P-Well and N^+ regions form a reverse bias junction, when an ESD event occurs, the ESD current flows into the deeper path via the HVNW and BNW layers to avoid device failure due to the current crowding effect at the drain-side LOCOS/N^+ edge. The cell names of the embedded P-Well length modulation are shown in Table 3.

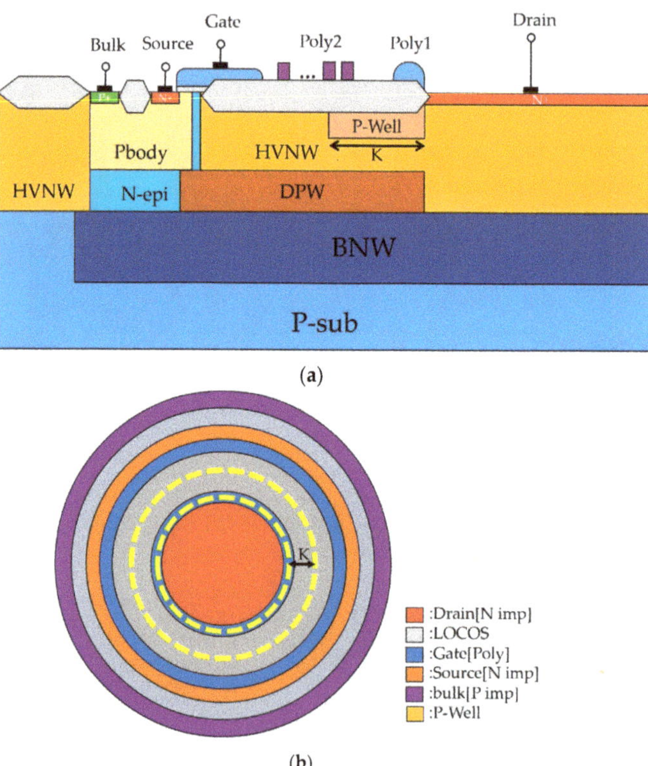

Figure 6. (**a**) Cross-sectional view and (**b**) layout top view of a circular nLDMOS with an embedded P-Well length modulation in the drift region.

Table 3. Cell names of the embedded P-Well length modulation.

Samples Name	P-Well Length (μm)
Ref.	0
PW_5	5
PW_7	7
PW_9	9
PW_11	11

3. Testing Machine

The related electronic instruments in a TLP testing system achieve an automated measurement process via the LabVIEW interface. This TLP machine provides a continuous rising square wave to get the I-V curve data of the DUTs. This testing system uses a square wave with 100 ns pulse width and has a short rising/falling time of <10 ns to obtain the voltage and current responses through the DUTs. This short transient pulse is used to simulate the human body model (HBM) waveform of an ESD event. Eventually, the I-V characteristics of the DUTs, such as the trigger voltage, holding voltage, and secondary breakdown current, can be measured.

4. Test Results and Discussion

4.1. UHV Circular nLDMOS—Polysilicon-Stripe Modulation above the Drift Region

The experimental results of the UHV nLDMOS-related DUTs with polysilicon-stripe modulation above the drift region obtained from the breakdown voltage measurement and TLP testing are shown in Table 4. These experiment results demonstrate that the floating poly2 improves the electric field distribution under the field-oxide layer, smoothing the electric field distribution and reducing the peak value of the electric field to enhance the breakdown voltage of a device. The secondary breakdown current is strongly related to the breakdown voltage that a device can withstand. Therefore, in the same device geometries, we find that as the breakdown voltage increases, the secondary breakdown current also increases.

Table 4. Snapback parameters of ultra-high-voltage (UHV) nLDMOS-related devices under test (DUTs).

Samples	V_{t1}(V)	V_h(V)	I_{t2}(A)	V_{BK}(V)
Spiral type	375.13	58.69	3.20	395.12
Concentric circle type	375.71	41.66	5.09	411.20

4.2. UHV Circular nLDMOS—Field-Oxide Width Modulation in the Drift Region

The experimental results of the UHV nLDMOS-related DUTs with the field-oxide width modulation in the drift region obtained from the breakdown voltage measurement and TLP testing are shown in Figures 7–9. As the field-oxide width decreases, the the equivalent series resistance, the breakdown voltage, and the trigger voltage are significantly reduced, and the holding voltage also lowers. Interestingly, the secondary breakdown currents of the ESD capability were higher than that of the reference group. It appears that the higher the operation voltage is, the lower I_{t2} value it has. The test results of the UHV nLDMOS with field-oxide width modulation in the drift region are shown in Table 5.

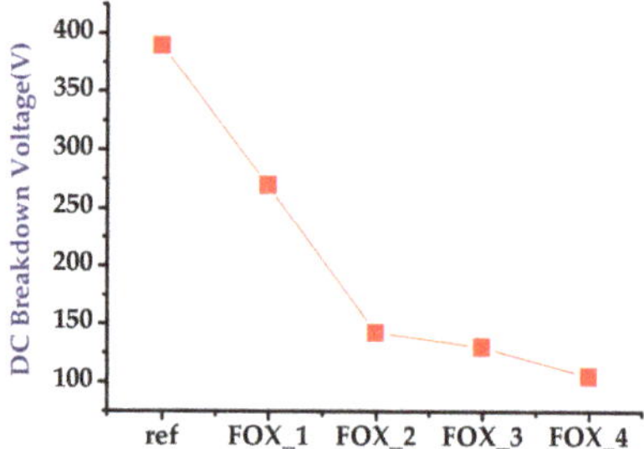

Figure 7. Breakdown voltage trend chart of nLDMOSs with the field-oxide width modulation in the drift region.

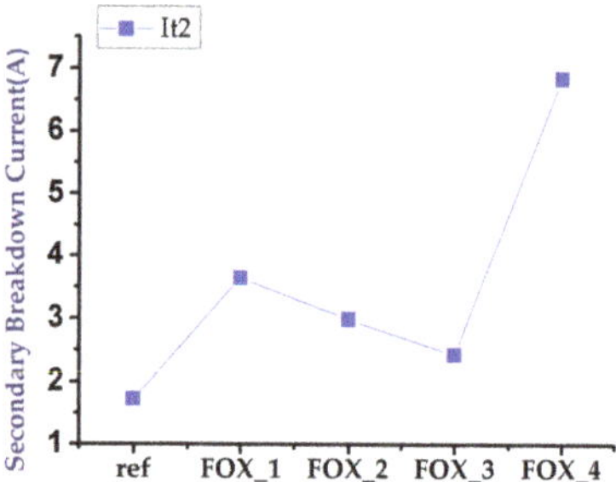

Figure 8. Secondary breakdown current trend chart of nLDMOSs with the field-oxide width modulation in the drift region.

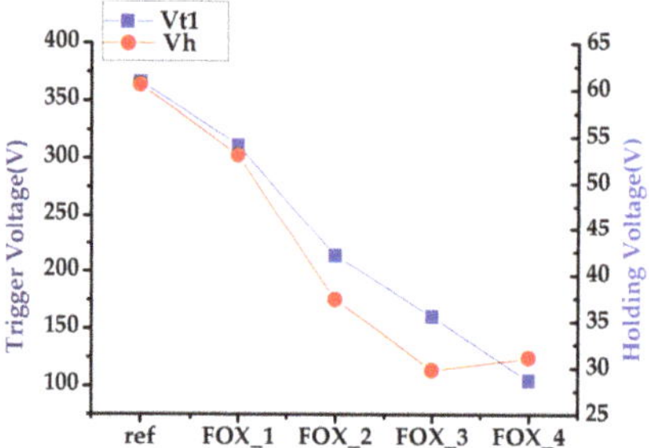

Figure 9. Trigger voltage and holding voltage trend charts of nLDMOSs with the field-oxide width modulation in the drift region.

Table 5. Snapback parameters of field-oxide width modulation in the drift region.

Samples		$V_{t1}(V)$	$V_h(V)$	$I_{t2}(A)$	$V_{BK}(V)$
Ref. nLDMOS		364.44	60.49	2.46	389.59
FOX width	FOX_1	311.40	52.97	3.65	269.85
	FOX_2	213.88	37.34	2.99	142.25
	FOX_3	160.61	29.74	2.43	130.34
	FOX_4	104.89	31.09	6.84	105.43

4.3. UHV Circular nLDMOS—Discrete HVNW Layer Modulation in the Drift Region

Similarly, the experimental results of the UHV nLDMOS-related DUTs with discrete HVNW layer modulation in the drift region obtained from using the breakdown voltage measurement and TLP testing are shown in Figures 10–12. In Figure 10, the breakdown voltage decreases when the

HVNW layer is discrete (or experiences an increase in unit cell spacing). Due to the continuous depletion region formed by HVNW/PBody/DPW in the reference device, it can withstand a breakdown voltage of more than 400 V. However, when the HVNW layer was discrete, the breakdown voltage decreased due to the discontinuous depletion layer, which cause the maximum electric breakdown decreased. Figures 11 and 12 demonstrate that since the concentration of the n-epi layer is lower than the HVNW layer, the holding voltages of these HVNW discrete devices are higher than the reference group voltages, because the concentration is inversely proportional to resistivity, which means that the equivalent resistance of the n-epi is indeed higher than that of the HVNW layer. The test results of the UHV nLDMOS with discrete HVNW layer modulation in the drift region are shown in Table 6.

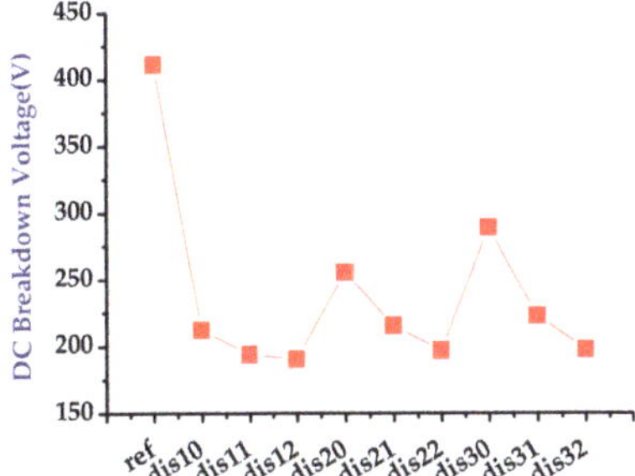

Figure 10. Breakdown voltage trend chart of nLDMOSs with the discrete HVNW layer modulation in the drift region.

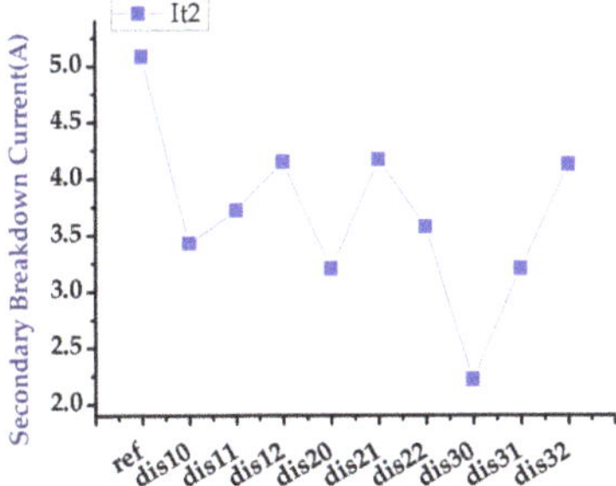

Figure 11. Secondary breakdown current trend chart of nLDMOSs with the discrete HVNW layer modulation in the drift region.

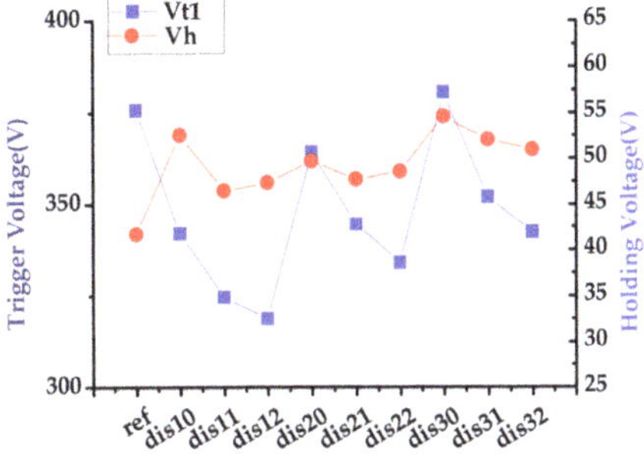

Figure 12. Trigger voltage and holding voltage trend charts of nLDMOSs with the discrete HVNW layer modulation in the drift region.

Table 6. Snapback parameters of discrete HVNW layer modulation in the drift region.

Samples		$V_{t1}(V)$	$V_h(V)$	$I_{t2}(A)$	$V_{BK}(V)$
Ref. nLDMOS		375.71	41.66	5.09	411.20
HVNW discrete	dis 10	341.94	52.52	3.43	212.32
	dis 11	324.67	46.48	3.72	193.79
	dis 12	318.69	47.31	4.15	190.29
	dis 20	364.15	49.65	3.20	255.48
	dis 21	344.48	47.72	4.17	215.27
	dis 22	333.98	48.52	3.57	196.76
	dis 30	380.50	54.60	2.22	289.29
	dis 31	351.92	51.99	3.20	222.75
	dis 32	342.27	50.95	4.12	197.62

4.4. UHV Circular nLDMOS—Embedded P-Well Length Modulation in the Drift Region

Finally, the experimental results of the UHV nLDMOS-related DUTs with embedded P-Well length modulation in the drift region obtained from the breakdown voltage measurement and TLP testing are shown in Figures 13–15. The trigger voltage and the holding voltage increase when the current flow path is blocked by the P-Well, due to the N^+/P-Well reverse bias junction, which results in an increase in the turn-on resistance. Nevertheless, even when the trigger voltage slightly decreases, the holding voltage increases related to the increase in P-Well length. When the P-Well length is 9 μm, the trigger voltage is the lowest and the holding voltage (65.5 V) is the highest. Meanwhile, its secondary breakdown current can be reached at 2.47 A, which is the best among the modulation samples. The test results of the modulations of UHV nLDMOS with embedded P-Well in the drift region are shown in Table 7.

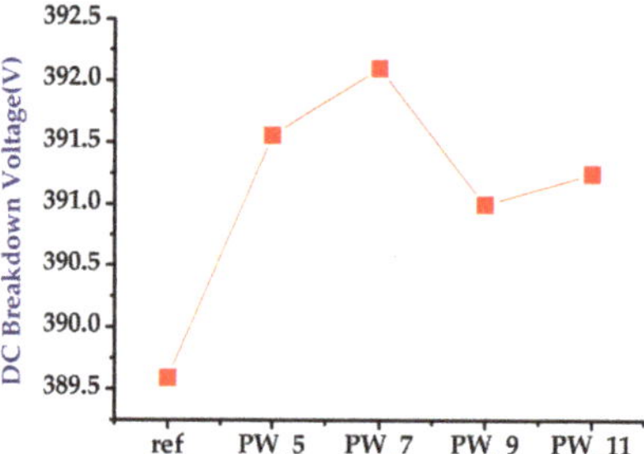

Figure 13. Breakdown voltage trend chart of nLDMOSs with the embedded P-Well length modulation in the drift region.

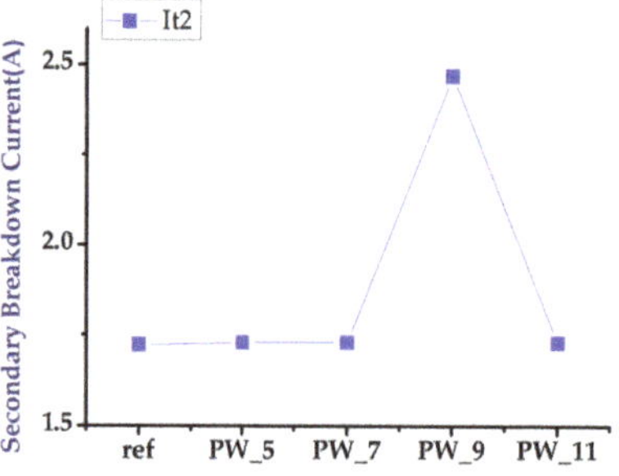

Figure 14. Secondary breakdown current trend chart of nLDMOSs with the embedded P-Well length modulation in the drift region.

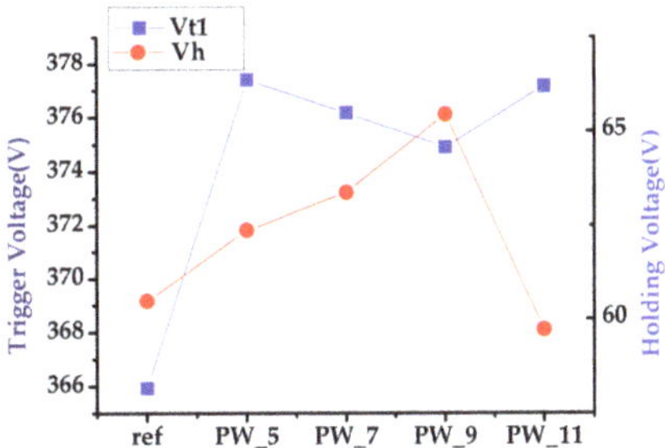

Figure 15. Trigger voltage and holding voltage trend chart of nLDMOSs with the embedded P-Well length modulation in the drift region.

Table 7. Snapback parameters of embedded P-Well length modulation in the drift region.

Samples		$V_{t1}(V)$	$V_h(V)$	$I_{t2}(A)$	$V_{BK}(V)$
Ref. nLDMOS		364.44	60.49	1.72	389.59
	PW_5	377.42	62.39	1.73	391.56
P-Well	PW_7	376.195	63.39	1.73	392.10
	PW_9	374.92	65.46	2.47	391.00
	PW_11	377.19	59.72	1.73	391.25

5. TCAD Simulation

To verify the differences, the impact generation rate profiles of the UHV nLDMOS transistors with (a) a reference device and (b) an embedded 9 μm P-Well in the drift region structures under the $V_G = V_S = V_{Bulk} = 0$ V and $V_D = 310$ V conditions are shown in Figure 16a,b. According to these three-dimensional (3-D) TCAD simulations, the general impact ionization process is described by Equation (1) [33]. Here, G represents the generation rate of the electron-hole pairs, and a device will fail if the G value is too high as in the reference DUT in Figure 16a The ionization coefficients for electrons and holes are $\alpha_{n,p}$, and these coefficients describe the number of electron-hole pairs generated per unit distance traveled by a solitary carrier between two collisions. Their current densities are represented by $J_{n,p}$. The impact generation rate profile of the reference device is higher than that of the embedded P-Well modulation. The minority carriers contribute to the drain current, and the majority of carriers are attracted and collected by the bulk electrode, thereby generating the bulk current. As the bulk current continues to increase, the additional carriers increase the forward currents in the transistors. Increasing the currents leads to massive growth heat generation that can lead to a device failure. Therefore, for ESD, latch-up immunities, and breakdown voltage performance, the embedded 9 μm P-Well in the drift region is the most suitable structure for drain-end modulated engineering.

$$G = \alpha_n |\vec{J}|_n + \alpha_p |\vec{J}|_p \tag{1}$$

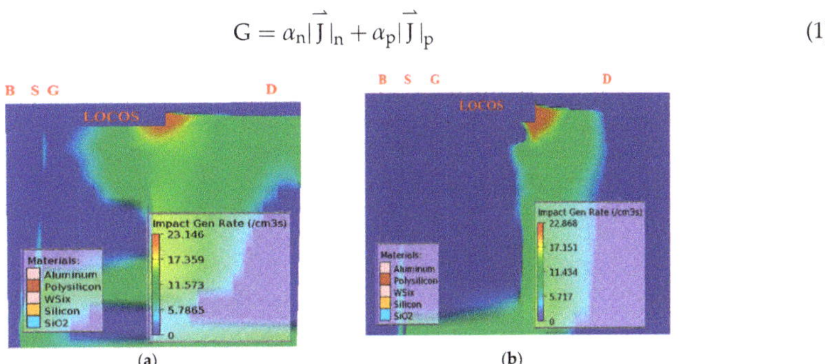

Figure 16. Impact generation rate diagrams of (**a**) reference device and (**b**) embedded 9 μm P-Well in the drift region (full scale: 1×10^{-7} A/cm^2) as the $V_G = V_S = V_{bulk} = 0$ V, $V_D = 310$ V bias condition.

6. Conclusions

Four kinds of modulations are used in circular UHV nLDMOS drift-region engineering: (1) changing the layout of the poly2 layer, (2) field-oxide width modulation, (3) discrete HVNW layer, and (4) embedded P-Well in the drift region. In the first type of modulation, the breakdown voltage increased more than 400 V due to the reduction of the peak electric field and an increase in the secondary breakdown current up to 5 A also occurred. In the second modulation, the breakdown voltage of the drain region reached 105 V, which meant that the operating voltage of the high-voltage circuits could be adjusted by the modulation length of the drift region. For the third modulation, the trigger voltage, holding voltage, and breakdown voltage were adjusted for different application voltages by adjusting the discrete unit cell size and spacing. Finally, for the embedded P-Well of different lengths in the drift region, when the embedded P-Well length was 9 μm, it had the best ESD ability due to the reduction of the impact ionization.

Author Contributions: Conceptualization, P.-L.L.; Formal analysis, P.-L.L.; Investigation, S.-K.F.; Project administration, S.-L.C.; Supervision, S.-L.C.; Validation, P.-L.L.; Writing—original draft, P.-L.L.; Writing—review and editing, S.-L.C.

Funding: This research was no external funding.

Acknowledgments: In this work, authors would like to thank the Taiwan Semiconductor Research Institute in Taiwan for providing the process information and fabrication platform.

Conflicts of Interest: The authors declare no conflict of interest.

References

1. Qiao, M.; Zhang, K.; Zhou, X.; Zou, J.; Zhang, B.; Li, Z. 250 V Thin-Layer SOI Technology with Field pLDMOS for High-Voltage Switching IC. *IEEE Trans. Electron Devices* **2015**, *62*, 1970–1976. [CrossRef]
2. Chen, Y.; Chang, C.; Yang, P. A Novel Primary-Side Controlled Universal-Input AC–DC LED Driver Based on a Source-Driving Control Scheme. *IEEE Trans. Power Electron.* **2015**, *30*, 4327–4335. [CrossRef]
3. Chen, Z.; Salman, A.; Mathur, G.; Boselli, G. Design and Optimization on ESD Self-Protection Schemes for 700V LDMOS in High Voltage Power IC. In Proceedings of the in 37th Electrical Overstress/Electrostatic Discharge Symposium, Reno, NV, USA, 27 September–2 October 2015.
4. Cong, L.; Lee, H. A 110-250V 2MHz Isolated DC-DC Converter with Integrated High-Speed Synchronous Three-Level Gate Drive. In Proceedings of the IEEE Energy Conversion Congress and Exposition, Montreal, QC, Canada, 20–24 September 2015; pp. 1479–1484.
5. Dai, S.; Knepper, R.; Horenstein, M. A 300 LDMOS analog-multiplexed drive for MEMS devices. *IEEE Trans. Circuits Syst. I. Reg. Pap.* **2015**, *62*, 2806–2815. [CrossRef]
6. Yi, B.; Chen, X. A 300-V Ultra-Low-Specific On-Resistance High-Side p-LDMOS With Auto-Biased n-LDMOS for SPIC. *IEEE Trans. Power Electron.* **2017**, *32*, 551–560. [CrossRef]
7. Chang, C.; Jiang, T.; Yang, P.; Xu, Y.; Xu, C.; Chen, Y. Adaptive line voltage compensation scheme for a source-driving controlled AC–DC LED driver. *IET Circuits Devices Syst.* **2017**, *11*, 21–28. [CrossRef]
8. Kim, S.; LaFonteese, D.; Zhu, D.; Sridhar, D.S.; Pendharkar, S.; Endoh, H.; Boku, K. A new ESD self-protection structure for 700V high side gate drive IC. In Proceedings of the 29th International Symposium on Power Semiconductor Devices and IC's, Sapporo, Japan, 28 May–1 June 2017; pp. 467–470.
9. Ker, M.; Lin, C.; Wu, Y.; Wang, W. ESD Protection Design with Low-Leakage Consideration for Silicon Chips of IoT Applications. In Proceedings of the 7th Annual IEEE International Conference on Cyber Technology in Automation, Control and Intelligent Systems, Honolulu, HI, USA, 31 July–4 August 2017; pp. 1496–1499.
10. Wu, J.; Lyu, X.; Kong, M.; Yi, B.; Chen, X. A novel level-shifter integrated on the edge termination region of the high voltage device. In Proceedings of the TENCON 2017 IEEE Region 10 Conference, Penang, Malaysia, 5–8 November 2017; pp. 2683–2686.
11. Li, W.; Makuuchi, M.; Chujo, N. Design of High-Voltage and High-Speed Driver. In Proceedings of the IEEE 12th International Conference on Power Electronics and Drive Systems, Honolulu, HI, USA, 12–15 December 2017; pp. 448–452.

12. Sun, W.; Ye, R.; Liu, S.; Wei, J.; Su, W.; Lin, F.; Sun, G.; Lin, Z. Layout Arrangement Concern for Lateral DMOS With Large Geometric Array Used as Output Device. *IEEE Trans. Device Mater. Reliab.* **2017**, *17*, 450–457. [CrossRef]

13. Yi, B.; Cheng, J.; Chen, X. A High-Voltage Quasi-p-LDMOS Using Electrons as Carriers in Drift Region Applied for SPIC. *IEEE Trans. Power Electron.* **2018**, *33*, 3363–3374. [CrossRef]

14. Adila, A.; Husam, A.; Husi, G. Towards the Self-Powered Internet of Things (IoT) by Energy Harvesting: Trends and Technologies for Green IoT. In Proceedings of the 2nd International Symposium on Small-scale Intelligent Manufacturing Systems, Cavan, Ireland, 16–18 April 2018.

15. Wang, H.; Qiao, M.; Jin, F.; Yu, Y.; Yuan, Z.Y.; Miao, B.; Yang, W.; Wu, J.; Qian, W.; Deng, T.; et al. A 0.35 μm 600 V Ultra-Thin Epitaxial BCD Technology for High Voltage Gate Driver IC. In Proceedings of the 30th International Symposium on Power Semiconductor Devices and ICs, Chicago, IL, USA, 13–17 May 2018; pp. 311–314.

16. Guo, S.; Chen, X. A Novel p-LDMOS Additionally Conducting Electrons by Control ICs. *IEEE J. Electron Devices Soc.* **2019**, *7*, 710–716. [CrossRef]

17. Mohammadi, F. Operation and Analysis of Float Cum Boost Charger in High-Voltage Switchgear Backup System. In Proceedings of the 1st International Conference on Modern Approaches in Engineering Science (ICMAES), Tbilisi, Georgia, 21 November 2018.

18. Ker, M. Whole-Chip ESD Protection Design with Efficient VDD-to-VSS ESD Clamp Circuits for Submicron CMOS VLSI. *IEEE Trans. Electron Devices* **1999**, *46*, 173–183.

19. Ker, M.; Chang, H.; Wu, C. A gate-coupled PTLSCR/NTLSCR ESD protection circuit for deep-submicron low-voltage CMOS ICs. *IEEE J. Solid-State Circuits* **1997**, *32*, 38–51.

20. Chen, T.; Ker, M. Investigation of the Gate-Driven Effect and Substrate-Triggered Effect on ESD Robustness of CMOS Devices. *IEEE Trans. Device Mater. Reliab.* **2001**, *1*, 190–203. [CrossRef]

21. Wang, C.; Ker, M. ESD Protection Design with Lateral DMOS Transistor in 40-V BCD Technology. *IEEE Trans. Electron Devices* **2010**, *57*, 3395–3404. [CrossRef]

22. Ker, M.; Hsu, K. Latchup-Free ESD Protection Design with Complementary Substrate-Triggered SCR Devices. *IEEE J. Solid-State Circuits* **2003**, *38*, 1380–1392.

23. Ker, M.; Hsu, K. Overview of On-Chip Electrostatic Discharge Protection Design With SCR-Based Devices in CMOS Integrated Circuits. *IEEE Trans. Device Mater. Reliab.* **2005**, *5*, 235–249.

24. Keppens, B.; Mergens, M.P.J.; Trinh, C.S.; Russ, C.C.; Camp, B.V.; Verhaege, K.G. ESD protection solutions for high voltage technologies. In Proceedings of the Electrical Overstress/Electrostatic Discharge Symposium, Grapevine, TX, USA, 19–23 September 2004; pp. 289–298.

25. Ker, M.; Lin, K. The Impact of Low-Holding-Voltage Issue in High-Voltage CMOS Technology and the Design of Latchup-Free Power-Rail ESD Clamp Circuit for LCD Driver ICs. *IEEE J. Solid-State Circuits* **2005**, *40*, 1751–1759.

26. Lin, C.; Wu, P.; Ker, M. Area-Efficient and Low-Leakage Diode String for On-Chip ESD Protection. *IEEE Trans. Electron Devices* **2016**, *63*, 531–536. [CrossRef]

27. Jiang, L.; Fan, H.; He, C.; Zhang, B. A reduced surface current LDMOS with stronger ESD robustness. In Proceedings of the 11th IEEE International Conference on Solid-State and Integrated Circuit Technology, Xi'an, China, 29 October–1 November 2012.

28. Wu, C.; Lee, J.; Lien, C. A Novel Drain Design for ESD Improvement of UHV-LDMOS. *IEEE Trans. Electron Devices* **2015**, *62*, 4135–4138. [CrossRef]

29. Yang, F.; Chen, H.; Tian, X.; Bai, Y.; Zhu, Y. Investigation on Current Crowding Effect in IGBTs. *IEEE Trans. Electron Devices* **2018**, *65*, 636–640. [CrossRef]

30. Parpia, Z.; Salama, C.A.T. Optimization of RESURF LDMOS Transistors: An Analytical Approach. *IEEE Trans. Electron Devices* **1990**, *37*, 789–796. [CrossRef]

31. Yang, K.; Guo, Y.; Pan, D.; Zhang, J.; Li, M.; Tong, Y.; He, L.; Yao, J. A Novel Variation of Lateral Doping Technique in SOI LDMOS With Circular Layout. *IEEE Trans. Electron Devices* **2018**, *65*, 1447–1452. [CrossRef]

32. Pjenčák, J.; Agam, M.; Šeliga, L.; Yao, T.; Suwhanov, A. Novel Approach for NLDMOS Performance Enhancement by Critical Electric Field Engineering. In Proceedings of the 30th International Symposium on Power Semiconductor Devices & ICs, Chicago, IL, USA, 13–17 May 2018; pp. 307–310.

33. *Atlas User's Manual*; Silvaco Group Inc.: Santa Clara, CA, USA, 2018.

 *electronics*

Article

Generative Noise Reduction in Dental Cone-Beam CT by a Selective Anatomy Analytic Iteration Reconstruction Algorithm

Lam Dao-Ngoc and Yi-Chun Du *

Department of Electrical Engineering, Southern Taiwan University of Science and Technology, No. 1, Nan-Tai Street, Yung-Kang district, Tainan 71005, Taiwan; da62b202@stust.edu.tw
* Correspondence: terrydu@stust.edu.tw; Tel.: +886-6-253-3131 (ext. 3321)

Received: 23 October 2019; Accepted: 17 November 2019; Published: 21 November 2019

Abstract: Dental cone-beam computed tomography (CBCT) is a powerful tool in clinical treatment planning, especially in a digital dentistry platform. Currently, the "as low as diagnostically acceptable" (ALADA) principle and diagnostic ability are a trade-off in most of the 3D integrated applications, especially in the low radio-opaque densified tissue structure. The CBCT benefits in comprehensive diagnosis and its treatment prognosis for post-operation predictability are clinically known in modern dentistry. In this paper, we propose a new algorithm called the selective anatomy analytic iteration reconstruction (SA2IR) algorithm for the sparse-projection set. The algorithm was simulated on a phantom structure analogous to a patient's head for geometric similarity. The proposed algorithm is projection-based. Interpolated set enrichment and trio-subset enhancement were used to reduce the generative noise and maintain the scan's clinical diagnostic ability. The results show that proposed method was highly applicable in medico-dental imaging diagnostics fusion for the computer-aided treatment planning, because it had significant generative noise reduction and lowered computational cost when compared to the other common contemporary algorithms for sparse projection, which generate a low-dosed CBCT reconstruction.

Keywords: cone-beam computerized tomography (CBCT); as low as diagnostically acceptable (ALADA); selective anatomy analytic iteration reconstruction (SA2IR); low-dosed; sparse projections; diagnostic ability

---

1. Introduction

Digital dentistry offers a comprehensive workflow in oral healthcare treatment and monitoring for most of the daily clinical practical protocols [1,2]. Oral and dentition anatomy are key to the image diagnostic efficiency, which would reduce complications in planning. This makes the operations more accurate, predictable, and safer. This also benefits the clinical expectation, which would help improve patient comfort. DICOM-based (Digital Imaging and Communications in Medicine) virtual planning is the golden standard in dental implantology, orthodontics, maxillofacial surgery, and comprehensive cosmetic dentistry. This method is used for systematic and specific computer-aided functionality. Cone-beam computed tomography was first introduction in the 1990s for dental implantology. Cone-beam computed tomography (CBCT) has been integrated into the current workflow of digital dentistry and has been improved upon by computer-aided design and manufacturing (CAD/CAM) solutions. These advancements have improved the clinical predictability and success rate; this is evidence-based and has been studied in dental literature. It is currently recommended in the dental professional community to use CBCT as a standard protocol [3–6]. Dental 3D imaging equipment allows clinicians to view enhanced visual information to aid in the correct diagnosis of the patient.

Newer imaging equipment has improved upon previous clinical results by producing reliable and consistent treatment planning. Soft tissue surface morphology has been digitally reproduced through 3D scanned modeling via CBCT or oral 3D scanners, which includes intra-oral and desktop scanners. Therefore, three-dimensional multi-modality fusion diagnosis is recommended to be applied for digital treatment planning standardization in modern dentistry [5–7].

According to the latest radioactive safety directives, dental CBCT reconstruction algorithms are being developed for dental hard and soft tissues attenuation density, and must be designed to avoid excessive radiation exposure. Those are issued as limitations to the number of projections and the exposure dosage, per a single revolution, or a couple revolutions. Recently, the standard for dosage optimization has changed from "As Low As Reasonably Achievable" (ALARA) to "As Low As Diagnostically Acceptable" (ALADA). This has caused a need for refined algorithms that lie within the range of ALADA for the new acceptable dosage optimizations. At the 2014 NCRP (National Council on Radiation Protection of United States) Annual Meeting, the term "ALADA" was first proposed by Dr. Jerrold T. Bushberg as a variation of the acronym ALARA to emphasize the importance of optimization in medical imaging [8]. ALADA began initially in pediatric imaging and then expanded into dental imaging, in regard to the Image Gently® Campaign (for the pediatric population) [9] and Image Wisely® Campaign (for the adult patients) to promote the clinician's responsibility in both medico-dental imaging indications and practices [8,10,11]. The "trade-off" constraints are between diagnostically valuable and radioactive safe practices. Currently, these are coined as dose reduction and essential informativeness enhancement. This makes it possible for three main approaches in 3D reconstruction that have high computational efficiency. The first approach is based on the approximation and its regularization. This is called the "regularized approach". The second approach is based on statistical modeling, in regard to the objects' anatomy and the radiation exposure physics; this is called the "statistical approach". The last one is based on the efficiency of parallel computation for sparse representation in the inverse problems, which is integrated with the graphical processing unit (GPU), which is called the "GPU-based approach". Over the past decade, difficulties in dental low-dosed CBCT and sparse-projection 3D reconstruction have been studied extensively. The first issue, inefficient input quality, is caused by generative blurring and defects, which are termed "generative noises". The second issue is ill-posed computation due to the sparsity of input projections. The aforementioned issues are being worked on intensively. These issues are addressed by approximate solutions and are iteratively solved by using compressed sensing theory for the under-defined sampling, or one can use the sparse representation theory for general sparsity computation. This is most efficiently implemented with GPU-based paralleling computation, during image post-processing [12]. As result, the fusion solutions of both estimation and filtering have been used globally, locally, and adaptively. These solutions are trended in technically prior-interpolated dictionary learning, and supervised and unsupervised machine learning. Afterwards, it will be clinically verified by professionals or expertized end-users in the field.

On the engineering side, it is not intuitive to understand clinical diagnostic uncertainties. This is due to the lack of anatomical and pathologic knowledge, and also in part due to a lack of clinical experience or understanding [13,14]. Similarly, the same situation is concerned in dental image diagnostics, which has been subjectively evaluated by clinicians as the augmented tool for their individual visuality experience, in which any uncertainties may cause clinical failures or non-manageable complications, in the studies of Jacobs et al. (2018) [14] and Katsumata et al. (2007) [15]. The importance of application understanding and experience for both sides is de facto essential to improve the clinical user's professionality and in vivo practical confidence. Additionally, with the recent advances in visualization augmentation, multi-formatted data structural registering, computer-aided design, and manufacturing integration, digitalization has been applied in modern dentistry. In fact, digitalization has been proposed to be applied in most branches of dentistry such as implantology [2,16], maxillofacial surgery (orthognathic or dentofacial cosmetic surgeries) [17], oral surgery, orthodontics, endodontics [18,19], periodontics [20], and the viral digital smile design in some comprehensive applications [21]. However, comprehensive dental implantology and maxillofacial surgery have been pioneered and successfully developed. It is recommended that modern digital diagnostic dentistry is used for dental implantology and maxillofacial surgery [22].

In this work, the proposed algorithm focuses on projection-based processing prior to reconstruction. The algorithm is made up of two parts: (a) projection-based pre-processing and (b) three-dimensional reconstruction. Due to the sparsity and low-dosed effects, the initial input projections set is interpolated to make a "pseudo up-sampling"; then, the interpolated pseudo-set is used as a dictionary to reweight for the new set bilaterally, which is based on both prior and posterior projections. According to the quadrant-based specifications, the anatomic containment in each projection is statistically metric and verified in the orthogonal and complimentary paired sets per each quadrant. After enhancing, a new pseudo-set is updated. This new set of projections is used to reconstruct iteratively the final three-dimensional model, which is processed by the GPU for computational effectiveness. Therefore, this algorithm is a combination of the regularization and statistical approaches and is implemented by GPU-based computation. The proposed algorithm is named "selective anatomy analytic iteration reconstruction" (SA2IR). This study proposes an alternative reconstruction algorithm for CBCT using dental imaging diagnostics, which is used to prevent the clinical radiological imaging overexposure and thus improve the radiation safety and assure the clinical diagnostic quality for the digital planned treatments in dentistry. For the detailed factors, the algorithm will be clarified and demonstrated further in the Methodology, Results, Discussion, and Conclusion.

2. Methodology

2.1. Equipment Configuration of the Duplicated Simulation

In order to lower the dosage radiation in image diagnostics, we propose a new approach using the sparse set of the CBCT's projections; this corresponds to the specific anatomical similarity analysis of the orthogonal and complimentary paired sets. This is different from other inverse problems in CBCT 3D reconstruction. The key to our approach is focused on sparse projection reconstruction, using new re-generatively interpolated projections to create a pseudo-set. This is based on their orthogonal and complimentary constraints by making successive prior-et-posterior weight approximations. The assumptions of our approach are as follows:

(1) Use a flat-panel imaging detector (FPID) to determine the CBCT modality.
(2) Set the number of the received projections to sparse.
(3) Scan the region of interest in dental anatomy structures for various diagnoses such as head-and-neck, oral, and dentition anatomy.

To simulate a real CBCT system, the parameters of CBCT DCT100-0X0 (Taiwan Care Tech Corporation (TCT), Taiwan, Integrated Biomedical System laboratory, STUST) is duplicated for the experimental studies, as shown in Table 1.

Table 1. DCT100-0X0 cone-beam computed tomography (CBCT) technical configuration. FPID: flat-panel imaging detector.

Technical Parameters	Information			-
	Notation	Unit	Value	
Source-to-Detector Distance	SDD	cm	72	Term of "Patient"
Source-to-Patient Distance	SPD	cm	50	position means
Patient-to-Detector Distance	PDD	cm	22	"Rotation Axis" position
X-ray beam exposure size	WE × HE	cm × cm	12.8 × 12.8	Width × Height
Cone-beam opening of FPID	β_x β_z	(°)	9.15° 9.15°	Respected to xx and yy axis

Table 1. *Cont.*

| Technical Parameters | Information | | | - |
	Notation	Unit	Value	
Detector's size (Width × Length)	W × L	cm × cm	13 × 13	Amorphous Silicon Receptor
Voxel size	V_x	mm	0.125 and 0.200	-
Rotation degree	A	(°)	180° or 360°	Projection arch per scan
Field-of-View (Diameter × Height)	FOV (D × H)	cm × cm	9 × 9 and 15 × 9	-
Projection rotation angular step	α_{pst}	(°/step)	0.6	-
Number of projections	N_p	projections	300 or 600	np = α/pp

2.2. Selective Anatomy Analytic Iteration Reconstruction (SA2IR)

In the overview, the regularization in CBCT reconstructed algorithms is used primarily in two ways: as analytic and synthetic formulations. Generically, the regularization is to minimize the cost function, with respect to the successive forward or backward projections per each projection, which are defined in Equation (1) as

$$\hat{p} = argmin_p\{Y(p) \triangleq J(p) + \lambda * \psi(S * p)\} \tag{1}$$

wherein $J(\cdot)$ represents data fidelity, $\psi(\cdot)$ is a functional regularization operator, S represents the sparse transforming coefficient set, and λ is a regularization parameter to optimize the computing. Due to the sparse view conditions for the low-dosed reconstruction, the regularized algorithms have trade-offs between image quality, computational complexity, and diagnostic ability, as mentioned in the literature reviews and studies [23–28]. However, excessively generating noise and significantly reducing the diagnostic values of the reconstructed images, sparsity, and low-dose exposure per projection are highly desirable in the recent studies. One such study is the sinogram-based dictionary learning patched-based algorithms in reconstructed images denoising and enhancement, by Karimi and Ward (2016) [23]. Another study by Zhu et al. (2013) applied compressed sensing (CS) algorithms when violating the sampling theory in the discrete wavelet or discrete gradient transforms by using the total-variation-based (TV-) CS algorithm to suppress the streaking artifacts significantly without any image quality compromises [24]. Zhang et al. (2016) used the sinogram-based inpainting technique in metal artifact reducing (MAR) [25]. In a different study Zhang et al. (2015) [26] used the combination of compressed sensing and dictionary learning to regularize parameter determination via sparse constraint of the TV minimization to reduce the computational cost. Du et al. (2019) [27] also used compressed sensing image recovery through dictionary learning and thresholding shape-adaptiveness of the discrete cosine transform (DCT). The sinogram-based and regularization solutions have been proved practically efficient in sparse representative for low-dose CT reconstruction. Kim et al. (2018) [28] proposed the iteration algorithm for the compressed-sensing reconstruction method to reduce the computational cost regularized voxelization. Therefore, the proposed algorithm is implemented to generate a pseudo-set from the input projections and the "prior-et-posterior" bounded coefficients are reweighed in order to reconstruct the three-dimensional volume. The input projections set is implemented from a set of preset angular-position-wise events in order to capture the projection. By means of the sparse view conditions and in order to have a safer low-dose mode to approach the ALADA requirements, the continuity of the input sequence is not efficient enough to reconstruct the diagnostic-able images. To address this issue, we propose to enrich the optimum quantity of the projections set and enhance the projection set quality before reconstructing the final three-dimensional model.

Firstly, the interpolation enrichment is implemented by Algorithm 1. The initial projection set $\left\{p_i^{init}\right\}\Big|_{n_{projection}^{init}}$ is sparse viewed, so the proposal of input set enrichment is implemented by the interpolation as a denser pseudo-set of projections with the projection number $n_{projection}^{interp} = n_{projection}^{init} *$ $(k_{int} + 1) - k_{int}$ to adapt the completeness of the basic computed tomography (CT) reconstruction fundamentals. Secondly, to enhance the projections set quality, our algorithm does not use the prior regularization as the others, but a loop of an "estimating–filtering–verifying" process for the regenerative interpolation (pseudo-) set of input projections is proposed to be applied with the same projection number $n_{projection}^{regen} = n_{projection}^{interp}$. This is based on the pseudo-complete interpolated projection set before applying the final reconstruction. It is an enhancement process. The generative interpolation set of the initial projections set is implemented by updating the successive trio projection subset per each new projection. This is similar to a sliding window modulation. That regeneration is individualized per each normalized view of the full revolution, as shown in Figure 1. The prior projection, as p_{i-1}^{interp}, is used to estimate an in-flow overlapping of the successive one, as $\hat{p}_i^{new} = K_i^{in-flow} * p_{i-1}^{interp}$, in which the $K_i^{in-flow}$ maximization is focused, which is termed the "Bhattacharyya distance" (D_B), as shown in Equation (2). Based on the total similarity, proportional to $K_i^{in-flow}$, the updated projection for a new (pseudo-) set is estimated and pre-enhanced by the first two projections of a trio subset, respectively with the interpolated prior projection and the interpolated central projections, which is implemented via the discretized probability distribution.

$$D_B\left(p_i^{interp}, p_{i-1}^{interp}\right) = -\ln\left(BC\left(p_i^{interp}, p_{i-1}^{interp}\right)\right) \tag{2}$$

wherein $BC\left(p_i^{interp}, p_{i-1}^{interp}\right) = \sum_{p_{i-1}, p_i \in \{P\}}^{interp} \sqrt{p_i^{interp} * p_{i-1}^{interp}}$, $0 \leq BC\left(p_i^{interp}, p_{i-1}^{interp}\right) \leq 1$ is called a "total similarity coefficient" (TSC), which is defined by "Bhattacharyya coefficient" (BC) of the pair of interpolated projections, of p_i^{interp} and p_{i-1}^{interp}, while $K_i^{in-flow}$ is called the "estimated operator", which is defined as shown in Equation (3).

$$K_i^{in-flow} = 1 - D_B\left(p_i^{interp}, p_{i-1}^{interp}\right) \tag{3}$$

The central projection, as a kernel of (p_i^{interp}), is used to filter the result of $\hat{p}_i^{new}$, while the posterior projection, as p_{i+1}^{interp}, is used to verify the $\hat{p}_i^{new}$, the updated regenerative projection, as $p_i^{regen} = \hat{p}_i^{new} = K_i^{out-flow} * p_{i+1}^{interp}$, in which the minimization of $K_i^{out-flow}$ is focused and termed as Kullback–Leibler's divergence conditioning, as its distance (D_{KL}) is shown in Equation (4). Then, the verification is implemented by the informative discrimination between the $\hat{p}_i^{new}$ and the posterior one, p_{i+1}^{interp}, to update the regenerative projections as $p_i^{regen} = \hat{p}_i^{new}$, if the informative discrimination is not exceeded ξ (computational parameter). Otherwise, $p_i^{regen} = \left(\hat{p}_i^{new}\right)^*$. The $\left(\hat{p}_i^{new}\right)^* = (1 - \xi) * p_{i+1}^{interp}$ is the proper state of $\hat{p}_i^{new}$, which is modified by the L_2-norm of the relevant p_i^{interp}, $\hat{p}_i^{new}$ to p_{i+1}^{interp}, as shown in Equation (5).

$$D_{KL}\left(\hat{p}_i^{new} \| p_{i+1}^{interp}\right) = -\sum_{p_i, p_{i+1} \in \{P\}}^{interp-regen} (p_{i+1}^{interp} \log\left(\frac{\hat{p}_i^{new}}{p_{i+1}^{interp}}\right)) \tag{4}$$

wherein $0 \leq D_{KL}\left(\hat{p}_i^{new} \| p_{i+1}^{interp}\right) \leq D_{KL}\left(p_i^{interp} \| p_{i+1}^{interp}\right) \leq 1$, and $K_i^{out-flow}$ is called "verification operator".

Let $\Delta_{KL}^i(\cdot)$ be the "absolute informative discrimination" of the verification at the ith projection, which is shown in Equation (5), as:

$$0 \leq \Delta_{KL}^i\left(projection_i \Big| p_{i+1}^{interp}\right) \triangleq \| D_{KL}\left(\hat{p}_i^{new} \| p_{i+1}^{interp}\right) - D_{KL}\left(p_i^{interp} \| p_{i+1}^{interp}\right) \|_2 . \tag{5}$$

Let $0 \leq \sigma^i_{r.i.d.}\left(\hat{p}^{new}_i, p^{interp}_i\right) = \dfrac{D_{KL}\left(p^{interp}_i \| p^{interp}_{i+1}\right)}{D_{KL}\left(\hat{p}^{new}_i \| p^{interp}_{i+1}\right)}$, where $\sigma^i_{r.i.d.}$ is called the "relative informative

discrimination" between the estimated and the origin of the ith interpolated projection $p^{(\cdot)}_i$. Substituting in Equation (6), we have:

$$0 \leq \Delta^i_{KL}\left(projection_i \| p^{interp}_{i+1}\right) \triangleq \| D_{KL}\left(\hat{p}^{new}_i \| p^{interp}_{i+1}\right)\left(1 - \sigma^i_{r.i.d.}\left(\hat{p}^{new}_i, p^{interp}_i\right)\right) \|_2 . \tag{6}$$

So, the updated projection is assigned to the regenerated projections set by "verification operator", which is defined as shown in Equation (7):

$$0 \leq K^{out-flow}_i = \begin{cases} 1, & if \ \sigma^i_{r.i.d.}\left(\hat{p}^{new}_i, p^{interp}_i\right) \to 1 \\ \Delta^i_{KL}\left(projection_i \middle| p^{interp}_{i+1}\right) \end{cases} \tag{7}$$

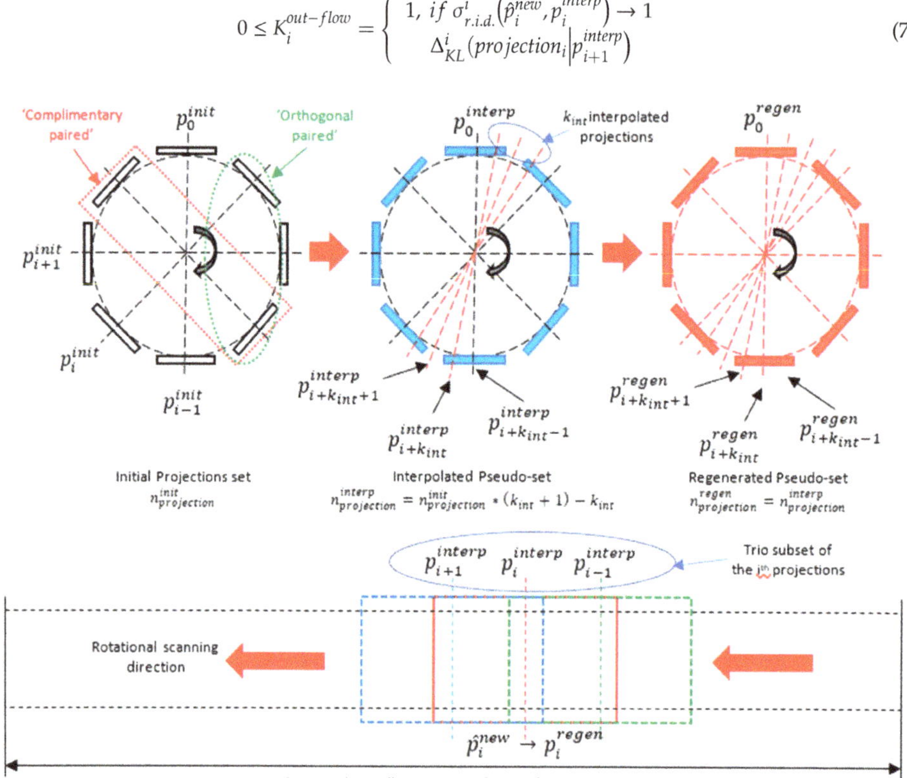

Figure 1. The illustrations of the projection set and the trio subset in selective anatomy analytic iteration reconstruction (SA2IR).

Algorithm 1: "Densification enrichment process" or Pseudo-Set Interpolation

---

Input: Initial projection set captured by FPID, as follows: $\left\{p_i^{init}\right\}\Big|_{n_{projection}^{init}} = \left\{p_0^{init}, p_1^{init}, \dots, p_{n_{projection}^{init}}^{init}\right\}$

Output: Interpolated pseudo-set for the reconstruction completeness availability, as

follows: $\left\{p_i^{interp}\right\}\Big|_{n_{projection}^{interp}} = \left\{p_0^{interp}, p_1^{interp}, \dots, p_{n_{projection}^{interp}}^{interp}\right\}$

Begin   Initialize necessary parameters. For $k = 1$ to $k = $ *maximum number of iterations*:

1: $p_0^{interp} = p_0^{init} \leftarrow p_0^{init}, n_{projection}^{init}, k_{int}$ and: $n_{projection}^{interp} = n_{projection}^{init} * (k_{int} + 1) - k_{int}$. Find the common

projection angle coincidences of two sets: $\left\{p_i^{init}\right\}\Big|_{n_{projection}^{init}}$ and $\left\{p_j^{interp}\right\}\Big|_{n_{projection}^{interp}}$

2: Assign a kernel to remove the non-attenuated outer-cored region of interest (ROI). Implement the iteration loop by $n_{projection}^{interp}$ times. Then, assign the targeted pseudo-set with a denser viewed set, implanted with k_{int} more interpolated projections in each interval of $\left\{p_i^{init}\right\}\Big|_{n_{projection}^{init}}$, between p_i^{init} and p_{i+1}^{init}.

3: Correct iteratively the implanted interpolated projections' sinograms quality by structural similarity and mean squared errors with the given number of iterations. Assign the corrected ones as the final interpolated pseudo-set $\left\{p_i^{interp}\right\}\Big|_{n_{projection}^{interp}}$.

End

Our proposed enhancement process of the pseudo-set is shown in Algorithm 2.

Algorithm 2: "Quality enhancement process" or Pseudo-Set Trio Subset Enhancement Operator

---

Input: Interpolated pseudo-set, as followed: $\left\{p_i^{interp}\right\}\Big|_{n_{projection}^{interp}} = \left\{p_0^{interp}, p_1^{interp}, \dots, p_{n_{projection}^{interp}}^{interp}\right\}$

Output: Regenerative pseudo-set for reconstruction, as follows:

$\left\{p_i^{regen}\right\}\Big|_{n_{projection}^{regen}} = \left\{p_0^{regen}, p_1^{regen}, \dots, p_{n_{projection}^{regen}}^{regen}\right\}$

Begin

1: $p_0^{regen} = p_0^{interp} \leftarrow p_0^{interp}, n_{projection}^{interp}$ and: $n_{projection}^{regen} = n_{projection}^{interp}$

2: Apply the trio subset's "estimating–filtering–verifying" processing loop per each projection in the interpolated pseudo-set iteratively.

1. Estimate the $\hat{p}_i^{new}$ by the prior projection p_{i-1}^{interp} and calculate the estimation operator $\left\{K_i^{in-flow}\right\}$, using Equation (3).

2. Apply the convolutional filter the $\hat{p}_i^{new}$ by the center projection p_i^{interp} and calculate the filtering operator K_i^{filt}. Update the $\left\{\hat{p}_i^{new}\right\}$ with the filtered $\hat{p}_i^{new}$.

3. Verify the $\hat{p}_i^{new}$ by the posterior projection p_{i+1}^{interp} by "verification operator" $\left\{K_i^{out-flow}\right\}$, using Equation (7).

3: Assign the corrected ones as the regenerative pseudo-set $\left\{p_i^{regen}\right\}\Big|_{n_{projection}^{regen}}$.

End

Finally, the output three-dimensional reconstruction is computed by our proposed variation of simultaneous iterative reconstruction technique (SIRT), which is called SA2IT (selective anatomy analytic iteration technique). The SIRT-based solution is chosen due to its advantage on the higher precision imaging quality reconstruction from the less accurate set of the appropriate noise-patterned projections, which is firstly introduced for medical computed tomography applications. However, its disadvantages are also mentioned in numerous studies, including (1) the expensive computational cost of the iteration, (2) the normalized blurring effect at the transition boundary of the less-intensity-discriminated levels, and (3) globalized intensity scale shifting, bandwidth truncating, and warping intensity distribution. Additionally, the effects of the background intensity levels affect the global contrast of an image, and the human-stimulated vision (as Weber's Law) will be ignored in this study. The proposed SA2IT is conceptualized as a SIRT variation with constraints regarding the post-reconstruction average volumetric stabilization coefficients, which is recalled from any view of the prime input projections. This is constructed by the complimentary paired and the orthogonal paired projections. The proposed workflow is shown in Algorithm 3. As an iterative reconstruction technique, the generalization of Equation (8) for SA2IT is applied in this algorithm.

$$F^{k+1} = F^k - \gamma * C * A^T * R * \left(A * F^k - P\right) \tag{8}$$

wherein:

$$\begin{cases} A_{M \times N} = \left. a_{i,j}\right|_{i=[1,M],\ j=[1,N]} \\ C_{i,i} = \frac{1}{\sum_{j=1}^{M} a_{i,j}}, \text{ and } R = \frac{1}{\sum_{j=1}^{N} a_{i,j}}, \text{ where } C \text{ and } R \text{ anddiagonal} \\ B_{i,j} = \frac{A_{i,j}}{\sqrt{A_{i+} * A_{j+}}}, \text{ converge when}: \ B^T * B < 2, \text{ and } \left(V_0^{calib} - V_{SIRT}^{GD}\right) < \varepsilon \end{cases}$$

where k is the iteration number, γ is a noise reduction relaxation parameter, $C = \{C_{i,i}\}$ and $R = \{R_{i,i}\}$ are weighted coefficients for the back-projection and correction computation of the projection, successively, and $\left(A * F^k - P\right)$ is the correction computation based on the forward projection. The $B = \{B_{i,i}\}$ is the convergence condition recommended by Byrne et al. [29]. Additionally, the orthogonality featured convergence condition, $\left(V_0^{calib} - V_{SIRT}^{GD}\right) < \varepsilon$, is proposed by our SA2IT, in which V_0^{calib} and V_{SIRT}^{GD} are known as the orthogonality-featured bounded volume and the gradient descent SIRT (GD-SIRT) reconstructed volume. The corresponded correction computation is indicated in the term of the denser pseudo-set $\left(\left\{p_i^{regen}\right\}\Big|_{n_{projection}^{regen}}\right)$, which is enriched, and enhanced by trio-subset principle as above.

Algorithm 3: SA2IT algorithm

Input: Regenerative pseudo-set, as follows: $\left\{p_i^{regen}\right\}\Big|_{n_{projection}^{regen}} = \left\{p_0^{regen}, p_1^{regen}, \ldots, p_{n_{projection}^{regen}}^{regen}\right\}$

Output: The SA2IT results

Begin   Initialize necessary parameters. For $k = 1$ to $k = maximum\ number\ of\ iterations$:

1: Apply the conventional SIRT algorithm to compute forward projection $\left(A * F^k\right)$, correction computation $(A * F^k - P)$, back-projection, and the reconstructed volume $(V_{SIRT}^{rec} = V\{F^{(k+1)}\})$, by Equations (8) and (9).

2: Apply the gradient descent to the reconstructed $V_{SIRT}^{rec} \rightarrow V_{SIRT}^{GD}$.

3: Assign the corrected ones, adapted $\left(V_0^{calib} - V_{SIRT}^{GD}\right) < \varepsilon$.

End

---

3. Experimental Setup

3.1. Phantom Reconstruction Experiments

As mentioned in dental literature and professional protocols, dental cone-beam computed tomography is used to give more clinical evidence for treatment planning and monitoring, which is anatomically relevant to hard tissues and various sinuses and air cavities of the maxillofacial region as well as dental structures. In ALADA's radiation safety standard, the exposure dose is accurately managed to maintain the quality of diagnostic ability while strictly conforming to the patient's radiation safety. Therefore, the experiments are designed as a sparse–viewed exposure of X-ray to lower the dosage per projection. This is different from the other algorithms; the proposed SA2IR enriches the input projection set with various sparsity, which is formulated with the densifying factorization or pseudo-up-sampling, to implant the possible slice into the sinogram of the original input. The specific number of input projections is set to minimum due to the trio-subset featured enhancement (a subset of three projections' enhancement) and the orthogonality (periodic quadrant orthogonality per revolution) of the algorithm. When then SA2IR algorithm is applied, the pseudo-set is quantitatively enriched and qualitatively enhanced. Its reconstruction is based on SA2IT and was introduced in Part 4. In this work, the modified Shepp–Logan phantom, a variant of the numerical phantom for computed tomography simulation suggested by Shepp and Logan (1974) [30], was used to simulate the studied experiment results. For the clinical simulation, the realistic digital three-dimensional head phantom for C-Arm computed tomography supported by Aichert et al. (2013) [31] was used to simulate the clinical applicability of the algorithm in dental diagnostic imaging.

3.2. Comparison with the Other Reconstruction Algorithms

The reconstructed result of the proposed SA2IR algorithm is compared to the results of other algorithms, such as the Feldkamp–Davis–Kress (FDK) Algorithm [32], simultaneous iterative reconstruction technique (SIRT) [33], simultaneous algebraic reconstruction technique algorithm (SART) [34], order-subset SART (OS-SART) [35], total variation SART (TV-SART), adaptive steepest descent projection onto convex sets (ASD-POCS) [36], order-subset ASD-POCS (OS-ASD-POCS) [36], and conjugated gradient least square algorithm (CGLS) [37], and fast iterative shrinkage-thresholding algorithm (FISTA). Furthermore, the application of the trio-subset enhancement process to the sparse input of those is also implemented and studied.

4. Results and Discussions

4.1. Results of the SA2IR Algorithm's Reconstruction Simulation

The simulated results of the proposed SA2IR are demonstrated into three parts: (a) the pseudo-set enrichment process, (b) the trio-subset enhancement process, and (c) the SA2IT reconstruction, respectively. The results are conducted on the modified Shepp–Logan phantom and the realistic patient's head data. The simulations are implemented in the MATLAB environment (MATLAB 2017b, The MathWorks Inc., Natick, Massachusetts, USA) and tomographic iterative GPU-based reconstruction framework [38]. The simulations are based on the configuration of DCT100-0x0 CBCT (shown in Table 1) and implemented with the compilation support of the Nvidia CUDA toolkit 10.1, accelerated by Nvidia GTX 1650 GPU. Corresponding to the experimental setup, the specific number of input projections is simulated at the minimum of 12 projections (conducted by the $k_{sparse} = 6$), which was demonstrated with the odd and even quadrant-bases number. The simulation results are conducted with $k_{sampling}$ values of 3, 8, and 14, which successively corresponded to the number of interpolated projections of 36, 96, and 168. The relevant sinograms of the enrichment process for the cases, including those sparse projections, denser projections, and enriched projections sets, are shown in Figure 2. Obviously, the lesser number of projection input is, the lower the reconstruction quality. Therefore, our proposed algorithm inserts the common between a sparse input to the denser expectation, with the given $k_{sampling}$ values of 3, 8, and 14, respectively as the number of projections in the enriched pseudo-sets. That made no changes in the histograms. However, it made a bounding enhancement. The generative noise is patterned as two parts, as external and internal patterned effects. The external patterned effect is shown as the bright streaked bounded frames and the aliased interfere, which is proportional to the times of overlapping, in terms of the common interaction of those, while the internal patterned effect is shown as a cumulative "edge-blurring". It causes an effect, which is similar to contrast diffusion from the internal patterned out to the external patterned part, in the meaning of gradient, while preserving the histogram. That is partly also discussed by Perona and Malik regarding the Perona–Malik diffusion, which is known as a nonlinear anisotropic diffusion and is equivalent to Gaussian blurring in the case of constant diffusion coefficient, as a conventional heat equation [39]. In the case of SA2IR enrichment, the noise pattern is disturbed by the spacing of the interpolation. The densification of those depends on the overlapping of the common projections of the initial and the pseudo-set, which would be normally distributed ($n_{interp} = 96$ and 144) or partly compressed ($n_{interp} = 168$). Due to the symmetricity of the phantom, the orthogonality repetition base on equilateral segments is studied, which shows the behavior of the pseudo-set's cumulative noise pattern. That orthogonality of the post-enrichment processing interpolated pseudo-set is also shown via eight equilateral segments of one scanning revolution (45 degrees per each segment), which are named successively as Ω_1, Ω_2, … , Ω_7 and Ω_8. The pattern and deterioration of the orthogonality per each segment are detected. The noise distribution changes per one scanning revolution; especially at the transition points at 0 degrees and 360 degrees, the flattened effect is observed. Otherwise, the shifting effects and blurring are differentiated per orthogonality pattern and the deterioration of each segment.

The parameters of trio-subset enhancement processing, such as the estimation operator ($\{K_i^{in-flow}\}$), filtering operator ($\{K_i^{filter}\}$), verification operator ($\{K_i^{out-flow}\}$), and their representation, are called 'total trio-subset enhancement operator' ($\{K_i^{total}\}$), which is shown quantitively in Table 2. The contribution of the operators on the total enhancement operator ($\{K_i^{total}\}$) is calculated, which corresponded to the overlapped indices of each projection with its prior and posterior projections. The results show the efficiency of the bilateral approximation to construct a new pseudo-set of the same interpolated projections. The results and the relationship between the estimation and verification processes of three cases A, B, and C, with the specific common views or projections in each segment Ω_i are shown in Table 2, between D_B^i, $K_i^{in-flow}$, Δ_{KL}^i, and $K_i^{out-flow}$. Those specific common views are categorized as the first, middle, and last projections groups; these describe the behavior of each process per each segment Ω_i and per different cases, which are set with 96, 144, and 168 projections to generate a new pseudo-set prior to the reconstruction.

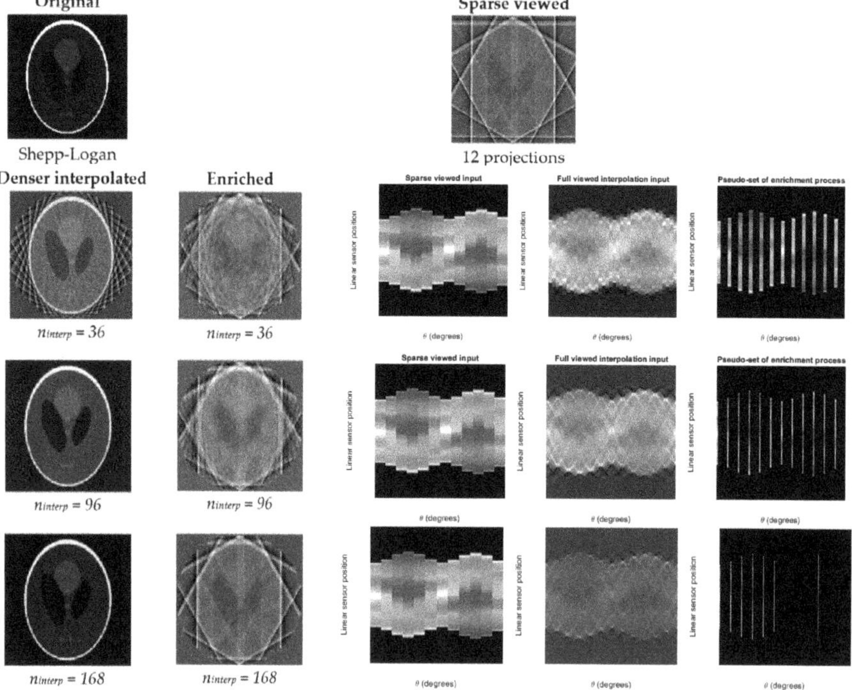

Figure 2. The reconstruction and sinograms illustration of the enrichment process of a sparse 12 projections (original, the first row), at the denser interpolated set of 36 (upper-left) (the second row), 96 (upper-right), and 168 projections (the third row).

Additionally, the results also show that the overlapping coefficient, $k_{overlapping}$, is meaningful in relation to the values regarding the number of projections and the operators in the trio sub-set enhancement. This coefficient is defined in Equation (9).

$$k_{overlapping} = DSO * \left(\frac{W_{detector}}{2*DSO} - \tan\left(\arctan\left(\frac{W_{detector}}{2*DSO} \right) - \frac{2*\pi}{2*n_{interp}} \right) \right) * $$
$$* \left(1 + \cos\left(2 * \arctan\left(\frac{W_{detector}}{2*DSO} \right) - \frac{2*\pi}{n_{interp}} \right) \right) \tag{9}$$

wherein DSO is the distance between the detector and subject, $W_{detector}$ is the width of the detector, and n_{interp} is the number of interpolated projections of the pseudo-set.

Table 2. The trio-subset enhancement operators of the simulated results.

	Simulated Results Based on the Modified Shepp–Logan Phantom				
Processes	**Estimation Process**		**Verification Process**		$k_{overlap}$
Studied Segments	D_B^i	$K_i^{in-flow}$	Δ_{KL}^i	$K_i^{out-flow}$	-
Ω_1 — A 1,6,12	0.1634 0.0050 0.0063	0.8366 0.9950 0.9937	2.4442 2.4575 2.4577	$2.4442 \pm 2.0901 \times 10^{-14}$ $2.4575 \pm 1.3341 \times 10^{-14}$ $2.4577 \pm 1.4675 \times 10^{-14}$	0.3215
B 1,9,18	0.1283 0.0052 0.0046	0.8717 0.9948 0.9954	2.4576 2.4575 2.4575	$2.4576 \pm 2.2235 \times 10^{-14}$ $2.4575 \pm 4.4470 \times 10^{-16}$ $2.4575 \pm 2.1790 \times 10^{-14}$	0.2139
C 1,11,21	0.1008 0.0028 0.0023	0.8992 0.9972 0.9977	2.4575 2.4575 2.4574	$2.4576 \pm 1.2451 \times 10^{-14}$ $2.4575 \pm 8.4492 \times 10^{-15}$ $2.4575 \pm 6.6704 \times 10^{-15}$	0.1833
Ω_2 — A 12,18,24	0.0063 0.0041 0.0019	0.9937 0.9959 0.9981	2.4577 2.4577 2.4577	$2.4577 \pm 1.4675 \times 10^{-14}$ $2.4577 \pm 2.3569 \times 10^{-14}$ $2.4577 \pm 2.0011 \times 10^{-14}$	0.3215
B 18,27,36	0.0046 0.9324 0.1033	0.9954 0.9991 0.9999	2.4575 2.4575 2.4575	$2.4575 \pm 2.1790 \times 10^{-14}$ $2.4575 \pm 7.5598 \times 10^{-15}$ $2.4575 \pm 8.4492 \times 10^{-15}$	0.2139
C 21,32,42	0.0023 0.0041 0.3650	0.9977 0.9959 1.0000	2.4575 2.4575 2.4575	$2.4575 \pm 6.6704 \times 10^{-15}$ $2.4575 \pm 1.4230 \times 10^{-14}$ $2.4575 \pm 1.5120 \times 10^{-14}$	0.1833
Ω_3 — A 24,30,36	0.0019 0.3053 0.9441	0.9981 0.9997 0.9999	2.4577 2.4576 2.4576	$2.4577 \pm 2.0011 \times 10^{-14}$ $2.4576 \pm 2.2401 \times 10^{-14}$ $2.4576 \pm 1.9122 \times 10^{-14}$	0.3215
B 36,45,54	0.1033 0.0016 0.0056	0.9999 0.9984 0.9944	2.4575 2.4575 2.4575	$2.4575 \pm 8.4492 \times 10^{-15}$ $2.4575 \pm 5.7810 \times 10^{-15}$ $2.4575 \pm 1.6009 \times 10^{-14}$	0.2139
C 42,53,63	0.3650 0.1337 0.0125	1.0000 0.9999 0.9875	2.4575 2.4575 2.4575	$2.4575 \pm 1.5120 \times 10^{-14}$ $2.4575 \pm 1.2451 \times 10^{-14}$ $2.4575 \pm 2.0901 \times 10^{-14}$	0.1833
Ω_4 — A 36,42,48	0.9441 0.0015 0.0036	0.9999 0.9985 0.9964	2.4576 2.4576 2.4576	$2.4576 \pm 1.9122 \times 10^{-14}$ $2.4576 \pm 1.2451 \times 10^{-14}$ $2.4576 \pm 2.3124 \times 10^{-14}$	0.3215
B 54,63,72	0.0056 0.0102 0.0060	0.9944 0.9898 0.9940	2.4575 2.4575 2.4575	$2.4575 \pm 1.6009 \times 10^{-14}$ $2.4575 \pm 1.9122 \times 10^{-14}$ $2.4575 \pm 1.0228 \times 10^{-14}$	0.2139
C 63,74,84	0.0125 0.6270 0.0022	0.9875 0.3730 0.9978	2.4575 2.4575 2.4575	$2.4575 \pm 2.0901 \times 10^{-14}$ $2.4575 \pm 1.1562 \times 10^{-14}$ $2.4575 \pm 1.7343 \times 10^{-14}$	0.1833

Table 2. *Cont.*

Simulated Results Based on the Modified Shepp–Logan Phantom					
Processes	**Estimation Process**		**Verification Process**		$k_{overlap}$
Studied Segments	D_B^i	$K_i^{in-flow}$	Δ_{KL}^i	$K_i^{out-flow}$	-
Ω_5 A 48,54,60	0.0036	0.9964	2.4576	$2.4576 \pm 2.3124 \times 10^{-14}$	
	0.0055	0.9945	2.4576	$2.4576 \pm 9.3386 \times 10^{-15}$	0.3215
	0.0050	0.9950	2.4576	$2.4576 \pm 1.3786 \times 10^{-14}$	
B 72,81,90	0.0060	0.9940	2.4575	$2.4575 \pm 1.0228 \times 10^{-14}$	
	0.0045	0.9955	2.4575	$2.4575 \pm 1.3341 \times 10^{-14}$	0.2139
	0.0013	0.9987	2.4575	$2.4575 \pm 1.5120 \times 10^{-14}$	
C 84,95,105	0.0022	0.9978	2.4575	$2.4575 \pm 1.7343 \times 10^{-14}$	
	0.0055	0.9945	2.4575	$2.4575 \pm 5.3363 \times 10^{-15}$	0.1833
	0.1044	0.9999	2.4575	$2.4575 \pm 1.0228 \times 10^{-14}$	
Ω_6 A 60,66,72	0.0050	0.9950	2.4576	$2.4576 \pm 1.3786 \times 10^{-14}$	
	0.0086	0.9914	2.4576	$2.4576 \pm 1.1562 \times 10^{-14}$	0.3215
	0.0054	0.9946	2.4576	$2.4576 \pm 1.2007 \times 10^{-14}$	
B 90,99,108	0.0013	0.9987	2.4575	$2.4575 \pm 1.5120 \times 10^{-14}$	
	0.8497	0.9999	2.4575	$2.4575 \pm 1.8677 \times 10^{-14}$	0.2139
	0.0022	0.9978	2.4575	$2.4575 \pm 9.3386 \times 10^{-15}$	
C 105,116,126	0.1044	0.9999	2.4575	$2.4575 \pm 1.0228 \times 10^{-14}$	
	0.3272	0.9997	2.4575	$2.4575 \pm 1.2007 \times 10^{-14}$	0.1833
	0.0108	0.9892	2.4575	$2.4575 \pm 1.8677 \times 10^{-14}$	
Ω_7 A 72,78,84	0.0054	0.9946	2.4576	$2.4576 \pm 1.2007 \times 10^{-14}$	
	0.0027	0.9973	2.4576	$2.4576 \pm 2.9456 \times 10^{-14}$	0.3215
	0.0019	0.9981	2.4576	$2.4576 \pm 8.8939 \times 10^{-15}$	
B 108,117,126	0.0022	0.9978	2.4575	$2.4575 \pm 9.3386 \times 10^{-15}$	
	0.0056	0.9944	2.4575	$2.4575 \pm 6.2257 \times 10^{-15}$	0.2139
	0.0077	0.9923	2.4575	$2.4575 \pm 6.2257 \times 10^{-15}$	
C 126,137,147	0.0108	0.9892	2.4575	$2.4575 \pm 1.8677 \times 10^{-14}$	
	0.0031	0.9969	2.4575	$2.4575 \pm 6.2257 \times 10^{-15}$	0.1833
	0.0057	0.9943	2.4575	$2.4575 \pm 3.5576 \times 10^{-15}$	
Ω_8 A 84,90,96	0.0019	0.9981	2.4576	$2.4576 \pm 8.8939 \times 10^{-15}$	
	0.0012	0.9988	2.4576	$2.4576 \pm 1.4657 \times 10^{-14}$	0.3215
	-	-	2.4532	$2.4532 \pm 6.2257 \times 10^{-15}$	
B 126, 135,144	0.0077	0.9923	2.4575	$2.4575 \pm 6.2257 \times 10^{-15}$	
	0.0060	0.9940	2.4575	$2.4575 \pm 4.8917 \times 10^{-15}$	0.2139
	-	-	2.4576	$2.4575 \pm 2.0011 \times 10^{-14}$	
C 147,158,168	0.0057	0.9943	2.4575	$2.4575 \pm 3.5576 \times 10^{-15}$	
	0.0054	0.9946	2.4575	$2.4575 \pm 1.6898 \times 10^{-14}$	0.1833
	-	-	2.4575	$2.4575 \pm 1.0228 \times 10^{-15}$	

Three cases A, B, and C, defined as sparse 12 projections' input, and the n_{interp} = 96, 144, and 168. The A, B, and C result sets are quoted at the first, middle, and last projections of each segment Ω_i (i = [1,8], $I \in \mathbb{N}$). The Ω_i is defined in degrees of (0,45), (45,90), (90,135), ... (270,315), (315,360). The notation explanation of the common first, middle, and last projections represented for each segment of those cases is shown in Table 3.

Table 3. The notation of the common projections for eight segments, notation of the common projections of three cases A, B, and C for eight segments Ω_i (sparse 12 projections input and the n_{interp} = 96, 144, and 168).

Segment Ω_i	Case A			Case B			Case C		
	First	Middle	Last	First	Middle	Last	First	Middle	Last
Ω_1	1	6	12	1	9	18	1	11	21
Ω_2	12	18	24	18	27	36	21	32	42
Ω_3	24	30	36	36	45	54	42	53	63
Ω_4	36	42	48	54	63	72	63	74	84
Ω_5	48	54	60	72	81	90	84	95	105
Ω_6	60	66	72	90	99	108	105	116	126
Ω_7	72	78	84	108	117	126	126	137	147
Ω_8	84	90	96	126	135	144	147	158	168

4.2. Comparison to the Other Reconstruction Algorithms

The reconstructed results are implemented on the sparse 12-projection set, the result of the proposed SA2IR is compared to different algorithms, as mentioned above, included FDK, SIRT, SART, OS-SART, TV-SART, ASD-POCS, OS-ASD-POCS, β-ASD-POCS, CGLS, and FISTA. Obviously, the sparse set is unable to be implemented with FDK; the other results are shown in Figure 3. The result of our proposed algorithm (Figure 3(k1)) for the sparse case shows the transition quality between the results of SIRT (Figure 3(b1)) and SART (Figure 3(c1)). The edge preservation of the proposed algorithm is better than that of the conventional SIRT and nearly the same as that of SART. The external generative noise is a little bit thinner than the result of SIRT, while the internal generative noise is likely homogenous and better than the result of SART. Furthermore, the proposed algorithm shows that the edge preservation and global homogeneity are significantly better than the other results of OS-SART (intensity reducing), TV-SART (informative loss of edge and intensity), ASD-POCS (structural similarity reducing), OS-ASD-POCS (blurring), β-ASD-POCS (edge and intensity deterioration), CGLS (edge blurring), and FISTA (total image quality reducing) (in Figure 3, from (d1) to (j1)). Consequently, in the case of sparse projection input, the proposed algorithm reduced the generative noise significantly, compared to the other algorithms, as mentioned above, while preserving the edge and intensity; thus, had diagnostic ability.

The reconstructed results are implemented on the denser projection set of 48, as shown in Figure 3 (from (a2) to (l2). The result of the SA2IR algorithm is compared to different algorithms, as mentioned above, included SIRT, SART, OS-SART, TV-SART, ASD-POCS, OS-ASD-POCS, β-ASD-POCS, CGLS, and FISTA. In this section, we implemented both results of the proposed algorithm, as the origin and the modified ones, successively shown in Figure 3(k2,l2). The results are shown to be the same as what was discussed in the case of the sparse projection input.

Due to the shape-ness characteristics, the noise distributions are shown patterned per segments. The boundary ripple causes the edge detail loss, and the shifting causes the intensity loss. The image quality metrics of the reconstructed results are qualitatively evaluated in Table 4. The qualitative metrics of image quality are proposed to be measured in root mean square error (RMSE), structural similarity (SSIM), peak signal-to-noise ratio (PSNR), and signal-to-noise (SNR), and the entropy of the results are shown in Figure 3. According to Table 4, the quantitative metric benefits of SA2IR are indicated as asterisked. Those metrics are divided into two main groups, included (1) the Referral metrics group, included RMSE, SSIM, PSNR, and SNR, which is referred to the original phantom, and (2) the Individual metrics group, as the entropy measurement. Corresponding to the results in Figure 3 and Table 4, the proposed SA2IR algorithm is proved to reduce the generative noise, both external and internal of the edges, or inner or outer of the edge, compared to the other algorithms.

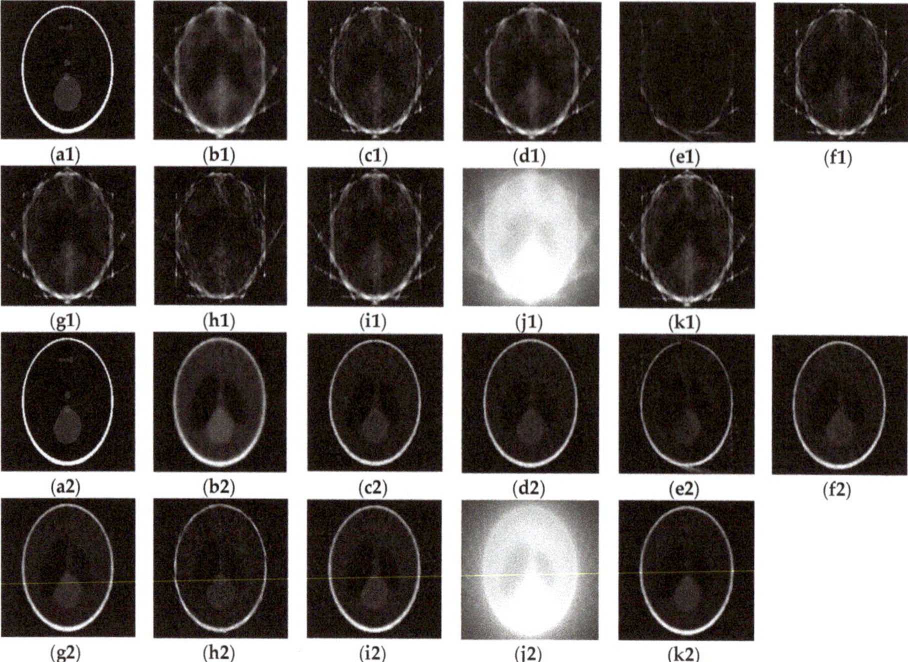

Figure 3. The Shepp–Logan phantom reconstructed results, based on sparse 12-projection input (from (**a1**) to (**k1**)) and sparse 48-projection input (from (**a2**)–(**l2**)), such as: (**a1,a2**)–Origin phantom, (**b1,b2**)–SIRT (simultaneous iterative reconstruction technique), (**c1,c2**)–SART (simultaneous algebraic reconstruction technique algorithm), (**d1,d2**)–OS-SART (order-subset SART), (**e1,e2**)–TV-SART (total variation SART), (**f1,f2**)–ASD-POCS (adaptive steepest descent projection onto convex sets), (**g1,g2**)–OS-ASD-POCS (order-subset ASD-POCS), (**h1,h2**)–β-ASD-POCS, (**i1,i2**)–CGLS (conjugated gradient least square algorithm), (**j1,j2**)–FISTA (fast iterative shrinkage-thresholding algorithm) and (**k1,k2**)–SA2IR.

Table 4. Qualitative image quality metric evaluation of the reconstructed results in Figure 3 (from (a2) to (l2)). RMSE: root mean square error, SSIM: structural similarity, PSNR: peak signal-to-noise ratio, SNR: signal-to-noise.

Algorithms	RMSE	SSIM	PSNR	SNR	Entropy
SIRT	0.0082 *	0.6401 *	20.8486 *	−1.7900	5.1575
SART	0.0057 *	0.6517 *	22.4720 *	−0.9387	5.3968 *
OS-SART	0.0055	0.7024	22.6087 *	−1.0994	5.1386 *
TV-SART	0.0156	0.5216	18.0633	−1.8817	5.8063
ASD-POCS	0.0063 *	0.5802	21.9976 *	−0.9116	5.3289 *
OS-ASD-POCS	0.0053	0.7150	22.7878 *	−0.91174	4.7335
β-ASD-POCS	0.0081	0.4037	20.8905	−0.5470	5.8491
CGLS	0.0058	0.6488	22.3796	−1.3787	5.6202
FISTA	0.0509	0.1795	12.9291	−25.0662	1.7108
SA2IR	0.0079 *	0.6452 *	21.0202 *	−1.9962	5.0969 *

The asterisks (*) mean the concerned values of the other algorithms in comparison to SA2IR.

For the patient head data clinical simulation, the reconstructed results are implemented and demonstrated into the different section of maxillofacial and mandibular structures (shown in Figure 4). The reconstruction was done with the initial sparse set of 68 projections. The enrichment ratio was chosen as $k_{sampling} = 8$ for doubling scan angular rotation. Those results are addressed as the upper jaw region (basal palatine bone), lower arch, mid-facial region, and temporomandibular joint region, respectively to the 68th, 52nd, 108th, and 112nd slices of the used patient head data (shown in Figure 5). The boxing volumetric size of the head data reconstruction is $360 \times 360 \times 360$ (in mm). According to the simulation results of the phantom in Table 4, the qualitative image quality metric evaluation of those is shown in Table 5. Regarding the meaning of image quality indexes and computational time "trade-off", the significant benefit of using the proposed algorithm compared to the other common contemporary algorithms for sparse projection reconstruction was observed. According to the results in Figure 3, the reconstructed results of the clinical simulation is to compare SA2IR to SIRT, SART, OS-SART, ASD-POCS, OS-ASD-POCS, and CGLS algorithms. The remainder were omitted, because of their disadvantages when compared to the SA2IR. To clarify the trade-off beneficiary of SA2IR to the others, the differentiation beneficiary is shown in Table 5.

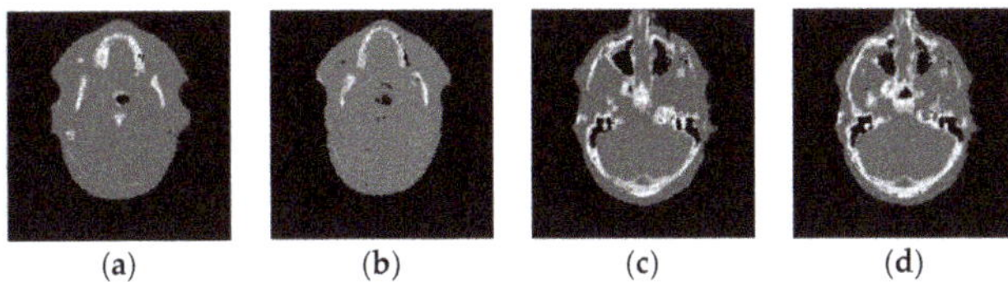

(a)   (b)   (c)   (d)

Figure 4. The patient head data referred to dental anatomical regions (from left to right), such as: (**a**)–Upper jaw region (at slice #68), (**b**)–Lower arch (at slice #52), (**c**)–Mid-facial region (at slice #108), (**d**)–Temporomandibular joint region (at slice #112).

According to Table 5, the results of the proposed algorithm are significantly beneficiary in balancing between the ALADA adaptation and computation time (in seconds) in comparison to the other referred algorithms. To evaluate the "clinical diagnostic-ability", the mean square error (MSE), structural similarity (SSIM), entropy (E), and the signal-to-noise ratios (PSNR and SNR) are used. In the complicative structures of the mid-facial and temporomandibular joint regions, the results of SA2IR are shown to be better than those of SART and ASD-POCS. The results are shown to have no meaningful diagnostic-able visual differences compared to OS-SART and OS-ASD-POCS. Except for the computation time of the SIRT and CGLS algorithms, SA2IR took less time to implement than the others. The computation time that the SA2IR algorithm took was 4.8, 5.0 and 5.3 s, which are significantly less when compared to SART (30.2, 30.2 and 35.1 s) and ASD-POCS (32.0, 35.9 and 37.1 s), while arithmetically less than that of OS-SART (5.4, 5.3 and 5.8 s) and OS-ASD-POCS (5.3, 5.3 and 5.6 s). However, the image quality of SIRT and CLGS is lower and has more blurring than the other algorithms, as shown in Figure 5. The SA2IR result has similar effects to the ordered-subset regularizations of both ASD-POCS and SART in generative noise reduction, contrast equalization, and cost-effective computation. Furthermore, the generative noise reduction in SA2IR is less than all the other algorithms shown in Figure 5. This is due to the "chirp soft ablation" effect of the trio subsets enhancement of SA2IR.

The proposed algorithm shows good agreement with the trade-off problem when using sparse projections to design low-dosed CBCT reconstructions. The results are promising when compared to other algorithms due to the balanced image quality, diagnostic potential, along with more cost-effective computation, and lowered exposure. The generative noise reduction effect is significantly reduced due to the initial projects set densification (enrichment) and trio subset enhancement. The structural features and contrast are maintained as able to be diagnostic. The computation time has been reduced significantly in comparison to the other algorithms such as SART, OS-SART, ASD-POCS, and OS-ASD-POCS.

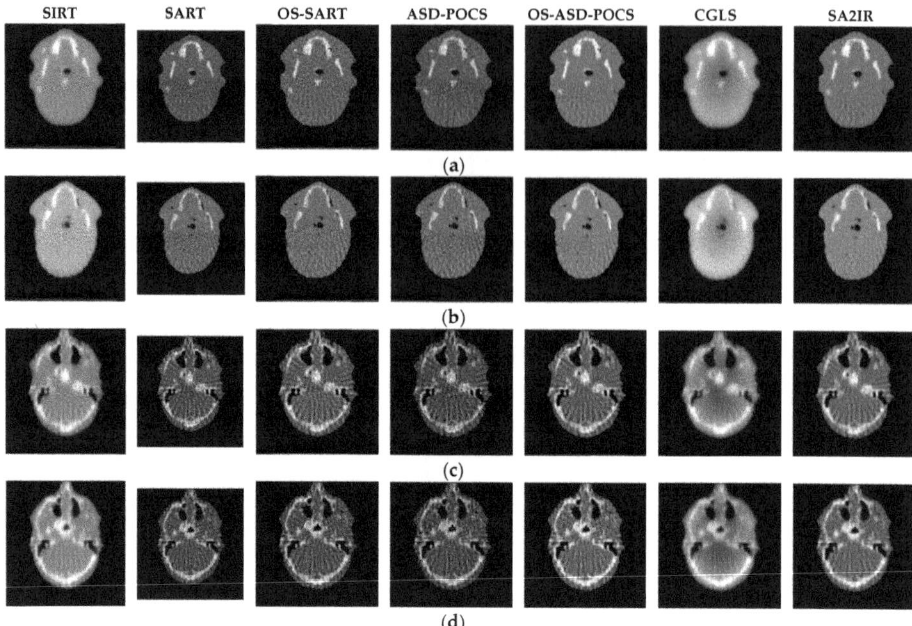

Figure 5. The reconstructed results of others algorithms compared to SA2IR, with respect to the different anatomical regions, as: (**a**)–Upper jaw region (the first row), (**b**)–Lower arch (the second row), (**c**)–Mid-facial region (the third row), and (**d**)–Temporomandibular region (the fourth row).

Table 5. Qualitative image quality metric evaluation of the patient data simulation results.

Algorithms	RMSE	SSIM	PSNR	SNR	Entropy	Computation Time (secs.)
			Upper Jaw Arch Region–Head			
SIRT	0.0013 *	0.6139	29.0292	−0.2467	4.1147	3.00
SART	0.0013 *	0.5423	28.9282	−0.0218	4.2795 *	30.20 *
OS-SART	0.0009 *	0.6005 *	30.5914 *	−0.0514	4.1238	5.40
ASD-POCS	0.0011 *	0.5625 *	29.6087 *	−0.0906	4.1928 *	32.00 *
OS-ASD-POCS	0.0006	0.6747	32.4155	−0.0827	3.712	5.30 *
CGLS	0.0022 *	0.4164	26.5544	−0.0671	5.6606	1.80
SA2IR	0.0009 *	0.5988 *	30.2784 *	−0.1543	4.2824 *	4.80 *
			Lower Jaw Arch Region–Head			
SIRT	0.0012 *	0.6003	29.254	−0.2533	3.8695	2.80
SART	0.0012 *	0.5412	29.22	−0.0383	4.0492 *	30.20 *
OS-SART	0.0008 *	0.5976 *	30.7944 *	−0.0614	3.9003 *	5.30 *
ASD-POCS	0.0010 *	0.5557	29.8809 *	−0.1041	4.072 *	35.90 *
OS-ASD-POCS	0.0006	0.6596	32.2566	−0.0931	3.5123	5.30 *
CGLS	0.0020 *	0.4042	26.8865	−0.0846	5.4896	1.90
SA2IR	0.0009 *	0.5865 *	30.3070 *	−0.1638	4.0506 *	5.00 *
			Mid-Facial Region–Head			
SIRT	0.0011 *	0.6161	29.7068	−0.2359	3.885	2.80
SART	0.0011 *	0.5509	29.4622 *	−0.0267	4.0768 *	29.40 *
OS-SART	0.0008 *	0.6062 *	31.108 *	−0.0526	3.9013 *	5.40 *
ASD-POCS	0.0010 *	0.5662	30.1793 *	−0.0919	3.9817 *	31.60 *
OS-ASD-POCS	0.0005	0.6764	32.8669	−0.0823	3.5258	5.20 *
CGLS	0.0019 *	0.4185	27.162	−0.0628	5.5153	1.80
SA2IR	0.0008 *	0.6037 *	30.8624 *	−0.1472	4.0744 *	4.80 *

Table 5. *Cont.*

Algorithms	RMSE	SSIM	PSNR	SNR	Entropy	Computation Time (secs.)
			Temporomandibular joint Region–Head			
SIRT	0.0040 *	0.5173	23.9879	−0.5294	4.7749	3.00
SART	0.0027	0.5377 *	25.6691 *	−0.0972	4.9691 *	35.10 *
OS-SART	0.0021	0.5889	26.8414	−0.1208	4.7041	5.80 *
ASD-POCS	0.0026 *	0.5494 *	25.8782 *	−0.1777	4.9102 *	37.10 *
OS-ASD-POCS	0.0018	0.6166	27.4522	−0.1851	4.6107	5.60 *
CGLS	0.0053 *	0.3314	22.7355	−0.2227	6.1786	1.90
SA2IR	0.0030 *	0.5151 *	25.1978 *	−0.3472	5.0687 *	5.30 *

The asterisks (*) mean the concerned values of the other algorithms in comparison to SA2IR.

5. Conclusions

In this paper, the SA2IR is proposed as a specific iterative method for dental CBCT reconstruction problems. This algorithm is made up of two parts: the pre-reconstruction processing and the 3D reconstruction of a CBCT projection set. The pre-reconstruction process is composed of the pseudo-set enrichment and the trio-subset enhancement processes. Experimental simulation was implemented with (1) a minimum projection set of 12 projections and (2) the sparse cases of 48 projections, on a phantom. Clinical simulation was implemented with the sparse projections set of 68 projections, with respect to the diagnostic anatomy differentiation of four regions. These regions are the upper jaw arch, lower jaw arch, mid-facial, and temporomandibular. The results have shown that when compared to previous conventional algorithms (SIRT, SART, OS-SART, ASD-POCS, OS-ASD-POCS, and CGLS), the SA2IR method produces comparable diagnostic ability image quality compared to the other conventional methods. SA2IR reduces generative noise, can preserve structural features, and has enhanced global contrast when compared to SART, OS-SART, ASD-POCS, and OS-ASD-POCS. The SA2IR's computation time is significantly better than all of the other methods, especially when compared to SART and ASD-POCS. The clinical simulation has shown the compatibility of the algorithm with many applied dental image diagnostics, especially for dental implantology and maxillofacial surgery.

The proposed algorithm is shown to be highly applicable in the sparse projection and low-dose CBCT reconstruction, adapting the ALADA concept, as "radiation safer and imaging diagnostic-ability optimization". That is more humane for the patients and clinicians, regarding both heath and responsibility. This approach is also available to be applied in other medical CBCT reconstruction applications. This selective anatomical analytic iteration reconstruction (SA2IR) algorithm uses a combination of the regularization approach and statistical approach, and utilizes the robustness of GPU-based computation. To improve upon this study, further research will be conducted on human head and body anatomy structures and properties. That study will help to enhance the SA2IR's trio subset enhancement process. The bio-physical relationship will aid in modeling between exposure–direction and various other anatomical structures. The outcome of this experiment is expected to reduce generative noise as well as computational uncertainties, and enhance the imaging of the soft tissue structure.

Author Contributions: Y.-C.D. and L.D.-N. developed the methodology. Y.-C.D. conceived, supervised, and coordinated the investigations as well as checked the manuscript's logical structure. L.D.-N. performed the simulation as well as analyzed the experimental data.

Funding: The authors would like to thank the Ministry of Education, Taiwan, for financially supporting this research in the Allied Advanced Intelligent Biomedical Research Center (A2IBRC) under the Higher Education Sprout Project.

Conflicts of Interest: The authors declare no conflict of interest.

Abbreviations

The following abbreviations are used in this manuscript:

ALADA	As Low As Diagnostically Acceptable
ALARA	As Low As Reasonably Achievable
(OS)-ASD-POCS	(Order-Subset) Adaptive Steepest Descent Projection On Convex Sets
CAD/CAM	Computer-Aided Design/ Computer -Aided Manufacturing
CBCT	Cone-Beam Computed Tomography
CGLS	Conjugated Gradient Least Square
CT	Computed Tomography
DICOM	Digital Imaging and Communication in Medicine
FISTA	Fast Iterative Shrinkage-Thresholding Algorithm
FPID	Flat Panel Imaging Detector
FDK	Feldkamp - Davis - Kress
GPU	Graphical Processing Unit
NCRP	National Council on Radiation Protection (United States)
RMSE	Root Mean Square Error
SA2IR	Selective Anatomy Analytic Iteration Reconstruction
SA2IT	Selective Anatomy Analytic Iteration Technique
(OS)-SART	(Order-Subset) Simultaneous Algebraic Reconstruction Technique
SIRT	Simultaneous Iterative Reconstruction Technique
SNR	Signal-To-Noise Ratio
PSNR	Peak Signal-To-Noise Ratio
SSIM	Structural Similarity
TSC	Total Similarity Coefficient

References

1. Fokas, G.; Vaughn, V.N.; Scarfe, W.C.; Bornstein, M.M. Acccuracy of linear measurement on CBCT images related to presurgical implant treatment: A systematic review. *Clin. Oral Implant. Res* **2018**, *29*, 393–415. [CrossRef] [PubMed]

2. Bornstein, M.M.; Jorner, K.; Jacobs, R. Use of cone beam computed tomography in implant dentistry: Current concepts, indications, and limitations for clinical practice and research. *Periodontology* **2017**, *73*, 51–72. [CrossRef] [PubMed]

3. Kim, I.H.; Singer, S.R.; Mupparapu, M. Review of cone beam computed tomography guidelines in North America. *Quintessence Int.* **2019**, *50*, 136–145.

4. Hayashi, T.; Arai, Y.; Chikui, T.; Hayashi-Sakai, S.; Honda, K.; Indo, H.; Kawai, T.; Kobayashi, K.; Murakami, S.; Nagasawa, M.; et al. Committee on Clinical Practice Guidelines Japanese Society for, Oral Maxillofacial, Radiology. Clinical guidelines for dental cone-beam computed tomography. *Oral Radiol.* **2018**, *34*, 89–104. [CrossRef]

5. Alghazzawi, T.F. Advancements in CAD/CAM technology: Options for practical implementation. *J. Prosthodont. Res.* **2016**, *60*, 72–84. [CrossRef]

6. Dao-Ngoc, L. The review of RP (Rapid Prototyping application in maxillofacial surgeries in Vietnam from 2010 to 2016: In the manufacturing engineer's view. *Cập nhật nha khoa–Tài liệu tham khảo và đào tạo liên tục* **2017**, *22*, 121–142.

7. Qin, Z.; Zhang, X.; Li, Y.; Wang, P.; Li, J. One-stage treatment for maxillofacial asymmetry with orthognathic and contouring surgery using virtual surgical planning and 3D-printed surgical templates. *J. Plast. Reconstr. Aesthet. Surg.* **2019**, *72*, 97–106. [CrossRef]

8. Bushberg, J.T. Eleventh annual Warren K. Sinclair keynote address-science, radiation protection and NCRP: Building on the past, looking to the future. *Health Phys.* **2015**, *108*, 115–123. [CrossRef]

9. White, S.C.; Scarfe, W.C.; Schulze, R.K.W.; Lurie, A.G.; Douglass, J.M.; Farman, A.G.; Law, C.S.; Levin, M.D.; Sauer, R.A.; Valachovic, R.W.; et al. The Image Gently in Dentistry campaign: Promotion of responsible use of maxillofacial radiology in dentistry for children. *Oral Surg. Oral Med. Oral Pathol. Oral Radiol.* **2014**, *118*, 257–261. [CrossRef]

10. Fernandes, K.; Levin, T.L.; Miller, T.; Schoenfeld, A.H.; Amis, E.S., Jr. Evaluating an Image Gently and Image Wisely Campaign in a Multihospital Health Care System. *J. Am. Coll. Radiol.* **2016**, *13*, 1010–1017. [CrossRef]

11. Jaju, P.P.; Jaju, S.P. Cone-beam computed tomography: Time to move from ALARA to ALADA. *Imaging Sci. Dent.* **2015**, *45*, 263–265. [CrossRef]

12. Matenine, D.; Goussard, Y.; Despres, P. GPU-accelerated regularized iterative reconstruction for few-view cone beam CT. *Med. Phys.* **2015**, *42*, 1505–1517. [CrossRef]

13. Zhang, H.; Wang, J.; Zeng, D.; Tao, X.; Ma, J. Regularization strategies in statistical image reconstruction of low-dose X-ray CT: A review. *Med. Phys.* **2018**, *45*, e886–e907. [CrossRef]

14. Jacobs, R.; Vranckx, M.; Vanderstuyft, T.; Quirynen, M.; Salmon, B. CBCT vs. other imaging modalities to assess peri-implant bone and diagnose complications: A systematic review. *Eur. J. Oral Implant.* **2018**, *11*, 77–92.

15. Katsumata, A.; Hirukawa, A.; Okumura, S.; Naitoh, M.; Fujishita, M.; Ariji, E.; Langlais, R.P. Effects of image artifacts on gray-value density in limited-volume cone-beam computerized tomography. *Oral Surg. Oral Med. Oral Pathol. Oral Radiol. Endodontol.* **2007**, *104*, 829–836. [CrossRef]

16. Stefanelli, L.V.; DeGroot, B.S.; Lipton, D.I.; Mandelaris, G.A. Accuracy of Dynamic Dental Implant Navigation System in a Private Practice. *Int. J. Oral Maxillofac. Implant.* **2019**, *34*, 205–2013. [CrossRef]

17. Harris, B.T.; Montero, D.; Grant, G.T.; Morton, D.; Llop, D.R.; Lin, W.S. Creation of a 3D dimensional virtual dental patient for computer-guided surgery and CAD-CAM interim complete removable and fixed dental prostheses: A clinical report. *J. Prosthet. Dent.* **2017**, *117*, 197–204. [CrossRef]

18. Patel, S.; Brown, J.; Pimentel, T.; Kelly, R.D.; Abella, F.; Durack, C. Cone beam computed tomography in Endodontics—A review of the literature. *Int. Endod. J.* **2019**, *52*, 1138–1152. [CrossRef]

19. Tchorz, J.P. 3D Endo: Three-dimensional endodontic treatment planning. *Int. J. Comput. Dent.* **2017**, *20*, 87–92.

20. Woelber, J.P.; Fleiner, J.; Rau, J.; Ratka-Kruger, P.; Hannig, C. Accuracy and Usefulness of CBCT in Periodontology: A Systematic Review of the Literature. *Int. J. Periodontics Restor. Dent.* **2018**, *38*, 289–297. [CrossRef]

21. Pozzi, A.; Arcuri, L.; Moy, P.K. The smiling scan technique: Facially driven guided surgery and prosthetics. *J. Prosthodont. Res.* **2018**, *62*, 514–517. [CrossRef] [PubMed]

22. McGuigan, M.B.; Duncan, H.F.; Horner, K. An analysis of effective dose optimization and its impact on image quality and diagnostic efficacy relating to dental cone beam computed tomography (CBCT). *Swiss Dent. J.* **2018**, *128*, 297–316.

23. Karimi, D.; Ward, R.K. Sinogram denoising via simultaneous sparse representation in learned dictionaries. *Phys. Med. Biol.* **2016**, *61*, 3536–3553. [CrossRef]

24. Zhu, Z.G.; Wahid, K.; Babyn, P.; Cooper, D.; Pratt, I.; Carter, Y. Improved Compressed Sensing Based Algorithm for Sparse—View CT Image Reconstruction. *Comput. Math. Methods Med.* **2013**, *2013*, 185750. [CrossRef]

25. Zhang, H.M.; Li, L.; Wang, L.Y.; Sun, Y.M.; Yan, B.; Cai, A.L.; Hu, G.E. Computed Tomography Sinogram Inpainting With Compound Prior Modelling Both Sinogram and Image Sparsity. *IEEE Trans. Nucl. Sci.* **2016**, *63*, 2567–2576. [CrossRef]

26. Zhang, C.; Zhang, T.; Zheng, J.; Li, M.; Lu, Y.; You, J.; Guan, Y. A Model of Regularization Parameter Determination in Low-Dose X-Ray CT Reconstruction Based on Dictionary Learning. *Comput. Math. Methods Med.* **2015**, *2015*, 831790. [CrossRef]

27. Du, D.; Pan, Z.; Zhang, P.; Li, Y.; Ku, W. Compressive sensing image recovery using dictionary learning and shape-adaptive DCT thresholding. *Magn. Reson. Imaging* **2019**, *55*, 60–71. [CrossRef]

28. Kim, G.; Park, S.; Je, U.; Cho, H.; Park, C.; Kim, K.; Lim, H.; Lee, D.; Lee, H.; Park, Y.; et al. A New Voxelization Strategy in Compressed-Sensing (CS)-Based Iterative CT Reconstruction for Reducing Computational Cost: Simulation and Experimental Studies. *J. Med. Biol. Eng.* **2018**, *38*, 129–137. [CrossRef]

29. Byrne, C. A unified treatment of some iterative algorithms in signal processing and image reconstruction. *Inverse Probl.* **2003**, *20*, 103–120. [CrossRef]

30. Shepp, L.A.; Logan, B.F. The Fourier reconstruction of a head section. *IEEE Trans. Nucl. Sci.* **1974**, *21*, 21–43. [CrossRef]

31. Aichert, A.; Manhart, M.T.; Navalpakkam, B.K.; Grimm, R.; Hutter, J.; Maier, A.; Hornegger, J.; Doerfler, A. A realistic digital phantom for perfusion C-arm CT based on MRI data. In Proceedings of the 2013 IEEE Nuclear Science Symposium and Medical Imaging Conference (2013 NSS/MIC), Seoul, Korea, 27 October–2 November 2013.

32. Feldkamp, L.A.; Davis, L.C.; Kress, J.W. Practical Cone-Beam Algorithm. *J. Opt. Soc. Am. A* **1984**, *1*, 612–619. [CrossRef]

33. Kak, A.C.; Slaney, M. *Principles of Computerized Tomographic Imaging, Classics in Applied Mathematics*; Society for Industrial and Applied Mathematics: Philadelphia, PA, USA, 2001. [CrossRef]

34. Andersen, A.H.; Kak, A.C. Simultaneous Algebraic Reconstruction Technique (SART)—A Superior Implementation of the ART Algorithm. *Ultrason. Imaging* **1984**, *6*, 81–94. [CrossRef]

35. Censor, Y.; Elfving, T. Block-iterative algorithms with diagonally scaled oblique projections for the linear feasibility problem. *SIAM J. Matrix Anal. Appl.* **2002**, *24*, 40–58. [CrossRef]

36. Sidky, E.Y.; Pan, X. Image reconstruction in circular cone-beam computed tomography by constrained, total-variation minimization. *Phys. Med. Biol.* **2008**, *53*, 47–77. [CrossRef]

37. Björck, Å. *Numerical Methods for Least Squares Problems*; Society for Industrial and Applied Mathematics: Philadelphia, PA, USA, 1996. [CrossRef]

38. Biguri, A.; Dosanjh, M.; Hancock, S.; Soleimani, M. TIGRE: A MATLAB-GPU toolbox for CBCT image reconstruction. *Biomed. Phys. Eng. Express* **2016**, *2*, 055010. [CrossRef]

39. Perona, P.; Malik, J. Scale-space and edge detection using anisotropic diffusion. *IEEE Trans. Pattern Anal. Mach. Intell.* **1990**, *16*, 629–639. [CrossRef]

Article

Dual-Input Isolated DC-DC Converter with Ultra-High Step-Up Ability Based on Sheppard Taylor Circuit

Chih-Lung Shen *, Li-Zhong Chen and Hong-Yu Chen

Department of Electronic Engineering, National Kaohsiung University of Science and Technology, Kaohsiung City 82445, Taiwan; 0652807@nkust.edu.tw (L.-Z.C.); 0352030@nkust.edu.tw (H.-Y.C.)
* Correspondence: clshen@nkust.edu.tw; Tel.: +886-925-871-685

Received: 8 September 2019; Accepted: 5 October 2019; Published: 7 October 2019

Abstract: A dual-input high step-up isolated converter (DHSIC) is proposed in this paper, which incorporates Sheppard Taylor circuit into power stage design so as to step up voltage gain. In addition, the main circuit adopts boosting capacitors and switched capacitors, based on which the converter voltage gain can further be improved significantly. Since the proposed converter possesses an inherently ultra-high step-up feature, it is capable of processing low input voltages. The DHSIC also has the important features of leakage energy recycling, switch voltage clamping, and continuous input-current obtaining. These characteristics advantage converter efficiency and benefit the DHSIC for high power applications. The structure of the proposed converter is concise. That is, it can lower cost and simplifies control approach. The operation principle and theoretical derivation of the proposed converter are discussed thoroughly in this paper. Simulations and hardware implementation are carried out to verify the correctness of theoretical analysis and to validate feasibility of the converter as well.

Keywords: dual-input converter; high voltage gain; leakage energy recycling; galvanic isolation; voltage clamping

---

1. Introduction

Nowadays, electricity is mostly generated by fossil fuels and nuclear fuel. Although nuclear power plants can generate considerable power by utilizing a little amount of nuclear fuel, nuclear waste influencing the environment is inevitable. Fossil fuels have been overused and become in shortage. Therefore, human beings attempt to discover more alternatives for maintaining global economic development and environment protection. To alleviate the problems of global temperature rising and serious greenhouse gas emission, many kinds of clean-energy power generation, such as photovoltaic (PV) panel, wind turbine, and fuel-cell stack, are developed imperatively [1,2].

In general, renewable power systems need a DC-bus voltage in the range of 380 V–400 V for grid-tied connection or in high power applications. Unfortunately, the output voltage of a PV module, battery or fuel cell is much lower than the dc-bus voltage and thus conventional DC/DC converters cannot be utilized directly to serve as an energy interface for dealing with power control. In addition, if the power supplied from two input ports has to be processed simultaneously, dual-input converter with high voltage gain is essentially anticipated. Figure 1 illustrates a hybrid generation system that includes two sources, a high step-up converter, and an inverter for AC application, which reveals that a dual-input converter with high step-up voltage gain is urgent in such system.

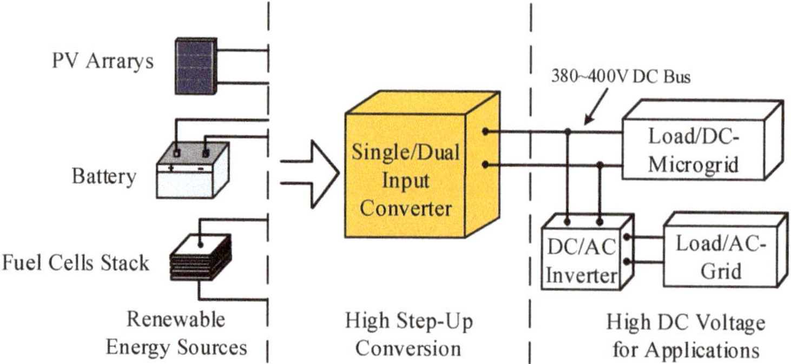

Figure 1. An illustration of the hybrid renewable-energy generation system.

Theoretically, conventional step-up converters, like Boost and Flyback [3–6], can achieve a high voltage gain under the operating with extremely high duty ratio or the design of a much higher turns ratio. However, extreme high duty cycle or turns ratio will dramatically degrade the conversion efficiency due to large conduction loss and copper loss of windings. In order to improve conversion efficiency and voltage gain, many transformerless high step-up converters are proposed [7–10]. Although these converters are designed to achieve high step-up characteristics with reasonable duty ratio, here still exits some problems, for instance, large transient current and limited voltage gain, confining converter flexibility. To mitigate the mentioned drawbacks, converters incorporating coupled inductor and/or switched capacitor are proposed [11–14], however, which are only able to process single-input source. In order to deal with two different kinds of inputs, dual-input converters (DIC) are proposed. In comparison with single input, a dual generation system is capable of providing higher reliability, durability and power rating. The structure of DICs can be simply classified as series type and parallel one. The conventional series-type DICs construct string connection at input ports [15,16]. Such a DIC will malfunction in case that either of the two inputs fails. The parallel-type DICs collocate different sources in parallel so that even if one of the inputs is out of commitment, it still can accomplish voltage stepping to meet DC-bus level [17–23]. Some of them are controlled with time-sharing scheme. That is, only one DC source is permitted to delivery its energy to the load at a time. Compared with series-type DICs, parallel ones possess much better features from the aspects of reliability and controllability. Nevertheless, limitation on voltage gain is unavoidable, which confines converter applications in the field of high DC-bus voltage requirement.

In order to convert a lower DC voltage to a much higher level and to provide consecutive power even under the situation that one input source shuts down, this paper proposes a DHSIC, which is developed by means of boosting capacitor, switched capacitor and Sheppard Taylor circuit. The proposed dual-input converter can achieve the following important features: ultra-high step-up ability, continuous input current, and galvanic isolation. Furthermore, the proposed converter possesses the competence of inherent voltage clamping without any additional devices and recycling leakage energy stored in transformers. Therefore, the voltage spike on the power switch can be suppressed and converter efficiency is also improved.

The structure of this paper is organized as follows. Following the introduction, the operation principle of the proposed dual-input high step-up isolated converter is described in Section 2. The steady-state analysis is discussed in Section 3, which covers voltage gain of the converter, voltage and current stresses of the semiconductor device, and magnetizing inductance design in continuous conduction mode (CCM). To verify the correctness of proposed converter, experimental results from a 200-W prototype are illustrated in Section 4. Finally, Section 5 summarizes the conclusions of this paper.

2. Operation Principle of Proposed Dual-Input Converter

The equivalent of the proposed DHSIC is shown in Figure 2. Parameters in Figure 2 are represented in the following. V_{in1} and V_{in2} are input voltages, while i_{in1} and i_{in2} denote input currents. The practical model of coupled inductor includes magnetizing inductance, leakage inductance, and an ideal transformer. The L_{m1} and L_{m2} are the magnetizing inductances of T_1 and T_2, respectively, meanwhile, leakage inductances are expressed as L_{k1} and L_{k2}. The S_1–S_4 represent the four main power switches. The C_1 and C_2 function as boosting capacitors, and C_3 and C_4 serve as switched capacitors. These capacitors can elevate converter voltage gain effectively. The D_1–D_6 are rectifier diodes. In addition, D_o and C_o are the output diode and filter capacitor, respectively. Output voltage and current of DHSIC are in turn described as V_o and I_o. Finally, the output equivalent resistance is presented as R_o. Even though the proposed DHSIC works normally in dual input operation (DIO), it still possesses the ability to be in single-input operation (SIO) while either input source fails. In Section 2, DIO will be first discussed followed by SIO.

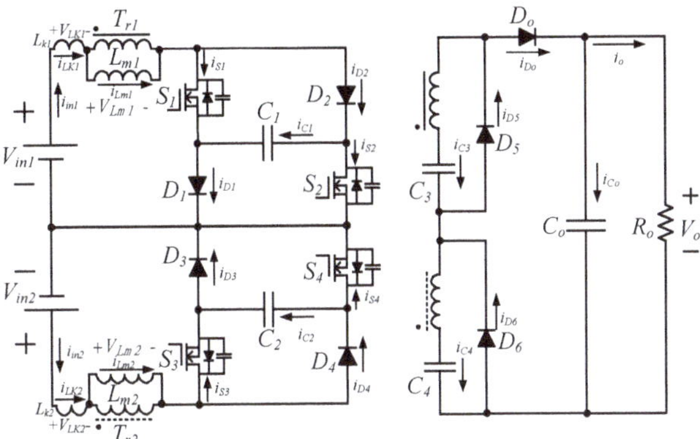

Figure 2. Equivalent circuit of the proposed converter.

2.1. Dual-Input Operation

The switches S_1 and S_2 are turned on/off simultaneously, so do the switches S_3 and S_4. Assume that the turn-on period of S_1 and S_2 is $D_1 T_s$ and $D_2 T_s$ is for S_3 and S_4. In addition, the magnitude of V_{in2} is twice that of V_{in1}. While the proposed DHSIC operates at DIO and in continuous conduction mode (CCM), converter operation over one switching cycle can be divided into six states. Converter operation will be described state by state as the following proceeds with. In addition, the equivalent of each state and conceptual waveforms are depicted in Figures 3 and 4, respectively.

- State 1 [t_0~t_1]: Converter operation begins at this state, in which all switches S_1–S_4 are turned on at $t = t_0$. All diodes are reverse except the output diode D_o. The currents of leakage inductances, i_{Lk1} and i_{Lk2}, increase linearly and steeply. Meanwhile, the energy stored in magnetizing inductances L_{m1} and L_{m2} is released to the output via transformers and diode D_o. When leakage inductance current rises to be equal to magnetizing current, the diode current flowing through D_o will drop to zero and then this state ends. The diode D_o turns OFF under zero current transition. That is, the reverse-recovery problem at D_o can be therefore overcome.
- State 2 [t_1~t_2]: In State 2, the switches S_1–S_4 remain ON. The diodes D_1–D_4 and D_o are reversely biased, but diodes D_5 and D_6 are forwarded. In this time interval, leakage inductance and magnetizing inductance of the coupled inductor T_1 absorb energy from V_{in1} and C_1, similarly, L_{k2} and L_{m2} of T_2 from V_{in1} and C_1. The voltage across T_{r1} is $V_{in1} + V_{C1}$ and T_{r2} is $V_{in2} + V_{C2}$. At the secondary of the DHSIC, switched capacitors C_3 and C_4 are charged by the energy from coupled

inductors T_{r1} and T_{r2}, respectively. This state lasts for a time interval much longer than that of State 2 and is a major state in the converter operation.

- State 3 [t_2~t_3]: During the period from t_2 to t_3, switches S_1 and S_2 continue conducting. On the contrary, S_3 and S_4 are turned off at t_2. The diodes D_1–D_4 and D_o are reversely biased, but D_5 and D_6 are in a forward state. The parasitic capacitances of S_3 and S_4 are charged and current i_{Lk2} decreases dramatically. As the increasing voltages blocked by S_3 and S_4 reach V_{C2}, diodes D_3 and D_4 become forwarded and then the operation state enters State 4.

- State 4 [t_3~t_4]: All active switches remain the same on-off conditions as in State 3. That is, S_1 and S_2 are closed but S_3 and S_4 open. The voltage $V_{in1} + V_{C1}$ will supply T_{r1} and forwards the energy to charge switched capacitor C_3. Meanwhile, the capacitor C_2 absorbs energy from V_{in2} and T_{r2}. Leakage energy of L_{k2} is recycled to C_2. The voltage stress of S_3 and S_4 will be clamped to V_{C2}. This operation state ends at the time both switches S_1 and S_2 are turned off.

- State 5 [t_4~t_5]: After S_1 and S_2 are turned off, the voltage across both switches increases. At the same time, their parasitic capacitances are charging toward the value of V_{C1}. With respect to the other switches, S_3 and S_4 are still in the OFF state. Once parasitic capacitance-voltage approaches to V_{C1}, State 5 ends and blocking voltage of S_1 and S_2 will be clamped at V_{C1}. The switched capacitor C_3 is still charging. During the time interval of State 5, the current flowing through L_{k1} drops steeply. The diodes D_1 and D_2 will be forwarded at $t = t_5$ and then this state ends.

- State 6 [t_5~t_6]: From t_5 to t_6, all switches remain OFF. The L_{m1} pumps its stored energy to charge C_1 and to the output as well. With respect to L_{m2}, it is also in energy-releasing but charges toward C_2, meanwhile, part of its energy will be transformed to the secondary of T_{r2} to power the output. The leakage energy stored in T_{r1} and T_{r2} will be recycled to capacitors C_1 and C_2, respectively. During State 6, the series voltage of T_{r1}, C_3, T_{r2}, and C_4 is connected to the output. That is, the output can accordingly obtain a high voltage level. Like State 2 and 4, State 6 also plays a major role in the converter operation. While switches S_1–S_4 are turned on again at $t = t_6$, this state ends and converter operation over one switching cycle is completed.

2.2. Single-Input Operation

Once one of the inputs fails to supply power, the proposed converter still can function as a high step-up feature. Suppose that only V_{in1} powers the DHSIC and in CCM condition. The converter will have four operation states over switching cycle. The corresponding key waveforms and equivalent circuits are illustrated in Figures 5 and 6 in turn.

- State 1 [t_0~t_1]: If only V_{in1} supplies the converter, power processing is controlled by switches S_1 and S_2. Both switches are turned on at $t = t_0$ and converter operation begins. As shown in Figure 6a, the voltage across T_{r1} will be equal to the series voltage of C_1 and V_{in1}. Magnetizing inductance L_{m1} pumps its stored energy to the secondary of T_{r1}. Therefore, the current flowing through L_{m1} decreases. The current i_{Lk1} increases quickly. When i_{Lk1} is equal to i_{Lm1}, this state ends. At this time, the current flowing through the diodes D_o and D_6 also drops to zero. That is, the reverse-recovery problem of both diodes is therefore resolved.

- State 2 [t_1~t_2]: The equivalent of State 2 is illustrated in Figure 6b, in which the switches S_1 and S_2 remain closed. The L_{m1} absorbs energy form V_{in1} and C_1 and thus i_{Lm1} increases linearly. At the secondary of T_{r1}, the switched capacitor C_3 is charging continuously over this stage. State 2 is a major state in the converter operation at SIO. At time $t = t_2$, the switches S_1 and S_2 are turned off and then the converter operation enters the next stage.

- State 3 [t_2~t_3]: In State 2, all diodes are in reverse bias except D_5. The parasitic capacitance of power switches S_1 and S_2 are charged and the current i_{Lm1} decreases. The blocking voltage of S_1 and S_2 is therefore increasing. The associated equivalent is shown in Figure 6c. When the voltage across S_1 and S_2 approaches to V_{C1}, diodes D_1 and D_2 become forwarded. Then, State 4 starts.

- State 4 [t_3~t_4]: The equivalent circuit refers to Figure 6d, in which the magnetizing inductance forwards its stored energy to charge capacitor C_1 and to the output via the ideal transformer. Meanwhile, the leakage energy of L_{k1} is recycled to C_1, which also suppresses the voltage spike on the active switches. Over the time interval from t_3 to t_4, switches S_1 and S_2 are open. With respect to diode, the D_5 is in reverse state but D_1, D_2, D_6 and D_0 are forward biased. State 4 is also a major state similar to State 2, dominating the converter operation. Both switches S_1 and S_2 will be turned on again at $t = t_4$ and then this state ends. Converter operation over one switching cycle is completed.

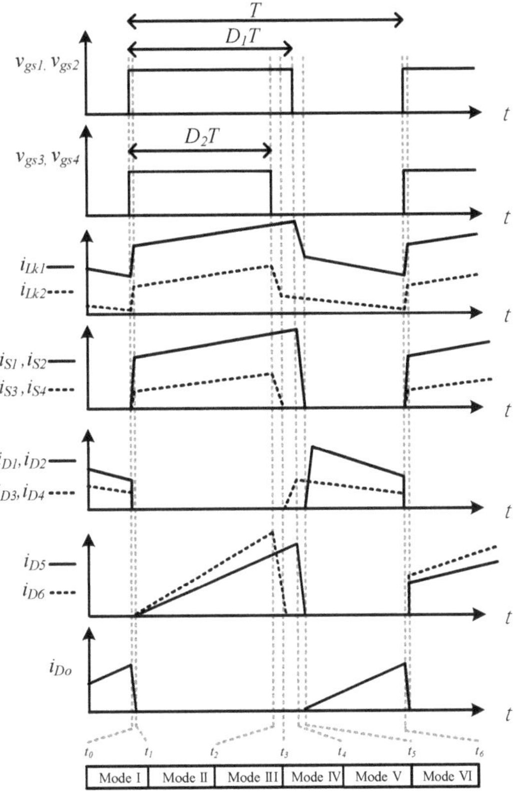

Figure 3. The conceptual key waveform of the proposed converter at DIO (dual input operation) and CCM (continuous conduction mode) operation.

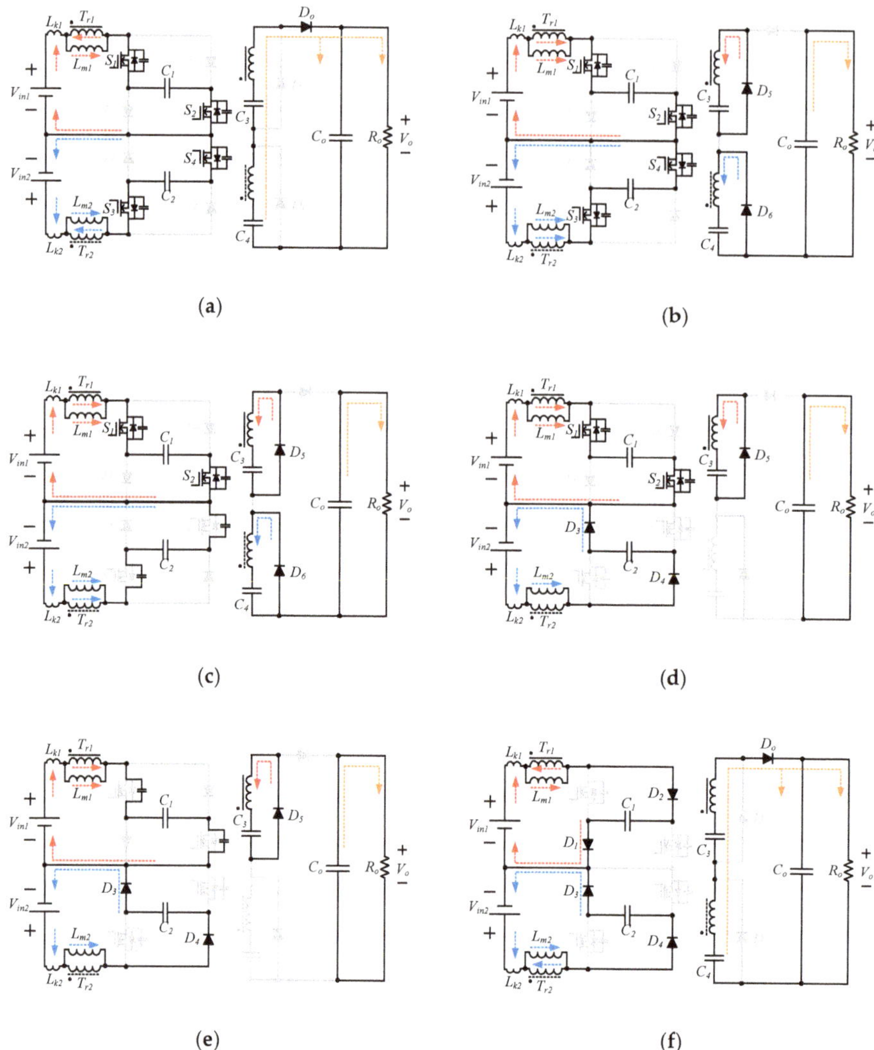

Figure 4. The equivalents of the proposed converter at DIO and CCM operation: (**a**) State 1, (**b**) State 2, (**c**) State 3, (**d**) State 4, (**e**) State 5, (**f**) State 6.

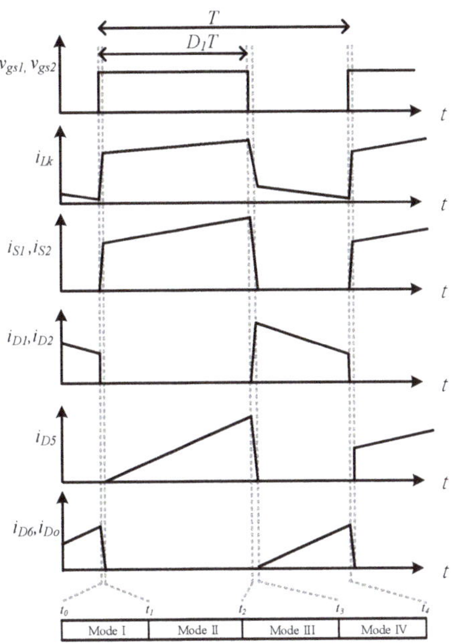

Figure 5. Conceptual key waveform of the proposed converter at SIO (single-input operation) and CCM operation.

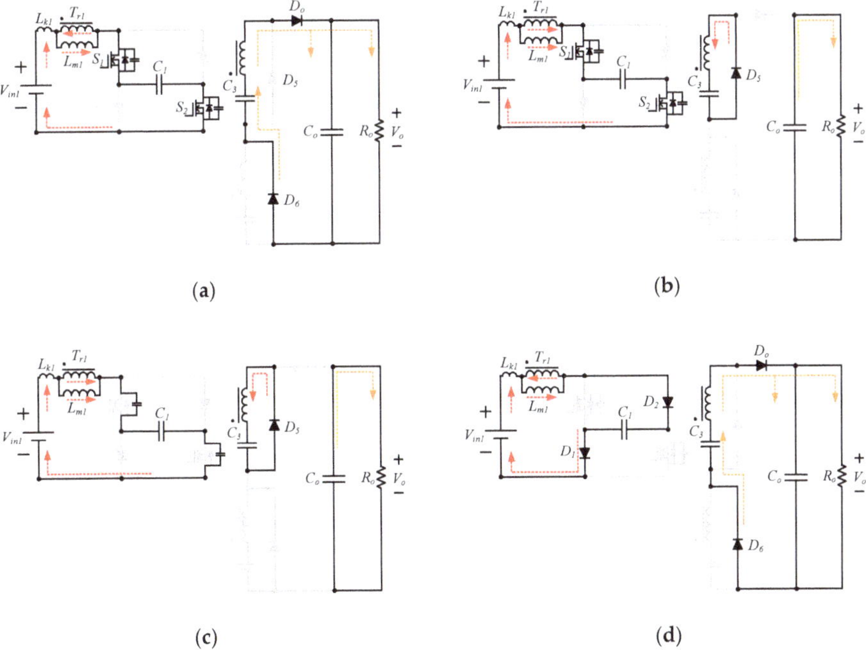

Figure 6. The equivalents of the proposed converter at SIO and CCM operation: (**a**) State 1, (**b**) State 2, (**c**) State 3, (**d**) State 4.

3. Steady-State Analysis of Proposed Converter

The voltage ratio of output to input, voltage stress and current stress of the semiconductor device, and magnetizing inductance design will be covered in this section. To simplify the steady-state analysis, the following assumptions are made.

1. The values of the boosting capacitors C_1 and C_2 are large enough so as to keep their across voltages invariant.
2. All diodes are regarded to be ideal. That is, forward drop voltage and ON-state resistance are neglected.
3. The magnetizing inductance of the coupled inductor is much more than leakage inductance so that influence of the leakage inductance can be ignored.
4. The turns ratios of the coupled inductors, N_{1s}/N_{1p} and N_{2s}/N_{2p}, are defined as n_1 and n_2, respectively.
5. The DHSIC is at CCM operation.

The driving pattern relating to the four switches is the same as that discussed in the previous section. The S_1 and S_2 are closed for D_1T and S_3 and S_4 for D_2T. In addition, the duty ratio of D_1 is greater than D_2. Based on the assumptions made at the beginning of this section, states 2, 4 and 6 in Figure 4 will dominate the converter operation.

3.1. Voltage Conversion Ratio

The proposed DHSIC can be regarded as a combination of two step-up converters which are symmetrical to common ground at the input side and in series connection at the output port. Furthermore, the two step-up converters are capable of operating individually. Therefore, the obtaining of voltage conversion ratio of the DHSIC can simply be derived from a single input situation and then to augment to dual-input situation. Suppose that only the V_{in1} powers the DIC and both switches S_1 and S_2 are closed for D_1T and open for $(1 - D_1)T$ over one switching period. The equivalents of switch ON and OFF are depicted in Figure 7. Applying voltage second balance criterion to L_{m1} can yield

$$V_{Lm1,on} \cdot D_1 T + V_{Lm2,off} \cdot (1 - D_1)T = 0 \tag{1}$$

From Figure 7b, the $V_{Lm1,on}$ and $V_{Lm1,off}$ can be obtained as follows:

$$V_{Lm1,on} = V_{in1} + V_{C1} \tag{2}$$

and

$$V_{Lm1,off} = V_{in1} - V_{C1}. \tag{3}$$

Substituting Equations (2) and (3) into Equation (1) can find the expression of V_{C1}:

$$V_{C1} = \frac{1}{1 - 2D_1} V_{in1}. \tag{4}$$

From Figure 7a, the voltage across capacitor C_3, V_{C3}, is found by the multiplication of turns ratio n_1 and $V_{Lm1,on}$, and then, from Equations (2) and (4) the V_{C3} can be written as

$$V_{C3} = \frac{2n_1(1 - D_1)}{1 - 2D_1} V_{in1}. \tag{5}$$

According to Figure 7b, the following relationship holds:

$$n_1 V_{Lm1,off} = V_{C3} - V_o. \tag{6}$$

Based on Equations (3)–(6), the output voltage of the converter at SIO can be given by

$$V_o = \frac{2n_1}{1-2D_1}V_{in1}.$$ (7)

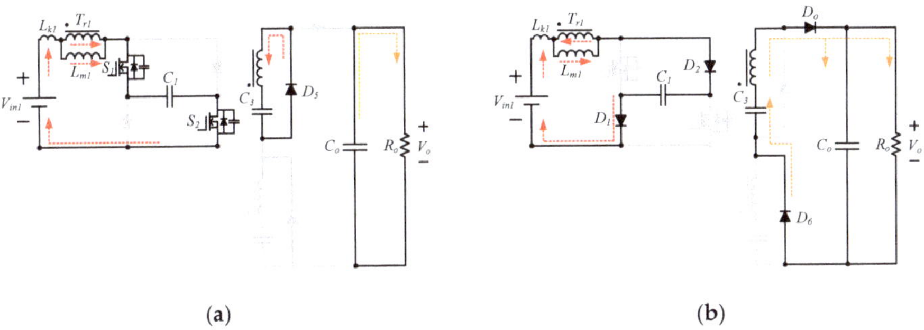

(**a**) (**b**)

Figure 7. The simplified equivalents while DHSIC (dual-input high step-up isolated converter) operates at SIO: during (**a**) switch-ON interval, (**b**) switch-OFF interval.

According to Equation (7), the switch duty ratio has to be less than 0.5, which is the converter limitation. Figure 8 depicts the relationship of voltage gain and duty ratio under different turns ratio.

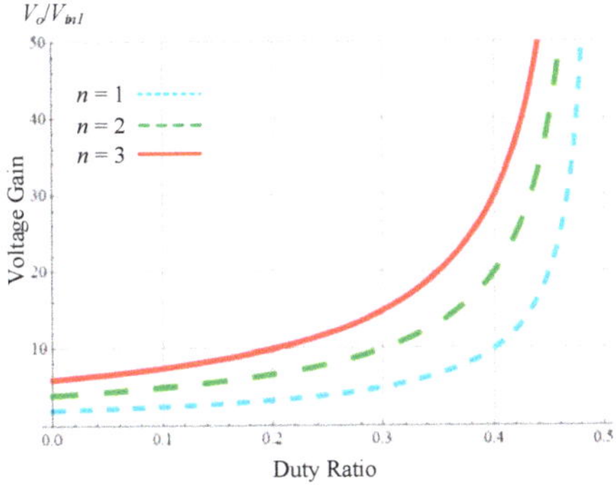

Figure 8. The voltage gain versus duty ratio under the SIO situation.

With respect to the DIO situation, the DHSIC has the same control scheme shown in Figure 3. The S_1 and S_2 are closed for $D_1 T$ and open for $(1 - D_1)T$, while S_3 and S_4 are closed for $D_2 T$ and open for $(1-D_2)T$ over one switching period. Equations (4) and (5) can be applied to the finding for the voltages of C_2 and C_4. That is, V_{C2} and V_{C4} are calculated by

$$V_{C2} = \frac{1}{1-2D_2}V_{in2}$$ (8)

and

$$V_{C4} = \frac{2n_2(1-D_2)}{1-2D_2}V_{in2},$$ (9)

respectively. Being similar to Equation (6), the following relationship will hold under a dual-input situation.

$$n_1 V_{Lm1,off} + n_2 V_{Lm2,off} = V_{C3} + V_{C4} - V_o.$$ (10)

Thus, the corresponding output voltage at DIO of the converter can be described as

$$V_o = \frac{2n_1}{1 - 2D_1} V_{in1} + \frac{2n_2}{1 - 2D_2} V_{in2}.$$ (11)

Assume that $V_{in2} = 2V_{in1}$, $D_1 = D_2 = D$, and $n_1 = n_2 = n$, according to Equation (11), Figure 9 represents the relationship of voltage gain versus turns ratio, while under DIO situation.

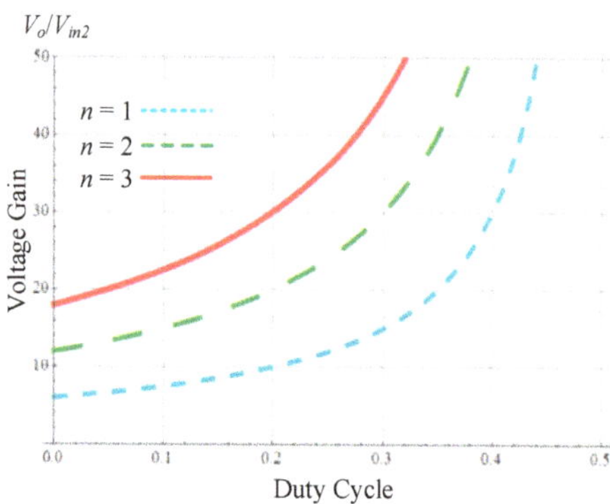

Figure 9. The voltage gain versus duty ratio under the DIO situation.

3.2. Voltage Stress of Semiconductor Device

According to the structure of the DHSIC, the semiconductor devices, S_1, S_2, D_1, and D_2, will have the same voltage stress. In addition, the power stage has inherently symmetrical configuration and is able to operate independently and individually at primary side. Therefore, the voltage stress of S_1, S_2, D_1, and D_2 can be determined from Figure 7, accordingly all of which will be equal to V_{C1}. From Equation (4), this voltage stress is expressed as

$$V_{S1,stress} = V_{S2,stress} = V_{D1,stress} = V_{D2,stress} = V_{C1} = \frac{1}{1 - 2D_1} V_{in1}.$$ (12)

Similarly, voltage stress of S_3, S_4, D_3, and D_4 can be determined by

$$V_{S3,stress} = V_{S4,stress} = V_{D3,stress} = V_{D4,stress} = V_{C2} = \frac{1}{1 - 2D_2} V_{in2}.$$ (13)

With attention to the semiconductor devices at the output port (the secondary side of the DHSIC), an associated determination is discussed in the following. While all active switches are in OFF-state, the blocking voltages of diodes D_5 and D_6 are obtained by

$$V_{D5,stress} = V_{C3} + n_1(V_{C1} - V_{in1})$$ (14)

and

$$V_{D6,stress} = V_{C4} + n_2(V_{C2} - V_{in2}),$$ (15)

respectively. The V_{C1} and V_{C3} can be founded by Equations (4) and (5) in turn, and V_{C2} and V_{C4} by Equations (8) and (9). As a result, the above Equations (14) and (15) can be rewritten as

$$V_{D5,stress} = \frac{2n_1}{1 - 2D_1} V_{in1}$$ (16)

and

$$V_{D6,stress} = \frac{2n_2}{1-2D_2} V_{in2},\tag{17}$$

respectively. Voltage stress of the output diode D_O is estimated at the state that all active switches are closed and thus it will be

$$V_{Do,stress} = V_{C3} + n_1(V_{C1} - V_{in1}) + V_{C4} + n_2(V_{C2} - V_{in2}).\tag{18}$$

Based on Equations (4), (5), (8) and (9), the expression of Equation (18) is rewritten as

$$V_{Do,stress} = \frac{2n_1}{1-2D_1} V_{in1} + \frac{2n_2}{1-2D_2} V_{in2}.\tag{19}$$

3.3. Magnetizing Inductance Design

The minimum currents of active switches S_1 and S_2, $i_{Lm1,min}$ and $i_{Lm2,min}$, can be expressed as

$$i_{Lm1,min} = I_{Lm1} - \frac{\Delta i_{Lm1}}{2},\tag{20}$$

and

$$i_{Lm2,min} = I_{Lm2} - \frac{\Delta i_{Lm2}}{2},\tag{21}$$

respectively. The Δi_{Lm1} and Δi_{Lm2} stand for the current change on L_{m1} and L_{m2} in turn, while I_{Lm1} and I_{Lm2} are the average currents of L_{m1} and L_{m2}. Assume that the converter is lossless. Then, The I_{Lm1} and I_{Lm2} can be determined as follows:

$$I_{Lm1} = \frac{2n_1}{(1-2D_1)} I_o,\tag{22}$$

and

$$I_{Lm2} = \frac{2n_2}{(1-2D_2)} I_o,\tag{23}$$

in which the I_o denotes output current. In addition, Δi_{Lm1} and Δi_{Lm2} can be calculated by

$$\Delta i_{Lm1} = \frac{V_{Lm1,on}}{L_{m1}} D_1 T_s\tag{24}$$

and

$$\Delta i_{Lm2} = \frac{V_{Lm2,on}}{L_{m2}} D_2 T_s.\tag{25}$$

At the boundary, Δi_{Lm1} and Δi_{Lm2} are equal to zero, that is,

$$\frac{2n_1 V_o}{(1-2D_1)R_o} = \frac{V_{Lm1.on}}{2L_{m1}} D_1 T_s\tag{26}$$

and

$$\frac{2n_2 V_o}{(1-2D_2)R_o} = \frac{V_{Lm2,on}}{2L_{m2}} D_2 T_s.\tag{27}$$

From Equations (26) and (27), the minimum magnetizing inductances for CCM operation have to meet the following inequality:

$$L_{m1} > \frac{(1-D_1)D_1 R_o V_{in1}}{2n_1 f_s \left(\frac{2n_1}{1-2D_1} V_{in1} + \frac{2n_2}{1-2D_2} V_{in2}\right)}\tag{28}$$

and

$$L_{m2} > \frac{(1-D_2)D_2 R_o V_{in2}}{2n_2 f_s \left(\frac{2n_1}{1-2D_1} V_{in1} + \frac{2n_2}{1-2D_2} V_{in2}\right)}.\tag{29}$$

Suppose that the turns ratio $n_1 = n_2 = n$, $V_{in2} = 2V_{in1} = 24$ V, switching frequency $f_s = 40$ kHz, and both switches have the same duty ratio denoted as D. In addition, the converter operates at boundary condition mode (BCM) at 25 W. Figure 10 illustrates the relationship between magnetizing inductance and duty ratio for L_{m1} and L_{m2}.

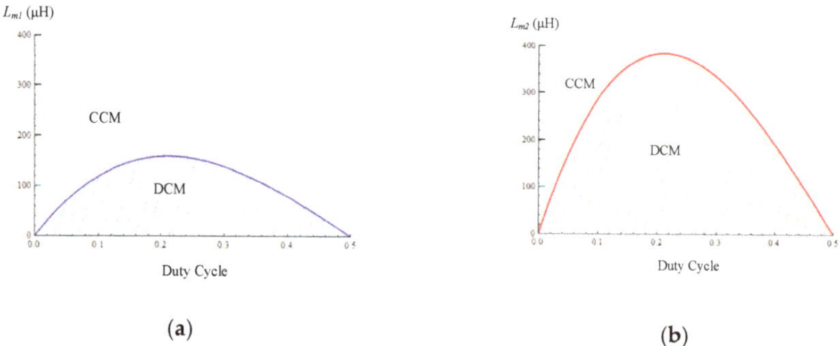

Figure 10. The relationship between the magnetizing inductance and the duty cycle. (**a**) L_{m1}, (**b**) L_{m2}.

The comparison with other similar converters is summarized in Table 1. It can be found that even though the DHSIC needs more power components, it achieves excellent voltage gain over other DICs in addition to that the features of galvanic isolation, continuous input current and switch voltage clamping still can be possessed.

Table 1. Performance Comparison among the Proposed and Other Multi-Input Converters.

Converter	Number of Switches	Number of Magnetic Components	Number of Capacitors	Input Current Ripple	Voltage Stress	Switch Voltage Camping	Galvanic Isolation	Output Voltage (V_o)
[15]	2	2	5	low	medium	no	no	$\frac{3}{1-D}V_{in1} = \frac{3}{1-D}V_{in2}$
[16]	4	2	2	high	low	yes	yes	$n_1 D_1 V_{in1} + n_2 D_2 V_{in2}$
[19]	3	1	5	high	medium	no	no	$\frac{3[D_1 V_{in1} + (D_1 - D_2)V_{in2}]}{(1-D_3)}$
[24]	6	2	2	low	high	no	no	$\frac{V_{in1}}{(1-D_1)} = \frac{V_{in2}}{(1-D_2)}$
DHSIC	4	2	5	medium	low	yes	yes	$\frac{2n_1}{(1-2D_1)}V_{in1} + \frac{2n_2}{(1-2D_2)}V_{in2}$

4. Experimental Results

To verify the proposed DIC, a 200-W prototype is built, simulated and measured. Converter parameters and components adopted are summarized in Table 2. The control block diagram of the prototype is depicted in Figure 11. Power of input port 1 is calculated according to the detected input voltage and current, that is, $V_{in1,fb}$, and $I_{in1,fb}$, and then it is compared to a reference P_{ref1}. The control signals for S_1 and S_2 are determined by the PI power controller and the carrier. Accordingly, the input power at port 1 can be readily controlled. With respect to the other part of the control block diagram, the output voltage is regulated by controlling the switches S_3 and S_4. With such voltage regulation, the input port 2 can accommodate the supplement to output power and thus power dispatch at both input ports is accomplished. The proposed converter adopts MCU dsPIC30F4011 to serve as system controller. Figure 12 shows the control signals and the corresponding input currents at full load under DIO situation. In Figure 12 the duty ratios of S_1 and S_2 are 0.32 and 0.23, respectively, which is consistent with Equation (11) for a 400-V output. Meanwhile, the switch voltages, v_{ds1} and v_{ds2}, are measured and shown in Figure 13, from which it can be observed that there is no voltage spike on active switches. That is, the boosting capacitors C_1 and C_2 are able to recycle leakage energy and can effectively clamp switch voltage. Additionally, withstood voltages on the switches at port 1 and port 2 are 34 V and 45 V, respectively, which verifies the theoretical analysis results of Equations (12) and (13). Figure 14 is the measurement of the voltages of boosting capacitors. This figure reveals that voltages of C_1 and C_2 are in turn 34 V and 45 V, both voltage magnitudes of which meet the theoretical results of Equations (4) and (8). While step change takes place from half load to full load and then drops back, Figure 15 shows the related output voltage and current. Figure 15 illustrates that constant 400-V output still can be kept with even under step change. The overshoots at step-up and step-down transitions are only 1 V and 0.7 V, respectively. With respect to SIO situation, once power failure occurs at input 1, Figure 16 shows the measured control signal, output voltage, and input current. Similarly, if at input 2, Figure 17 is the related measurement. Both figures demonstrate the operation ability of the converter at SIO situation. The measured waveform of output current is presented in Figure 18, from which it can be observed that the output current of the converter is near to be ripple-free. The efficiency of the proposed converter is measured and then shown in Figure 19, in which the peak value is about 91.4%. In the experiment, a very low level of voltage, $V_{in1} = 12$ V, is considered, therefore, which yields that higher current has to be demanded at a specific power, resulting in large conduction loss. This is the major reason why the converter efficiency is not as high as satisfied in the measurement. However, if input voltage is raised, the converter efficiency will be advanced. At single-input situation, if only the 24 V of V_{in2} suppled the DHSIC, measured converter efficiency is shown in Figure 20. This figure reveals that a higher voltage input can yield a better efficiency even under the operation of single input. A photo of the built-up DHSIC is shown in Figure 21.

Table 2. Parameters of the DHSIC for Simulations and Practical Measurements.

Parameter	Value
V_{in1} (PV arrays)	12–16 V
V_{in2} (fuel-cells stack)	24–30 V
f_s (switching frequency)	40 kHz
V_o (output voltage)	400 V
Power rating	200 W
L_{m1} and L_{m2} (magnetizing inductors)	176 µH and 302 µH
L_{Lk1} and L_{Lk2} (leakage inductors)	1.9 µH and 2.4 µH
S_1–S_4 (power MOSFET)	IRFP4668
D_1, D_2, D_3, and D_4	DSSK60-02AR
D_5, D_6 and D_o	BYR29-600
C_1	68 µF
C_2	33 µF
C_3 and C_4	22 µF
C_o	82 µF
n_1 and n_2 (transformer turns ratio)	3 and 2.5

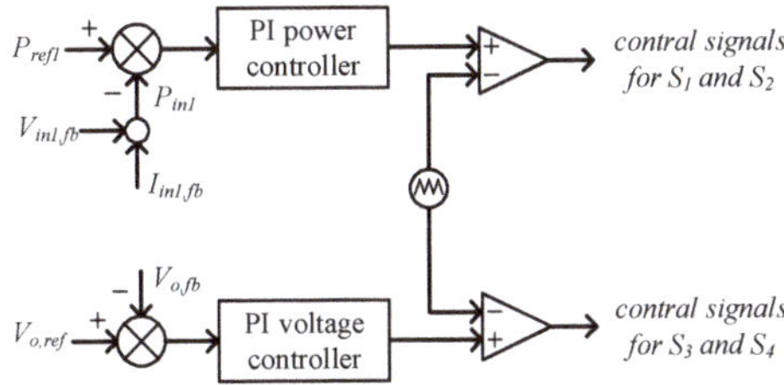

Figure 11. Control block diagram of the proposed converter.

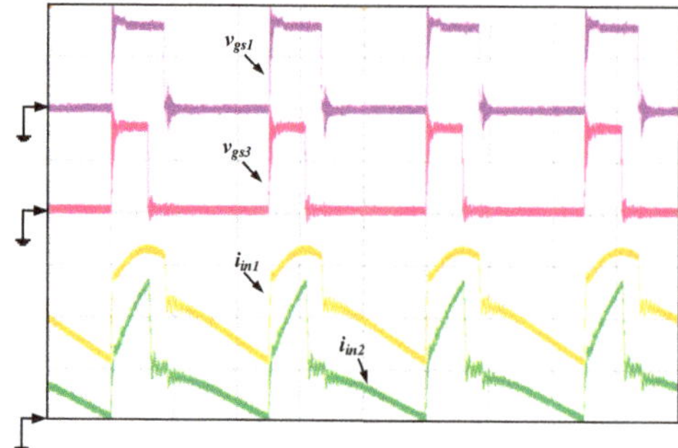

(control signals: 10 V/div, i_{in1} *and* i_{in2}: 5 A/div, time: 10 μs/div)

Figure 12. The waveforms of control signals and corresponding input currents at full load under the DIO situation.

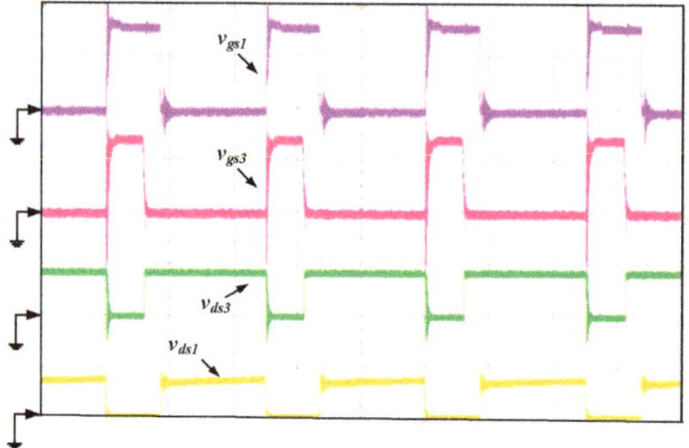

(control signal: 10 V/div, v_{ds1} and v_{ds2}: 50 V/div, time: 10 μs/div)

Figure 13. The practical waveforms of the voltages across active switches.

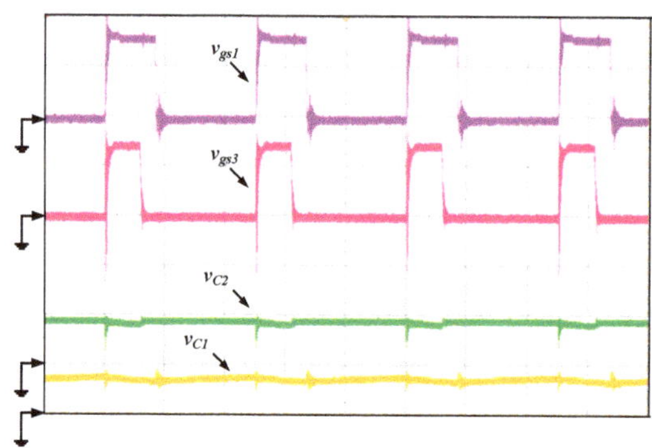

(control signals: 10 V/div, v_{C1} and v_{C2}: 50 V/div, time: 10 μs/div)

Figure 14. The voltage measurement of boosting capacitors C_1 and C_2.

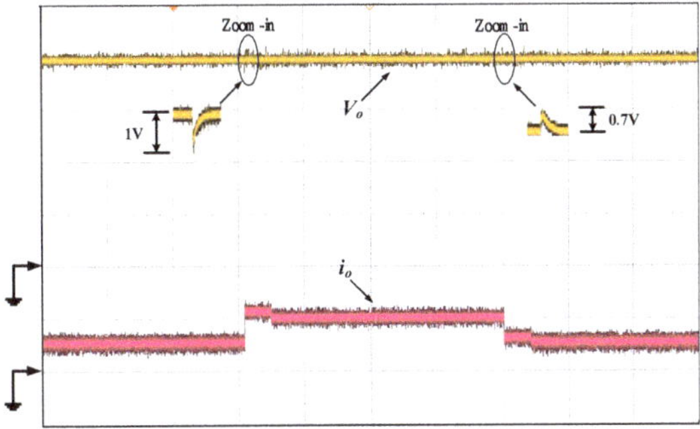

(*V$_o$*: 100 V/div, *i$_o$*: 0.5 A/div, time: 500 ms/div)

Figure 15. The waveforms of output voltage and current while step change occurs at load.

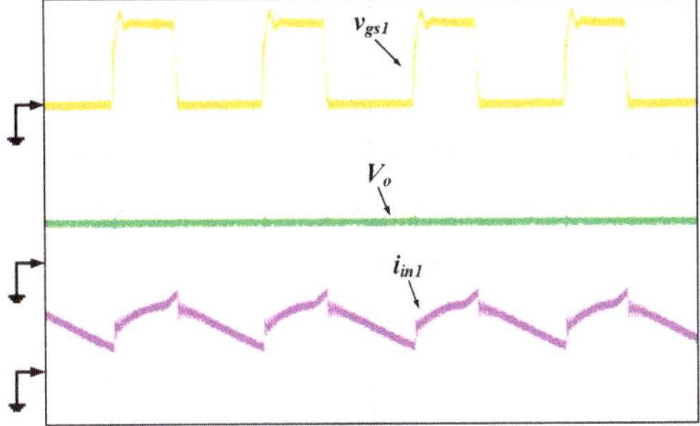

(*v$_{gs1}$*: 10 V/div, *V$_o$*: 500 V/div, *i$_{in1}$*: 10 A/div, time: 10 μs/div)

Figure 16. The measured waveforms of the control signal, output voltage, and input current while input 2 encounters power failure.

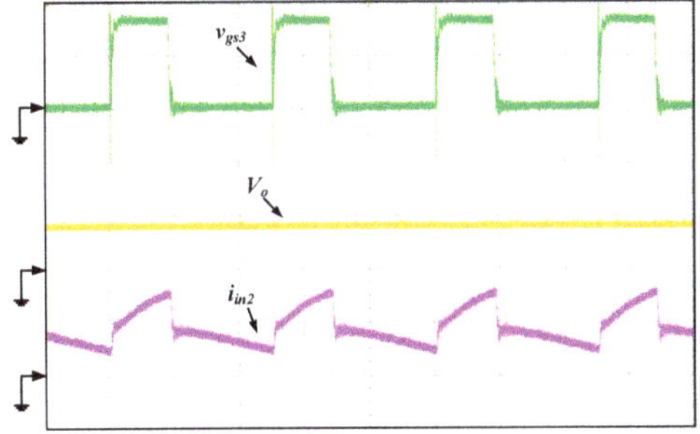

(*v$_{gs3}$*: 10 V/div, *V$_o$*: 500 V/div, *i$_{in2}$*: 10 A/div, time: 10 μs/div)

Figure 17. The measured waveforms of the control signal, output voltage, and input current while input 1 encounters power failure.

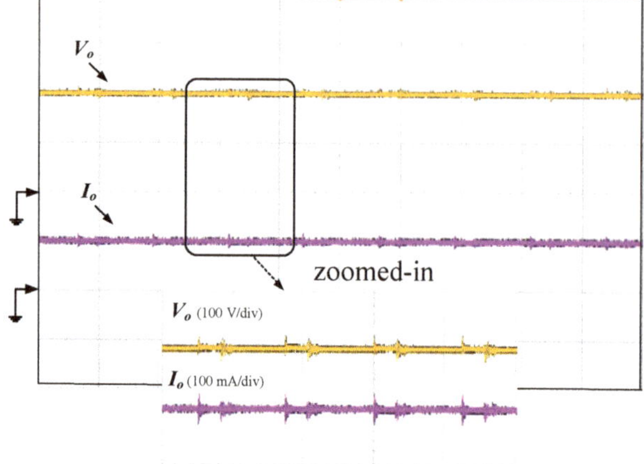

Figure 18. The practical measurement of output current I_o.

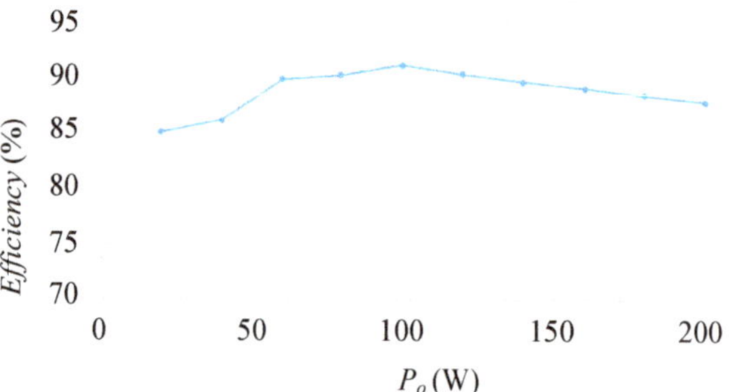

Figure 19. The conversion efficiency of the proposed converter.

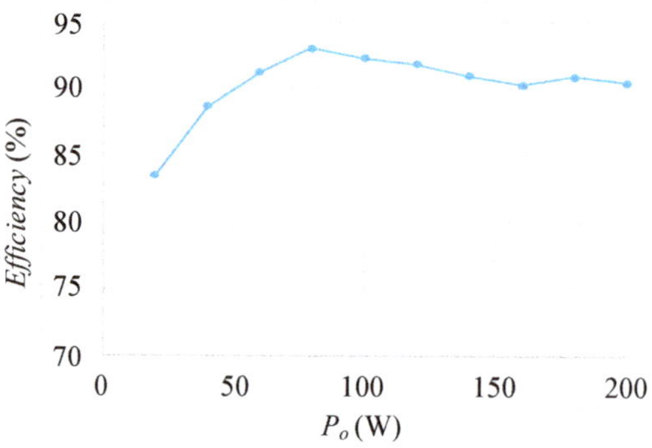

Figure 20. Measured converter efficiency while only input port 2 powers the DHSIC.

Figure 21. Photo of the prototype.

5. Conclusions

In this paper, a dual-input converter is proposed, which possesses the characteristics of ultra-high voltage gain, continuous input currents, galvanic isolation, inherent voltage-clamp feature, and recycling the energy stored in leakage inductance. This converter is capable of controlling the dual inputs independently and individually. Moreover, it still can accomplish all the mentioned features even under the case that either input encounters power failure. That is, the converter has operation flexibility to operate at dual-input situation or single-input situation for accommodating input conditions. A maximum of measured efficient is about 91.4% at dual-input operation.

Author Contributions: Conceptualization, C.-L.S., L.-Z.C. and H.-Y.C.; writing-original draft preparation, C.-L.S. and L.-Z.C.; writing-review and editing, C.-L.S. and L.-Z.C.; methodology, C.-L.S., L.-Z.C. and H.-Y.C.; validation, C.-L.S. and L.-Z.C.; supervision, C.-L.S.

Funding: This research received no external funding.

Conflicts of Interest: The authors declare no conflict of interest.

References

1. Wang, Z.; Li, H. Integrated MPPT and Bidirectional Battery Charge for PV Application Using One Multiphase Interleaved Three-Port DC-DC Converter. In Proceedings of the Applied Power Electronics Conference and Exposition (APEC 2011), Fort Worth, TX, USA, 6–11 March 2011; pp. 295–300.
2. Williams, M.C. Fuel Cells and the World Energy Future. In Proceedings of the Power Engineering Society Summer Meeting, Vancouver, BC, Canada, 15–19 July 2001; p. 725.
3. Bryant, B.; Kazimierczuk, M.K. Voltage-Loop Power Stage Transfer Functions with MOSFET Delay for Boost PWM Converter Operating in CCM. *IEEE Trans. Ind. Electron.* **2007**, *54*, 347–353. [CrossRef]
4. Medina-Garcia, A.; Schlenk, M.; Morales, D.P.; Rodriguez, N. Resonant Hybrid Flyback, a New Topology for High Density Power Adaptors. *Electronics* **2018**, *7*, 363. [CrossRef]
5. Lin, B.R.; Hsieh, F.Y. Soft-Switching Zeta-Flyback Converter with a Buck-Boost Type of Active Clamp. *IEEE Trans. Ind. Electron.* **2007**, *54*, 2813–2822.
6. Wang, C.M. A Novel ZCS-PWM Flyback Converter with a Simple ZCS-PWM Commutation Cell. *IEEE Trans. Ind. Electron.* **2008**, *55*, 749–757. [CrossRef]
7. Almalaq, Y.; Matin, M. Three Topologies of a Non-Isolated High Gain Switched-Inductor Switched-Capacitor Step-Up Cuk Converter for Renewable Energy Applications. *Electronics* **2018**, *7*, 94. [CrossRef]
8. Shu, L.J.; Liang, T.J.; Yang, L.S.; Lin, R.L. Transformerless High Step-Up DC-DC Converter Using Cascode Technique. In Proceedings of the Power Electronics Conference, Sapporo, Japan, 21–24 June 2010; pp. 63–67.

9. Young, C.M.; Chen, M.H.; Chang, T.A.; Ko, C.C. Transformerless High Step-Up DC-DC Converter with Cockcroft-Walton Voltage Multiplier. In Proceedings of the 2011 6th IEEE Conference on Industrial Electronics and Applications, Beijing, China, 21–23 June 2011; pp. 1599–1604.

10. Axelorod, B.; Berkovic, Y.; Ioinovici, A. Switched-Capacitor/Switched-Inductor Structures for Getting Transformerless Hybrid DC-DC PWM Converters. *IEEE Trans. Circuits Syst. I.* **2008**, *55*, 687–696. [CrossRef]

11. Changchien, S.K.; Liang, T.J.; Chen, J.F.; Yang, L.S. Novel high step-up dc-dc converter for fuel cell energy conversion system. *IEEE Trans. Ind. Electron.* **2010**, *57*, 2007–2017. [CrossRef]

12. Chen, S.M.; Liang, T.J.; Yang, L.S.; Chen, J.F. A Cascade high step-up dc-dc converter with single switch for microsource application. *IEEE Trans. Power Electron.* **2011**, *26*, 1146–1153. [CrossRef]

13. Hsieh, Y.P.; Chen, J.F.; Liang, T.J.; Yang, L.S. Novel high step-up dc-dc converter with coupled-inductor and switched –capacitor technique for a sustainable energy system. *IEEE Trans. Power Electron.* **2011**, *26*, 3481–3490. [CrossRef]

14. Cheng, X.-F.; Zhang, Y.; Yin, C. A ZVS Bidirectional Inverting Buck-Boost Converter Using Coupled Inductors. *Electronics* **2018**, *7*, 221. [CrossRef]

15. Hou, S.; Chen, J.; Sun, T.; Bi, X. Multi-input Step-Up Converters Based on the Switched-Diode-Capacitor Voltage Accumulator. *IEEE Trans. Power Electron.* **2016**, *31*, 381–393. [CrossRef]

16. Liu, F.; Wang, Z.; Mao, Y.; Ruan, X. Asymmetrical Half-Bridge Double-Input DC/DC Converters Adopting Pulsating Voltage Source Cells for Low Power Applications. *IEEE Trans. Power Electron.* **2014**, *29*, 4741–4751. [CrossRef]

17. Benavides, N.D.; Chapman, P.L. Power Budgeting of A Multiple-Input Buck-Boost Converter. *IEEE Trans. Power Electron.* **2005**, *20*, 1303–1309. [CrossRef]

18. Reddi, N.K.; Ramteke, M.R.; Suryawanshi, H.M. Dual-Input Single-Output Isolated Resonant Converter with Zero Voltage Switching. *Electronics* **2018**, *7*, 96. [CrossRef]

19. Babaei, E.; Abbasi, O. Structure for multi-input multi-output dc–dc boost converter. *IET Power Electron.* **2016**, *9*, 9–19. [CrossRef]

20. Matsuo, H.; Shigemizu, T.; Kurokawa, F.; Watanable, N. Characteristics of the Multiple-Input DC-DC Converter. In Proceedings of the Power Electronics Specialists Conference, Seattle, WA, USA, 20–24 June 1993; pp. 115–120.

21. Kobayashi, K.; Matsuo, H.; Sekine, Y. Novel Solar-Cell Power Supply System Using A Multiple-Input DC-DC Converter. *IEEE Trans. Ind. Electron.* **2006**, *53*, 281–286. [CrossRef]

22. Matsuo, H.; Lin, W.; Kurokawa, F.; Shigemizu, T.; Watanable, N. Characteristics of the Multiple-Input DC-DC Converter. *IEEE Trans. Ind. Electron.* **2004**, *51*, 625–631. [CrossRef]

23. Chen, Y.M.; Liu, Y.C.; Lin, S.H. Double-Input PWM DC-DC Converter for High/Low Voltage Sources. In Proceedings of the Telecommunications Energy Conference, Yokohama, Japan, 23 October 2003; pp. 27–32.

24. Danyali, S.; Hosseini, S.H.; Gharehpetian, G.B. New Extendable Single-Stage Multi-input DC–DC/AC Boost Converter. *IEEE Trans. Power Electron.* **2014**, *29*, 775–788. [CrossRef]

 *electronics*

Article

Secondary Freeform Lens Device Design with Stearic Acid for A Low-Glare Mosquito-Trapping System with Ultraviolet Light-Emitting Diodes

Wei-Hsiung Tseng [1], Wei-Cheng Hsiao [2], Diana Juan [2], Cheng-Han Chan [3], Wen-Sheng Hsiao [1], Hsin-Yi Ma [4] and Hsiao-Yi Lee [1,5,*]

[1] Department of Electrical Engineering, National Kaohsiung University of Science and Technology, Kaohsiung City 807, Taiwan; tly885@gmail.com (W.-H.T.); vic0920499311@gmail.com (W.-S.H.)
[2] Yuan General Hospital, Kaohsiung City 802, Taiwan; cheng_2034@yahoo.com.tw (W.-C.H.); dianajuan@yuanhosp.com.tw (D.J.)
[3] Department of Aviation and Communication Electronics, Air Force Institute of Technology, Kaohsiung City 820, Taiwan; errmatlab@gmail.com
[4] Department of Industrial Engineering and Management, Minghsin University of Science and Technology, Hsinchu County 30401, Taiwan; hsma@must.edu.tw
[5] Department of Graduate Institute of Clinical Medicine, Kaohsiung Medical University, Kaohsiung City 807, Taiwan
* Correspondence: leehy@nkust.edu.tw; Tel.: +886956-161-988

Received: 11 May 2019; Accepted: 31 May 2019; Published: 2 June 2019

Abstract: Dengue fever is a public health issue of global concern. Taiwan is located in the subtropical region featuring humid and warm weather, which is conducive to the breeding of mosquitoes and flies. Together with global warming and the increasing frequency of international exchanges, in addition to the outbreak of pandemics and dengue fever, the number of people infected has increased rapidly. This study is dedicated to the development of a new mosquito-trapping system. Research has shown that specific wavelengths, colors, and temperatures are highly attractive to both Aedes aegypti and Aedes albopictus. In this study, we create equipment which effectively improves the trapping capabilities of mosquitoes in a wider field. The design of the special Secondary Freeform Lens Device (SFLD) is used to expand the range for trapping mosquitoes and create illumination uniformity; it also directs light downward for the protection of users' eyes. In addition, we use the correct amount of stearic acid as a mosquito attractant to allow a better control effect against mosquitoes during the day. In summary, when the UV LED mosquito trapping system is combined with a quadratic free-form lens, the experimental results show that the system can extend the capture range to $100 \, \pi \, m^2$ in which the number of captured mosquitoes is increased by about 350%.

Keywords: mosquitoes; Aedes aegypti; Aedes albopictus; secondary freeform lens device (SFLD); stearic

---

1. Introduction

Dengue fever is a public health issue of global concern. Zika virus, dengue fever, [1,2] chikungunya, and yellow fever are all transmitted to humans by the Aedes aegypti mosquito. More than half of the world's population live in areas where this mosquito species is present. To date, there are still no dengue vaccines or therapeutic drugs. The worldwide incidence of dengue has risen 30-fold over the past 30 years. Vector-mosquito control is the main prevention and treatment method against this disease.

Taiwan is in the subtropical region and enters the season of facilitated dengue infection every summer, especially in southern Taiwan. According to the statistics of the Disease Control Agency, it is enters an epidemic period from July to November every year. Dengue fever is an arthropod-borne virus.

The main vectors are Aedes aegypti and Aedes albopictus. Aedes aegypti are mainly distributed in tropical and subtropical countries south of the Tropic of Cancer, including Southeast Asian countries, parts of the Pacific Ocean, Central and South America, Northern Australia, Africa, and the Eastern Mediterranean, etc. (Guzman et al., 2010; [3] Li et al. 2007 [4]).

There are still no effective vaccines or drugs for treating dengue, but there has been continuous research, development, and improvement. General mosquito repellent products use mosquito vision and phototaxis to attract the insects. Possible physical trapping factors for mosquitoes include light, color, and carbon dioxide concentration (see Brown et al. [5,6]). Mosquitoes have been found to be highly sensitive to wavelength, direction, light intensity, color, and contrast. A cross-sectional analysis of mosquito eyes by Kay [7] et al. revealed their sensitivity to ultraviolet (UV) light. A study by Shimoda et al. [8] explored pest control using low-energy and specific-wavelength light emitting diode (LED) sources.

The effectiveness of the luminescent traps used to guide mosquitoes has been supported by the above studies and experimental evidence. Moreover, such illuminating trap systems for guiding mosquitoes are now mass-produced and widely used. The Secondary Freeform Lens Device can be used to control the angle and range of illumination of UV LED strips to effectively enhance mosquito attraction. For example, to efficiently trap mosquitos, a secondary lens, which relies on total internal reflection (TIR) [9] and the bird-wing asymmetrical reflector [10], was proposed by Tseng et al. to accomplish a UV LED with wide-angled batwing light distribution. In this paper, a system is proposed with a novel TIR lens array with stearic acid can achieve 300% times the effectiveness of commercial products in terms of its mosquito-trapping efficiency.

2. Principles and Methods

2.1. Principles of the UV LED Mosquito-Trapping Device

The outgoing surface of the freeform lens is responsible for refracting its incident light to achieve the design targets. The light refracted by the outgoing surface is governed by Snell's law.

The vector equation of Snell's law is expressed as follows: [11]

$$O \cdot n_0 - I \cdot n_I = [n_0{}^2 + n_I{}^2 - 2n_0 n_I (O \cdot I)]^{1/2} \cdot N. \tag{1}$$

where O is the refraction unit vector; n_1 is the refractive index of incidence within the lens; n_0 is the refractive index of reflection within the lens; I denotes the incident unit vector; and N is the normal vector corresponding to the incident and refraction vectors.

To improve the optical efficiency of the mosquito trapping system, the UV LED lamp is combined with a quadratic free-form lens, and its reflecting surface is designed to totally and internally reflect the LED emitting light to the outgoing surface. To facilitate prototyping and optical testing, polymethylmethacrylate (PMMA) is used as the material of the Secondary Freeform Lens Device (SFLD) prototyping sample in the experiments.

2.2. Use of Stearic Acid as a Mosquito Attractant

Previous studies have demonstrated that the attraction of dengue fever mosquitos to humans is increased by the presence of acid. Moreover, humans have uniquely high levels of skin-borne lactic acid when compared with other mammals, which is similar to stearic acid in terms of smell. In previous studies, lactic acid and stearic acid have been shown to be attractants of the yellow fever mosquito Aedes aegypti L. (Acree et al. 1968 [12]; Smith et al. 1970 [13]; Eiras and Jepson 1991 [14]; Geier et al.

1996 [15]; Bernier et al. 2002 [16]). Knols et al. (1997) [17] demonstrated that a blend of 12 carboxylic acids affected the behavior of the malaria vector Anopheles gambiae Giles sensu stricto.

In summary, lactic acid and stearic acid may play a critical role in the host-seeking behavior of anthropophilic mosquitoes, which seem to have adapted to utilize this human compound (Dekker et al. 2002) [18].

2.3. Design Method

The design flow of the SFLD is shown in Figure 1.

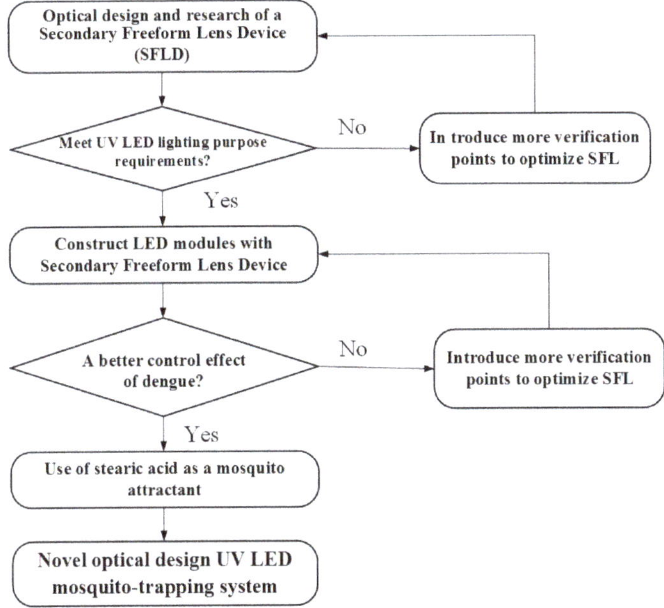

Figure 1. The design process of the Secondary Freeform Lens Device (SFLD).

The luminous intensity distribution curve (LIDC) and uniform illumination of the UV LED mosquito trap system was set as the main target of the SFLD design advantage function. In the SFLD design process, the total internal reflection surface and the output surface can be freely modified; the the trapping range of the UV LED mosquito trapping system can expand the range of trapping mosquitoes and, furthermore, better control the effects of dengue fever.

Additionally, Aedes aegypti plays a considerable role in the transmission of dengue fever. The lifespan of Aedes aegypti is about 30 days; in contrast, the lifespan of Aedes albopictus is about 14 days (Chen, 2006) [19]. After the vector mosquito bites the viremia patient, the virus takes about 10 days to travel from the gastrointestinal tract of the mosquito to the salivary gland. After about 8–12 days, the vector mosquito will have the ability to spread the virus for life. Therefore, the right amount of stearic acid as a mosquito attractant has a better control effect on active mosquitoes during the day (Aedes aegypti L.).

Next, we set the target light mode for the secondary freeform lens device (SFLD) and used a PMMA (PolyMethyl MethAcrylate) with a refractive index of 1.49 and a TIR critical angle of 42.15 degrees (10) as the material, as shown in Figure 2.

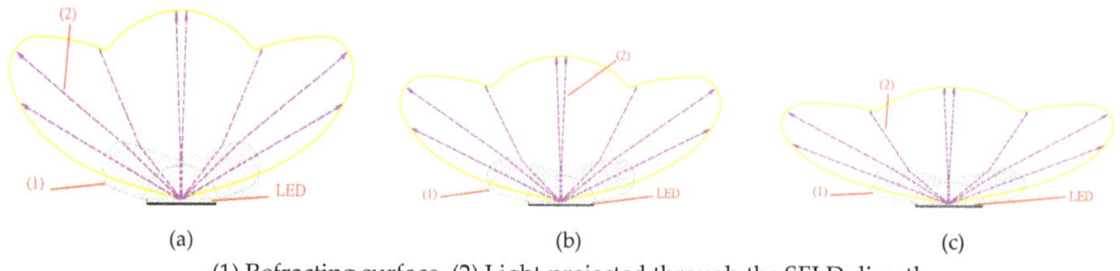

(a)   (b)   (c)

(1) Refracting surface. (2) Light projected through the SFLD directly.

Figure 2. Target light pattern for the SFLD. (**a**) Lens of angle 130; (**b**) Lens of angle 150°; (**c**) Lens of angle 170°.

3. Experiments and Results

3.1. Optical Model of a Secondary Freeform Lens Device

We designed and optimized the SFLD with Solidworks software, as shown in Figure 3.

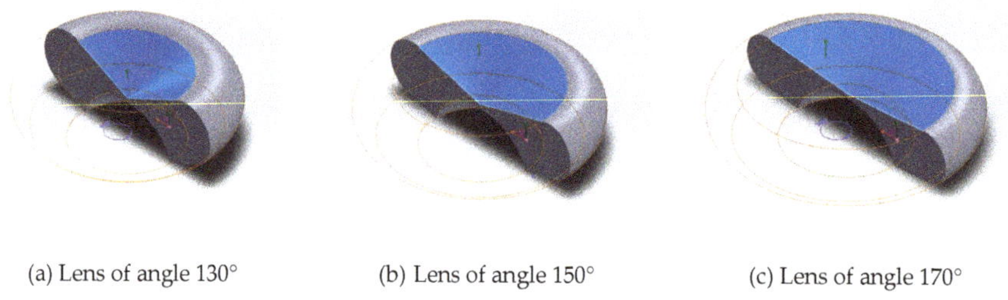

(a) Lens of angle 130°   (b) Lens of angle 150°   (c) Lens of angle 170°

Figure 3. Design and optimization of the SFLD by Solidworks software.

We simulated and analyzed the luminous intensity distribution curve (LIDC) in the space of the traditional LED mosquito lamp and the SFLD's designed UV LED strip module using the TracePro optical software, as shown in Figure 4.

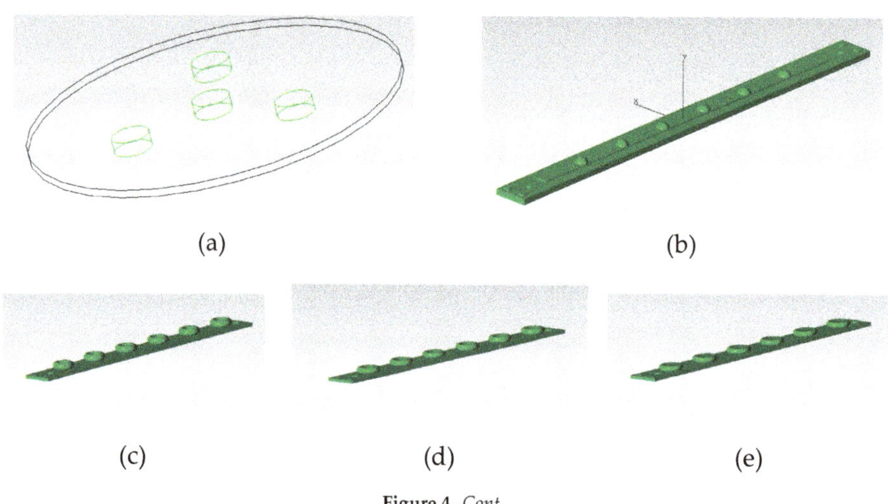

(a)   (b)

(c)   (d)   (e)

Figure 4. *Cont.*

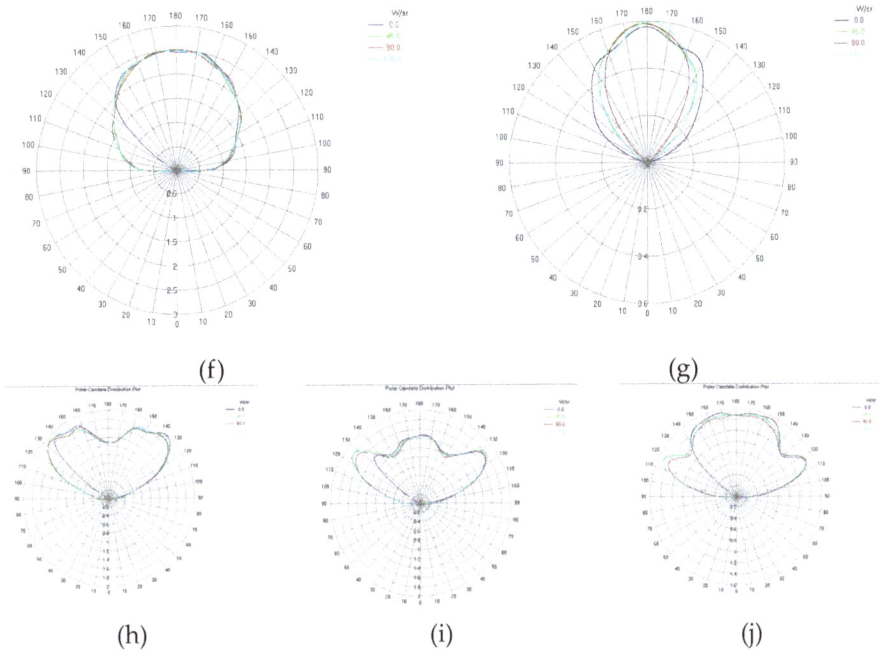

Figure 4. Simulated and analyzed luminous intensity distribution curves (LIDCs) in the space of the traditional LED mosquito lamp and the SFLD's designed UV LED strip module. (**a**) The traditional LED mosquito lamp; (**b**) the traditional designed UV LED strip module of the semicircular lens; (**c**) the SFLD's designed UV LED strip module with an angle of 130°; (**d**) the SFLD's designed UV LED strip module with an angle of 150°; (**e**) the SFLD's designed UV LED strip module with an angle of 170°; (**f**) the traditional LED mosquito lamp of LIDCs; (**g**) the UV LED strip designed semicircular lens module of LIDCs; (**h**) SFLD_UV LED module of LIDCs (130°); (**i**) SFLD_UV LED module of LIDCs (150°); (**j**) SFLD_UV LED module of LIDCs (170°).

The light-intensity distribution angle and the divergent angle centering the angle of the new trapper were targeted at 130°, 150°, and 170°, respectively, in the SFLD's designed UV LED strip module using the TracePro optical software optimization process. We found the SFLD_UV LED module can restrict upward UV ray generation to decrease the glare for the human eye and cover the trapping area as far as possible.

3.2. Process PMMA to Create a SFLD Model

We used a five-axis CNC (Computer Numerical Control) milling machine to process PMMA to create an SFLD model, as shown in Figure 5 [20].

3.3. Luminous Intensity Distribution Measurements

We used the ProMetric near field measurement system (PM-NFMS) developed by Radiant Vision Systems Co, and we measured the LIDC measurements of the LED lights as shown in Figure 6.

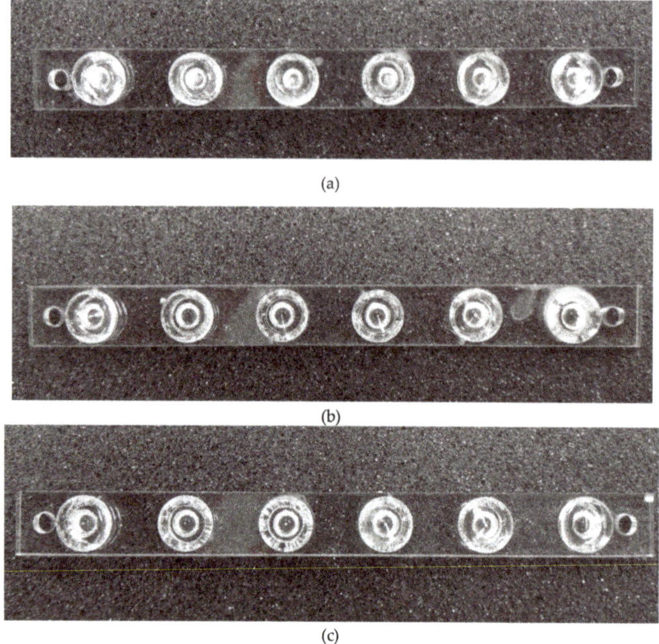

Figure 5. Polymethylmethacrylate (PMMA) model of SFLD. (**a**) Lens of angle 130°; (**b**) Lens of angle 150°; (**c**) Lens of angle 170°.

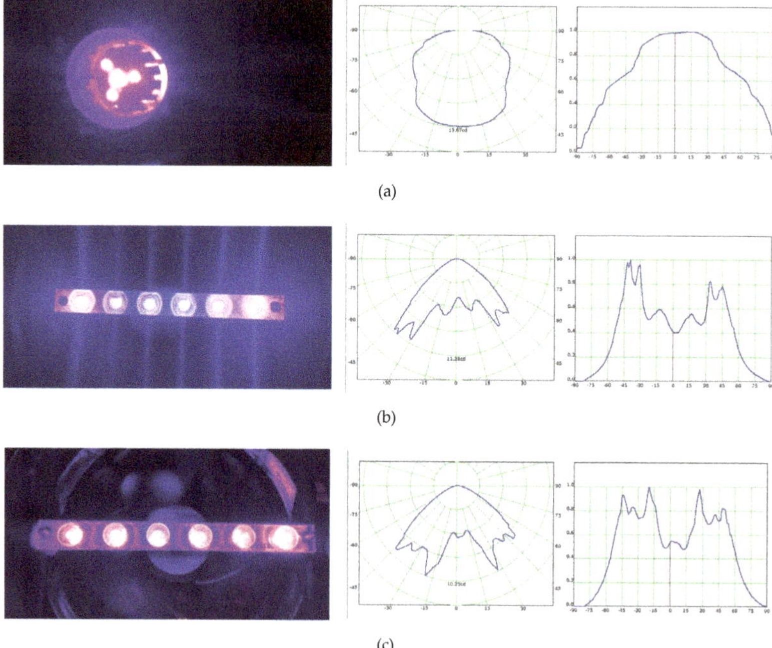

Figure 6. *Cont.*

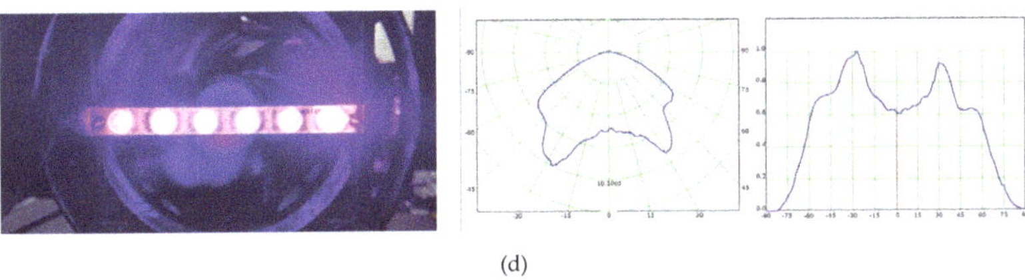

(d)

Figure 6. The 2D LIDC of the narrow UV LED strips and novel mosquito coil UV_LED strip modules (unit: mcd/Klm). (**a**) The 2D LIDC of the module of the traditional LED mosquito lamp; (**b**) The 2D LIDC of the module of the novel mosquito trap UV_LED light bar module (130°); (**c**) The 2D LIDC of the module of the novel mosquito trap UV_LED light bar module (150°); (**d**) The 2D LIDC of the module of the novel mosquito trap UV_LED light bar module (170°).

3.4. Real Machine Verification and Data Analysis

On the basis of the above three different central angles of the SFLD, three sets of the SFLD modules were prototyped and installed with inward air flow for the experiments, as shown in Figure 7. The 3W commercial mosquito trapper was also used in the experiments for comparison, as shown in Figure 8. After the experiments, the amounts of captured mosquitos were compared and are shown in Figure 9. On the basis of the results, it can be calculated that the trapped mosquito numbers are elevated by 350% by the proposed system at most in compared with the conventional method. It was found that the LIDC of the SFLD module with a central angle of 150° had the best result of all experimental systems.

Figure 7. Proposed mosquito trapper with Secondary Freeform Lens Device and its mosquito trapping results.

Figure 8. Traditional mosquito trapper and its mosquito trapping results.

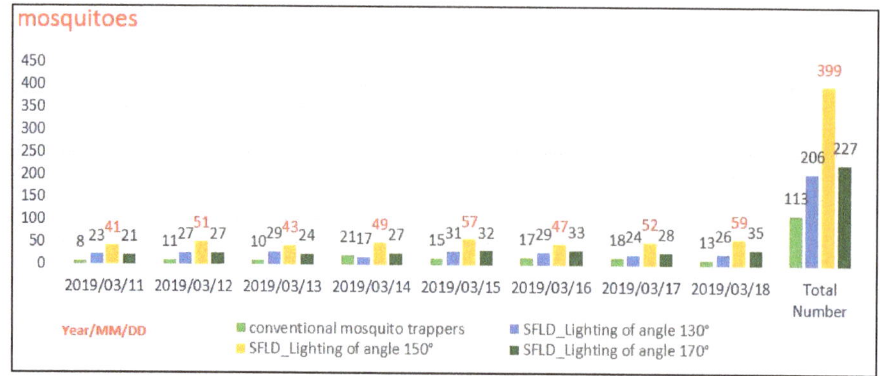

Figure 9. Comparison of the proposed mosquito trapper with Secondary Freeform Lens Device and the traditional one in terms of mosquito trapping number.

To verify the effect of use of stearic acid as a mosquito attractant on the proposed system, stearic acid was smeared on the reflective surface portion of the UV LED mosquito-trapping system, as shown in Figure 10.

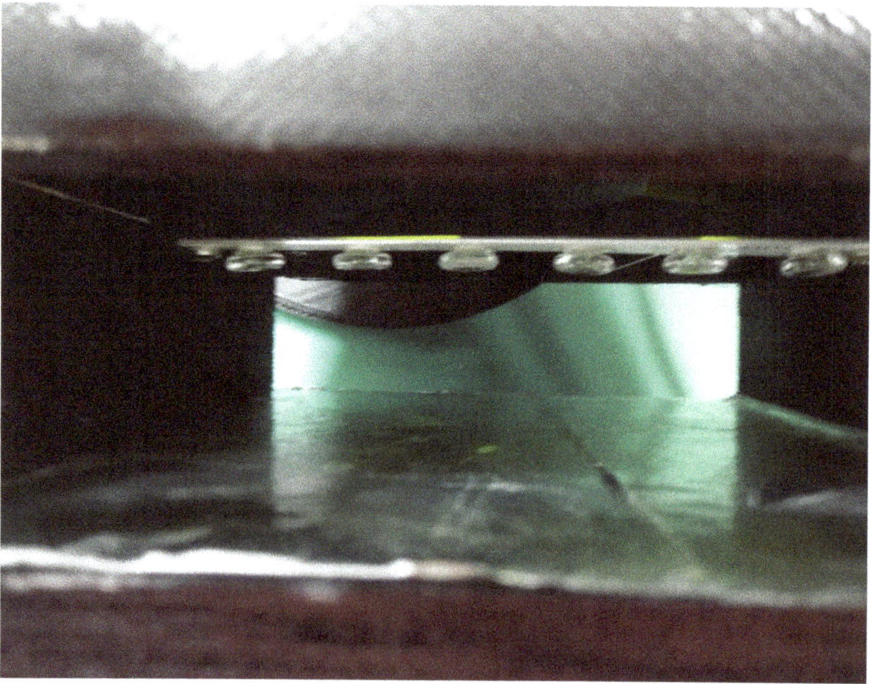

Figure 10. Use of the right amount of stearic acid as a mosquito attractant.

Because stearic acid can have a critical role in the host-seeking behavior of anthropophilic mosquitoes, the SFLD module with a central angle of 150° for a UV LED mosquito-trapper was used as the test sample to evaluate the difference before and after the coating of stearic acid in terms of mosquito trapping efficiency. According to our experiments, it was found that the trapped amount can be even elevated by 50% after applying the coating, as shown in Figure 11.

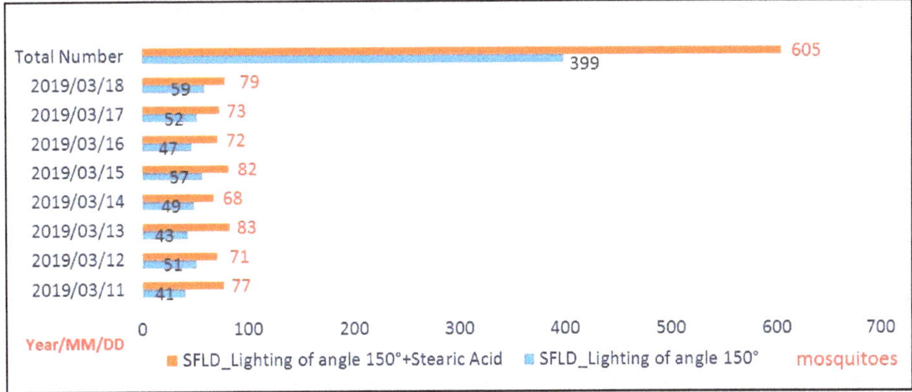

Figure 11. Comparison of the SFLD with a central angle of 150° for the UV LED mosquito-trapping system and that coated with stearic acid.

4. Conclusions

The experimental results showed that the specific secondary free-form lens device (SFLD) can be used to change the trapping range of the UV LED mosquito trapping system. The LIDC of the SFLD module with a central angle of 150° was the best of all experimental systems. Additionally, it can be extended to 100 πm^2 in terms of the outdoor trapping range. The number of mosquitoes captured by the UV LED mosquito trap system increased by about 350% in comparison with the general violet blue LED mosquito trap. In addition, when using the appropriate amount of stearic acid as a mosquito attractant, the UV LED mosquito trapping system of the SFLD module demonstrated a better control effect on mosquitoes during the day. According to our experiments, we found that, after coating, the amount of capture can be increased by 50%.

We can choose not to use chemical methods to fight mosquitoes to avoid harming people and the environment. We designed and optimized a mosquito trap, and, through the actual machine test, the UV LED mosquito trapping system of the SFLD module offers obvious advantages. The effectiveness of mosquito trapping is a disease vector control device that is worth promoting.

Author Contributions: Conceptualization, W.-H.T., H.-Y.M. and H.-Y.L.; Data curation, W.-H.T. and C.-H.C.; Formal analysis, W.-H.T. and H.-Y.L.; Funding acquisition, D.J., W.-C.H. and H.-Y.L.; Methodology, W.-H.T. and Y.C.C.; Project administration, W.-C.H. and H.-Y.M.; Resources, D.J., W.-C.H., W.-S.H., H.-Y.M. and H.-Y.L.; Supervision, D.J. and H.-Y.M.; Validation, W.-H.T. and W.-S.H.; Visualization, W.-H.T. and H.-Y.L.; Writing—original draft, W.-H.T.; Writing—review & editing, H.-Y.L. The authors of the present work made contributions to all of its parts.

Funding: This research received funding from National Kaohsiung University of Science and Technology and Yuan General Hospital.

Acknowledgments: This work was supported by the National Science Council of the Republic of China, project MOST 107-2622-E-992-002 -CC3.

Conflicts of Interest: The authors declare no conflict of interest.

References

1. Chang, C.; Ortiz, K.; Ansari, A.; Gershwin, M.E. The Zika outbreak of the 21st century. *J. Autoimmun.* **2016**, *68*, 1–13. [CrossRef] [PubMed]

2. Chen, H.L.; Tang, R.B. Why Zika virous infection has become a public health concern? *J. Chin. Med. Assoc.* **2016**, *79*, 174–178. [CrossRef] [PubMed]

3.   Guzman, M.G.; Halstead, S.B.; Artsob, H.; Buchy, P.; Farrar, J.; Gubler, D.J.; Hunsperger, E.; Kroeger, A.; Margolis, H.S.; Martínez, E.; et al. Dengue: A continuing global threat. *Nat. Rev. Microbiol.* **2018**, *8*, S7–S16. [CrossRef] [PubMed]

4.   Li, Y.C.; Liu, C.W.; Huang, K.P. Dengue Fever and Dengue Hemorrhagic Fever. *Infect. Control J.* **2007**, *17*, 307–315.

5.   Bidlingmayer, W.L. How mosquitoes see traps: Role of visual responses. *J. Am. Mosq. Assoc.* **1994**, *10*, 272–279.

6.   Browne, S.M.; Bennett, G.F. Response of Mosquitoes (Diptera: Culicidae) to Visual Stimuli. *J. Med. Entomol.* **1981**, *18*, 505–521. [CrossRef] [PubMed]

7.   Kay, R.E. Fluorescent materials in insect eyes and their possible relationship to ultra-violet sensitivity. *J. Insect Physiol.* **1969**, *15*, 2021–2038. [CrossRef]

8.   Shimoda, M.; Honda, K.-I. Insect reactions to light and its applications to pest management. *Appl. Entomol. Zool.* **2013**, *48*, 413–421. [CrossRef]

9.   Tseng, W.-H.; Juan, D.; Hsiao, W.-C.; Chan, C.-H.; Ma, H.-Y.; Lee, H.-Y. Design of a Secondary Freeform Lens of UV LED Mosquito-Trapping Lamp for Enhancing Trapping Efficiency. *Crystals* **2018**, *8*, 335. [CrossRef]

10.   Tseng, W.-H.; Juan, D.; Hsiao, W.-C.; Chen, Y.C.; Chan, C.-H.; Ma, H.-Y.; Lee, H.-Y. Bird-Wing Optical-Reflector Design with Photocatalyst for Low-Glare Mosquito Trapping System with Light-Emitting Diodes. *Crystals* **2019**, *9*, 139. [CrossRef]

11.   Lai, M.F.; Quoc Anh, N.D.; Gao, J.Z.; Ma, H.Y.; Lee, H.Y. Design of multi-segmented freeform lens for LED fishing/working lamp with high efficiency. *Appl. Opt.* **2015**, *54*, E69–E74. [CrossRef] [PubMed]

12.   Acree, F.; Turner, R.B.; Gouck, H.K.; Beroza, M.; Smith, N. L-Lactic Acid: A Mosquito Attractant Isolated from Humans. *Science* **1968**, *161*, 1346–1347. [CrossRef] [PubMed]

13.   Smith, C.N.; Gouck, H.K.; Weidhaas, D.E.; Gilbert, I.H.; Mayer, M.S.; Smittle, B.J.; Hofbauer, A. L-Lactic Acid as a Factor in the Attraction of Aedes aegypti (Diptera: Culicidae) to Human Hosts2. *Ann. Entomol. Soc.* **1970**, *63*, 760–770. [CrossRef] [PubMed]

14.   Eiras, A.E.; Jepson, P.C. Host location by Aedes aegypti (Diptera: Culicidae): A wind tunnel study of chemical cues. *Bull. Entomol. Res.* **1991**, *81*, 151–160. [CrossRef]

15.   Geier, M.; Sass, H.; Boeckh, J. A search for components in human body odour that attract females of Aedes aegypti. *Ciba Symp.* **1996**, *200*, 132–148.

16.   Bernier, U.R.; Kline, D.L.; Schreck, C.E.; Yost, R.A.; Barnard, D.R. Chemical analysis of human skin emanations: comparison of volatiles from humans that differ in attraction of Aedes aegypti (Diptera: Culicidae). *J. Am. Mosq. Assoc.* **2002**, *18*, 186–195.

17.   Knols, B.G.; Van Loon, J.J.; Cork, A.; Robinson, R.D.; Adam, W.; Meijerink, J.; De Jong, R.; Takken, W. Behavioural and electrophysiological responses of the female malaria mosquito Anopheles gambiae (Diptera: Culicidae) to Limburger cheese volatiles. *Bull. Entomol. Res.* **1997**, *87*, 151–159. [CrossRef]

18.   Dekker, T.; Steib, B.; Cardé, R.T.; Geier, M. l-Lactic acid, a human-signifying host cue for the anthropophilic mosquito Anopheles gambiae sensu stricto. *Med. Vet. Entomol.* **2002**, *16*, 91–98. [CrossRef] [PubMed]

19.   Chung, C.L. Talking about dengue prevention and control. *Taiwan Epidemiol. Bull.* **2006**, *22*, 589.

20.   Wei, Z.C.; Guo, M.L.; Wang, M.J.; Li, S.Q.; Liu, S.X. Prediction of cutting force in five-axis flat-end milling. *Int. J. Adv. Manuf. Technol.* **2018**, *96*, 137–152. [CrossRef]

Article

Using Different Ions in the Hydrothermal Method to Enhance the Photoluminescence Properties of Synthesized ZnO-Based Nanowires

Ya-Fen Wei [1,2], Wen-Yaw Chung [2,*], Cheng-Fu Yang [1,3,*], Jei-Ru Shen [3] and Chih-Cheng Chen [1]

[1] School of Information Engineering, Jimei University, Xiamen 361021, China; yafenwei@jmu.edu.cn (Y.-F.W.); 201761000018@jmu.edu.cn (C.-C.C.)

[2] Department of Electronic Engineering, Chung Yuan Christian University, Taoyuan City 320, Taiwan

[3] Department of Chemical and Material Engineering, National University of Kaohsiung, Kaohsiung 811, Taiwan; maltese1114@gmail.com

* Correspondence: eldanny@cycu.edu.tw (W.-Y.C.); cfyang@nuk.edu.tw (C.-F.Y.); Tel.: +886-3-2654602 (W.-Y.C.); +886-7-5919283 (C.-F.Y.)

Received: 18 March 2019; Accepted: 16 April 2019; Published: 18 April 2019

Abstract: ZnO films with a thickness of ~200 nm were deposited on SiO_2/Si substrates as the seed layer. Then $Zn(NO_3)_2$-$6H_2O$ and $C_6H_{12}N_4$ containing different concentrations of $Eu(NO_3)_2$-$6H_2O$ or $In(NO_3)_2$-$6H_2O$ were used as precursors, and a hydrothermal process was used to synthesize pure ZnO as well as Eu-doped and In-doped ZnO nanowires at different synthesis temperatures. X-ray diffraction (XRD) was used to analyze the crystallization properties of the pure ZnO and the Eu-doped and In-doped ZnO nanowires, and field emission scanning electronic microscopy (FESEM) was used to analyze their surface morphologies. The important novelty in our approach is that the ZnO-based nanowires with different concentrations of Eu^{3+} and In^{3+} ions could be easily synthesized using a hydrothermal process. In addition, the effect of different concentrations of Eu^{3+} and In^{3+} ions on the physical and optical properties of ZnO-based nanowires was well investigated. FESEM observations found that the undoped ZnO nanowires could be grown at 100 °C. The third novelty is that we could synthesize the Eu-doped and In-doped ZnO nanowires at temperatures lower than 100 °C. The temperatures required to grow the Eu-doped and In-doped ZnO nanowires decreased with increasing concentrations of Eu^{3+} and In^{3+} ions. XRD patterns showed that with the addition of Eu^{3+} (In^{3+}), the diffraction intensity of the (002) peak slightly increased with the concentration of Eu^{3+} (In^{3+}) ions and reached a maximum at 3 (0.4) at%. We show that the concentrations of Eu^{3+} and In^{3+} ions have considerable effects on the synthesis temperatures and photoluminescence properties of Eu^{3+}-doped and In^{3+}-doped ZnO nanowires.

Keywords: ZnO-based nanowires; hydrothermal method; Eu^{3+} and In^{3+} ions; photoluminescence properties

---

1. Introduction

Nanostructured semiconducting ZnO-based materials have been widely investigated, attracting significant attention due to their novel physical and chemical properties. Important applications include solar cells [1], light-emitting diodes [2], and super-hydrophobic surfaces [3]. The electron transport efficiency and mechanisms of semiconducting ZnO-based materials are dependent on surface states closely linked to the surface-to-volume ratio. One-dimensional (1D) and two-dimensional (2D) ZnO nanostructures are of great interest because they possess a large surface-to-volume ratio, enabling them to absorb more test molecules or accept more measured signals on their surface and, thereby, be highly efficient sensors. They are considered promising materials for various sensors because they have high electrochemical stability, are not toxic, are receptive to doping, and are inexpensive [4]. Many different methods of growing ZnO-based nanostructured materials have been investigated. For example, Lupan et al. used a successive ionic layer adsorption and reaction (SILAR) method to deposit undoped and Sn and Ni co-doped nanostructured ZnO thin films on glass [5]. Niarchos et al. investigated a reliable and low-cost method for large-scale ZnO nanorod production, using an alternative aqueous chemical growth (ACG) low-temperature process to grow ZnO nanorods on patterned Si substrates [6]. Kenanakis et al. used an aqueous solution to thoroughly investigate the growth of highly oriented ZnO nanowires on different substrates [7].

Recently, ZnO-based nanostructured materials, including nanotubes, nanowires (nanorods), and thin films, have been synthesized by various physical and chemical methods and used to fabricate sensors for a variety of applications. For example, Mondal et al. used a $ZnO-SnO_2$ composite material to fabricate a micro-electro–mechanical system (MEMS) microheater on silicon (Si) to develop a low-power gas sensor [8]. Zhao et al. grew a single ZnO nanowire on a flexible substrate using a custom-built nano-manipulation system and investigated it as an ultra-high-sensitivity strain sensor [9]. Thomas et al. used a microwave successive ionic layer adsorption reaction to synthesize pure and Al-doped photosensitive ZnO films and investigated their luminescence properties [10].

Various attempts have been made to grow pure ZnO and different ion-doped ZnO nanowires and investigate them as different applications. For example, Bai et al. used a seed-assisted hydrothermal method to grow the Al-doped ZnO nanowires on a silicon substrate [11]. Even though they could grow the ZnO nanowires at a low temperature of 95 °C, this method is very complicated to grow the Al-doped ZnO nanowires because they needed to deposit the Al on ZnO seed layer and annealed the samples at 550 °C. Chang et al. used a traditional thermal evaporation method to synthesize ZnO nanowires on a (100) Si substrate and then used a molecular beam epitaxy system to subsequently carry out the Mn doping process [12]. They constructed a back-gated Mn-doped ZnO nanowire field-effect transistor (FET) to demonstrate the electric-field control properties of ferromagnetism.

In-doped ZnO nanowires can be synthesized and grown using different methods. For example, the In-doped ZnO nanowires were synthesized via a thermal evaporation process, using metallic powders of zinc and indium as precursors and oxidizing them in the presence of oxygen [13,14]. Xu et al. also used a thermal evaporation method to synthesize the In-doped ZnO nanowires, but the used precursors were ZnO and In_2O_3 powders [15]. The Eu-doped ZnO nanowires could also be synthesized using different methods. Rifai and Kulnitskiy used chemical vapor deposition method to synthesize the single-crystal Eu^{3+}-doped wurtzite ZnO micro- and nanowires [16]. Lupan et al. investigated Eu-doped ZnO nanowire arrays that could be electrodeposited in a three electrode electrochemical cell using an aqueous solution containing $ZnCl_2$, KCl, and $EuCl_3$ as the supporting electrolyte [17]. Geburt et al. first used the vapor–liquid–solid mechanism to synthesize ZnO nanowires with diameters of about 100 to 300 nm. After that, they used ion implantation to dope the Eu^{3+} ions in ZnO nanowires [18]. Apparently, the concentrations of Eu^{3+} and In^{3+} ions in these researches are difficult to control well, and it is difficult to investigate the effects of different concentrations of Eu^{3+} and In^{3+} ions on their properties.

When all the synthesis methods are compared, using an aqueous solution to grow ZnO nanostructured materials is considered better, mainly due to low growth temperature and good potential for large-scale production. Previously, we found that changing the deposition parameters in our hydrothermal method (the concentrations of $Zn(NO_3)_2 \cdot 6H_2O$ and $C_6H_{12}N_4$; the face direction of $ZnO/SiO_2/Si$ substrates during hydrothermal deposition; and deposition time) resulted in ZnO films with three different morphologies on $ZnO/SiO_2/Si$ substrates [19]. Different deposition parameters yielded undoped ZnO in the shape of irregular plate-structured films, nanowires (nanorods), and chrysanthemum-like clusters (nanoflower films). We also found that with a synthesis temperature of 100 °C, nanowires could be grown using an undoped ZnO solution. ZnO-based nanowires can be used in many fields, for example, as ultraviolet (UV) photodetectors or sensors [20]. Trivalent lanthanide ion-doped wide-bandgap semiconducting ZnO nanowires could be the promising active materials in opto-electronic devices [17]. We believe that if the photoluminescence excitation (PLE) and photoluminescence emission (PL) properties are enhanced, the sensitivity of these fabricated UV photodetectors or sensors and the emission properties of opto-electronic devices will also be improved. We, therefore, used Eu^{3+} and In^{3+} ions and investigated the PLE and PL properties of Eu-doped and In-doped ZnO nanowires.

In the present study, the first important novelty is that we used $Eu(NO_3)_3 \cdot 6H_2O$ and $In(NO_3)_2 \cdot 6H_2O$ as the dopant sources of Eu^{3+} and In^{3+} ions because we could control their concentrations well. Next, we could synthesize the Eu-doped and In-doped ZnO nanowires using the hydrothermal method at low temperature (below 100 °C) and investigate the effects of different concentrations of Eu^{3+} and In^{3+} ions on their crystalline and photoluminescence properties. The last novelty is that the Eu^{3+} and In^{3+} ions form a compound with ZnO during the synthesis process of ZnO-based nanowires, for that the Eu^{3+} and In^{3+} ions can reach the whole region of the synthesized Eu-doped and In-doped ZnO nanowires. We found that at 100 °C, ZnO-based nanowires did not grow well when different concentrations of Eu^{3+} and In^{3+} ions were added. The needed synthesis temperatures drop increased with the concentrations of Eu^{3+} and In^{3+} ions. We will demonstrate how the designs of synthesis temperature and precursor materials can control the structures of pure ZnO and of Eu-doped and In-doped ZnO nanowires with hexagonal prismatic structures.

As compared with other researches, another important novelty is that no researches prove tht the concentrations of Eu^{3+} and In^{3+} ions will affect the growth temperatures of Eu-doped and In-doped ZnO nanowires and investigate the effect of the concentrations of Eu^{3+} and In^{3+} ions on the growth morphologies and PL properties. As the hydrothermal method was used to grow the Eu-doped and In-doped ZnO nanowires, their crystal qualities have been enhanced. The visible-light emission, which is caused by the defect states, was not found in Eu-doped and In-doped ZnO nanowires and the near-band-edge emission peak (located at around 395 nm) was enhanced because the number of defect states was reduced. We thoroughly investigate the effects of different concentrations of Eu^{3+} and In^{3+} ions on the growth properties of ZnO-based nanowires. We also show that the ions used and their concentrations have considerable effects on the synthesis temperatures and photoluminescence properties of ZnO-based nanowires.

2. Experimental

A ZnO seed layer is necessary to initialize the uniform growth of oriented nanowires using aqueous solutions. ZnO powder (Us Research Nanomaterials Inc., purity 99.99%, particle sizes small than 1 μm, Houston, TX, USA) was mixed with polyvinyl alcohol (PVA, ECHO Chemical Co., Ltd. Miaoli, Taiwan) as a binder, and the ZnO-PVA mixture was pressed into pellets 6 mm thick and 56 mm in diameter using a steel die. After debindering, each ZnO pellet was sintered at 1100 °C for 2 h to form a ceramic target. SiO_2/Si (Summit-Tech Resource Corp. Hsinchu, Taiwan) was used as the substrate to fabricate pure (undoped) as well as Eu^{3+}-doped and In^{3+}-doped ZnO nanowires. First, the SiO_2/Si substrates were cleaned with deionized (DI) water (office created), acetone (ECHO Chemical Co., Ltd.), and isopropyl alcohol (ECHO Chemical Co., Ltd.), then radio frequency (RF) magnetron sputtering was used to deposit the ZnO seed layers ($ZnO/SiO_2/Si$ substrates). Next, $Zn(NO_3)_2 \cdot 6H_2O$ (Alfa Aesar, MA, USA), $C_6H_{12}N_4$ (ECHO Chemical Co., Ltd.), and $Eu(NO_3)_3 \cdot 6H_2O$ (Alfa Aesar) or $In(NO_3)_2 \cdot 6H_2O$ (Alfa Aesar) were mixed with DI water with the designed compositions. The $Zn(NO_3)_2 \cdot 6H_2O$, $Eu(NO_3)_3 \cdot 6H_2O$, and $In(NO_3)_2 \cdot 6H_2O$ decomposed to form the mixed solutions of Zn^{2+} and In^{3+} ions or Zn^{2+} and Eu^{3+} ions in DI water. It was impossible for In^{3+} and Eu^{3+} ions to be separated with Zn^{2+} ions from the solutions during the growth processes of ZnO-based nanowires, the Eu-doped and In-doped ZnO nanowires could be grown from the mixed solutions. The Eu^{3+}-doped and In^{3+}-doped ZnO nanowires were synthesized at temperatures of 100 to 60 °C for 1 h using a hydrothermal method. Previously, we used a well-designed structure to grow ZnO-based nanostructured materials via a hydrothermal method [19]. We found that when $Zn(NO_3)_2 \cdot 6H_2O$ and $C_6H_{12}N_4$ were used as reagents to synthesize

ZnO-based nanostructured materials, the concentration of the diluted solution had important effects on the synthesis results. We, therefore, fixed the concentration of the diluted solution as a reference point [19].

We also found that the face direction of the ZnO seed layer and the synthesis time were two important factors affecting the synthesis of ZnO-based nanostructured materials. We found that when the face direction was down, nanowires grew on the ZnO seed layer, and when the direction was up, nanoflowers grew on the layer. We also determined that 1 h was the best synthesis duration for growing ZnO nanowires on the ZnO seed layer. Hence, facedown and 1 h were used as the parameters for growing pure, Eu^{3+}-doped, and In^{3+}-doped ZnO nanowires. The compositions for growing the Eu^{3+}-doped nanowires were ZnO + y Eu^{3+} ions, where y = 0, 1, 2, or 3 at%, abbreviated as ZnO-0-Eu (undoped-ZnO), ZnO-10-Eu, ZnO-20-Eu, ZnO-30-Eu, and ZnO-40-Eu. The compositions for growing the In^{3+}-doped ZnO nanowires were ZnO + x In^{3+} ions, where x = 0, 0.4, 0.8, or 1.2 at%, abbreviated as ZnO-4-In, ZnO-8-In, and ZnO-12-In. The morphologies of pure ZnO and Eu^{3+}-doped and In^{3+}-doped ZnO nanowires were observed by field-emission scanning electron microscopy (FESEM, Hitachi 4800, Tokyo, Japan) and used to determine the effects of synthesis temperature and the concentrations of Eu^{3+} and In^{3+} ions on the synthesis properties of the pure and doped ZnO nanowires. Crystalline phases were analyzed using X-ray diffraction (XRD, D8, Bruker, MA, USA) to determine the effects of concentrations of Eu^{3+} and In^{3+} ions on the crystalline and photoluminescence properties of the ZnO-based nanowires.

3. Results

In the case of undoped ZnO grown at 80 °C, ZnO nanowires were not synthesized on the ZnO/SiO$_2$/Si substrate and only irregular ZnO nano-particles were observed, as Figure 1a shows (presents a top-down image). As the temperature was 100 °C, ZnO nanowires were successfully synthesized on the ZnO/SiO$_2$/Si substrate, as Figure 1b shows. To investigate the effects of concentrations of Eu^{3+} and In^{3+} ions on the synthesis properties of hydrothermally grown Eu^{3+}-doped and In^{3+}-doped ZnO nanostructures, the synthesis temperature of all the ZnO + y Eu^{3+} and ZnO + x In^{3+} compositions was set at 100 °C. The general surface morphologies of the hydrothermally grown Eu^{3+}-doped ZnO nanostructures were examined by FESEM, and the results are also shown in Figure 1. The SEM image of this sample revealed that the undoped ZnO nanowires had a hexagonal wurtzite structure, with an easily discernable hexagonal prism arrangement. When the compositions were changed to ZnO-20-Eu and ZnO-40-Eu, as Figure 1c,d show, no nanowires grew on the ZnO/SiO$_2$/Si substrates and only irregular plate-structured grains were observed. When the compositions were changed to ZnO-60-Eu and ZnO-80-Eu (not shown here), only very large, irregular plate-structured grains were observed. These results prove that synthesis temperature is an important parameter to affect the synthesized results of ZnO-based nanowires.

Figure 1. Surface morphologies of hydrothermally grown Eu^{3+}-doped ZnO nanostructures, (**a**) undoped ZnO synthesized at 80 °C; synthesized at 100 °C: (**b**) undoped ZnO, (**c**) ZnO-20-Eu, and (**d**) ZnO-40-Eu.

However, when In^{3+} was used, the surface morphologies of the In^{3+}-doped ZnO nanostructures resembled those of the Eu^{3+}-doped ZnO nanostructures. When ZnO-4-In, ZnO-8-In, and ZnO-12-In were used to grow ZnO-based nanowires at 100 °C, their FESEM images were similar to those of ZnO-20-Eu and ZnO-40-Eu: no nanowires grew, and only irregular plate-structured grains were observed (not shown here). These results suggest that when the ZnO seed layer is deposited, the concentrations of Eu^{3+} or In^{3+} ions have an important effect on the synthesis results of the ZnO-based nanostructures. Increasing the concentrations of Eu^{3+} or In^{3+} ions lowers the temperature required to form the ZnO-based nanowires. We next demonstrated that the concentrations of Eu^{3+} or In^{3+} ions (or the synthesis temperature) are the most important factor in the syntheses of Eu^{3+}-doped and In^{3+}-doped ZnO nanowires on $ZnO/SiO_2/Si$ substrates.

The Eu^{3+}-doped ZnO nanomaterials were synthesized at 100 °C using various concentrations of Eu^{3+} ions. The XRD patterns of the Eu^{3+}-doped ZnO nanostructures are shown in Figure 2. Only the diffraction peak of the (002) plane was observed; no (004) plane diffraction peak appeared. Apparently, the diffraction intensity of the Eu^{3+}-doped ZnO nanostructures decreased as the concentration of Eu^{3+} ions increased, reaching a minimum at ZnO-30-Eu, then becoming saturated as the concentration of Eu^{3+} ions was further increased. Using Figure 2, we also measured the 2θ value and full width at half maximum (FWHM) value of the (002) plane of the Eu^{3+}-doped ZnO nanostructures. The 2θ values of the (002) plane of the 100 °C-synthesized Eu^{3+}-doped ZnO nanostructures were unchanged at 34.44 ± 0.02 as the concentration of Eu^{3+} ions increased. The FWHM value of the (002) plane initially increased, reaching saturation when the concentration of Eu^{3+} ions was 3 at%. These results suggest that the crystallinity of the Eu^{3+}-doped ZnO nanostructures degenerated as the concentration of Eu^{3+} ions increased. Comparison with the results in Figure 1 leads us to believe that the degeneration was caused by unformed nanostructures.

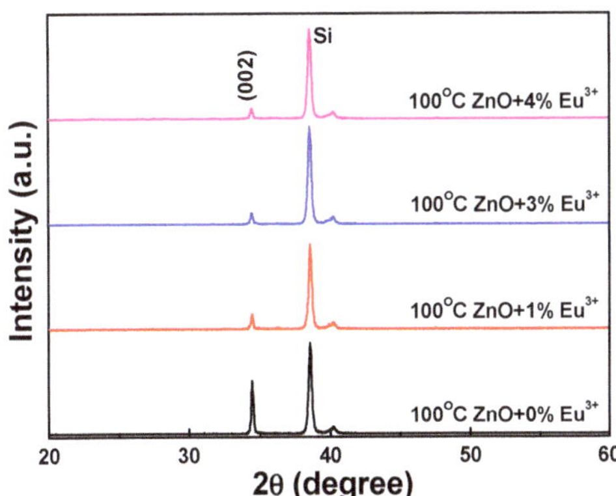

Figure 2. X-ray diffraction (XRD) patterns of synthesized Eu^{3+}-doped ZnO nanowires as a function of the concentration of Eu^{3+} ions; the synthesis temperature was 100 °C.

When In^{3+} was used as the ion, the XRD patterns of the In^{3+}-doped ZnO nanomaterials were similar to those of the Eu^{3+}-doped ZnO nanomaterials (not shown here). These results also suggest that increasing the concentration of In^{3+} ions led the In^{3+}-doped ZnO to form ZnO-based nanowires at lower temperatures. In addition, only the diffraction peak of the (002) plane was observed in the In^{3+}-doped ZnO nanostructures; no diffraction peak for the (004) plane was evident. The diffraction intensity of the In^{3+}-doped ZnO nanostructures decreased as the concentration of In^{3+} ions increased, reaching a minimum at ZnO-4-In, then it was unchanged as the concentration of In^{3+} ions was further increased.

The morphologies of the synthesized Eu^{3+}-doped ZnO nanowires are shown in Figure 3 for different concentrations of Eu^{3+} ions and synthesis temperatures. Figure 3a shows that when ZnO-10-Eu was used at 90 °C, nanowires with diameters in the range of 50 to 160 nm and hexagonal prism structures were readily observable. When the concentration of Eu^{3+} ions increased, the synthesis temperature could be lowered. As Figure 3b–d show, the synthesis temperatures of the ZnO-20-Eu, ZnO-30-Eu, and ZnO-40-Eu nanowires were 80, 70, and 60 °C, and their diameters were in the ranges of 50 to 85, 40 to 80, and 140 to 450 nm, respectively. Other than the nanowire structure changing from an equilateral hexagon to a non-equilateral hexagon, the surface morphologies experienced no apparent change, and all had a hexagonal prism structure.

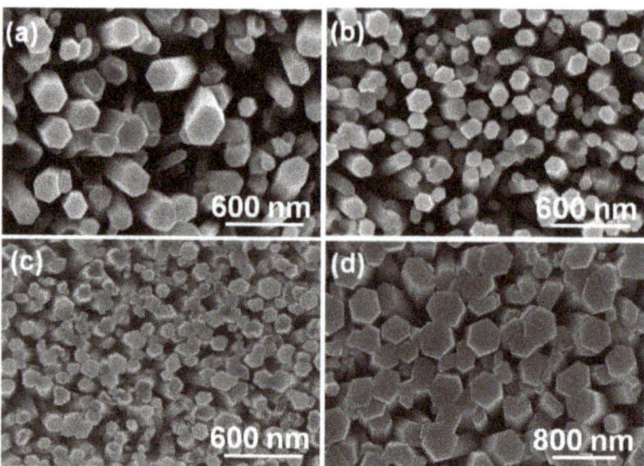

Figure 3. Surface morphologies of hydrothermally grown Eu^{3+}-doped ZnO. (**a**) ZnO-10-Eu nanowires synthesized at 90 °C, (**b**) ZnO-20-Eu nanowires synthesized at 80 °C, (**c**) ZnO-30-Eu nanowires synthesized at 70 °C, and (**d**) ZnO-40-Eu nanowires synthesized at 60 °C.

Figure 4 shows top-view SEM images of high-density In^{3+}-doped ZnO nanowires grown on $ZnO/SiO_2/Si$ substrates at different synthesis temperatures and with different concentrations of In^{3+} ions. Figure 4a shows that the 88 °C ZnO-4-In nanowires had the structure of hexagonal prisms with diameters in the range of 45 to 150 nm. When the concentration of In^{3+} ions was increased, the In^{3+}-doped ZnO nanowires could be synthesized at a low temperature. As Figure 4b,c shows, the synthesis temperatures of ZnO-8-In and ZnO-12-In nanowires were 75 and 60 °C, and their diameters were in the ranges of 130 to 280 and 70 to 150 nm, respectively. The top-view image shows that the In^{3+}-doped ZnO nanowires changed from an equilateral hexagon structure to a non-equilateral hexagon structure as the concentration of In^{3+} ions increased. When Eu^{3+} or In^{3+} ions were added, the nanowires still displayed a hexagonal wurtzite structure, providing strong evidence that the undoped ZnO and the Eu^{3+}-doped and In^{3+}-doped ZnO nanowires grew in the (002) direction, independent of the concentrations of Eu^{3+} or In^{3+} ions.

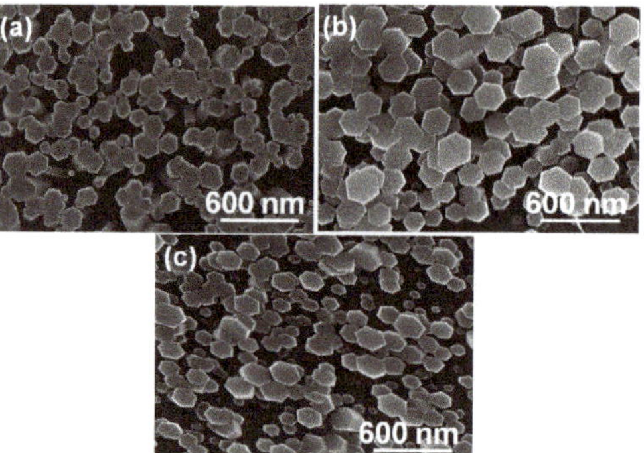

Figure 4. Surface morphologies of hydrothermally grown In^{3+}-doped ZnO: (**a**) ZnO-4-In nanowires synthesized at 88 °C, (**b**) ZnO-8-In nanowires synthesized at 75 °C, and (**c**) ZnO-12-In nanowires synthesized at 60 °C.

Table 1 shows the corresponding FESEM images of formed Eu-doped and In-doped ZnO nanowires (Figures 3 and 4) the elemental ratios obtained by FESEM equipped with energy dispersive X-ray spectroscopy (EDX) analyses for elemental Zn and Eu or Zn and In. Six different areas of Eu-doped and In-doped ZnO nanowires were depicted for analysis, and the range of the measured elemental ratio for Eu and In and the average value of the measured elemental ratio are shown in Table 1. The Eu and In elements were detected in Eu-doped and In-doped ZnO nanowires, respectively, and the measured elemental ratios increased with the concentrations of Eu^{3+} and In^{3+} ions.

Table 1. Energy dispersive X-ray spectroscopy (EDX) analyses for elemental Zn and Eu or Zn and In. Eu(In)-measured: the range of the measured elemental ratio; Eu(In)-average: the average value of the measured elemental ratio.

Composition	Eu-Measured	Eu-Average	Composition	In-Measured	In-Average
ZnO-10-Eu	0.56–0.78%	0.69%	ZnO-4-In	0.04–0.34%	0.21%
ZnO-20-Eu	1.32–1.77%	1.47%	ZnO-8-In	0.38–0.72%	0.59%
ZnO-30-Eu	2.02–2.51%	2.24%	ZnO-12-In	0.85–1.11%	1.02%
ZnO-40-Eu	3.25–3.56%	3.32%			

We investigated the crystallinity of Eu^{3+}-doped ZnO nanowires synthesized with different concentrations of Eu^{3+} ions and synthesis temperatures using XRD, and the results are shown in Figure 5. All of the patterns were in agreement with the diffraction data from the standard card (JCPDS 36-1451). The main diffraction peak of ZnO is (101) (JCPDS 36-1451), which is located around 2θ~36.25°. However, the stronger intensity of the (002) diffraction peak, which is located at 2θ = 34.44 ± 0.02~34.40 ± 0.02, was discernible for all of the Eu^{3+}-doped ZnO nanowires, suggesting that all of the Eu^{3+}-doped ZnO nanowires had a high c-axis orientation. The 2θ value of the c-orientation (200) peak decreased from 34.43 ± 0.02, 34.43 ± 0.02, 34.42 ± 0.02, 34.41 ± 0.02, to 34.40 ± 0.02 as the concentration of Eu^{3+} ions increased from 0 1, 2, 3, to 4 at%. The radius of Eu^{3+} ions larger than that of Zn^{2+} ions is the reason to cause an unapparent decrease in the 2θ value of the c-orientation (200) peak. The FWHM value of the (200) diffraction peak decreased from 2θ = 0.19 (34.37–34.55), 0.18 (34.35–34.52), 0.17 (34.34–34.50), to 0.15 (34.34–34.48) as the concentration of Eu^{3+} ions increased from 0, 1, 2, to 3 at%, indicating that the crystallization of the nanowires increased with the concentration of Eu^{3+} ions. When the concentration of Eu^{3+} ions increased from 3 or 4 at%, the (100) peak was observed, the diffraction intensity of the (200) diffraction peak increased, and the FWHM value increased from 2θ = 0.15 to 0.17 (34.34–34.50). Figure 5 also shows that the diffraction intensity of the (100) peak increased and the FWHM value decreased as the concentration of Eu^{3+} ions increased from 3 to 4 at%. These results suggest that the crystallization property changed when the concentration of Eu^{3+} ions was 3 at% or more.

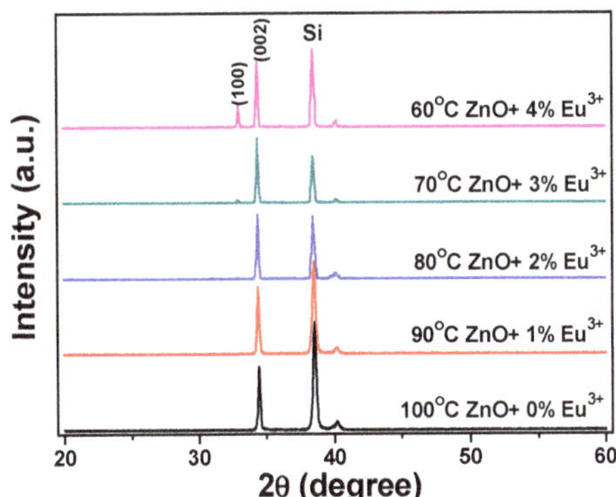

Figure 5. XRD patterns of the synthesized Eu^{3+}-doped ZnO nanowires as a function of the concentration of Eu^{3+} ions (or synthesis temperature).

The crystallinities of In^{3+}-doped ZnO nanowires synthesized with different concentrations of Eu^{3+} ions were investigated using XRD, and the results are shown in Figure 6. Comparison of these results showed differences when the concentration of In^{3+} ions was varied. As the concentration of In^{3+} ions was increased from 0 to 0.4 at%, the FWHM value of the (200) diffraction peak decreased from $2\theta = 0.18$ (34.37–34.55) to 0.17 (34.37–34.54) and the diffraction intensity increased. As the concentration of In^{3+} ions was further increased from 0.4, 0.8, to 1.2 at%, the FWHM value of the (200) diffraction peak decreased from $2\theta = 0.17$, 0.20 (34.36–34.55), to 0.25 (34.33–34.57) and the diffraction intensity decreased. Nevertheless, the synthesized In^{3+}-doped ZnO nanowires exhibited no (100) diffraction peak. Further analysis of the XRD data in Figure 6 showed that the radius of the In^{3+} ions (0.80 nm) was larger than that of the Zn^{2+} ions (0.74 nm), and all the (002) diffraction peaks were located at $2\theta = 34.44 \pm 0.02$ as the concentrations of In^{3+} ions were 0.0 and 0.4 at% and located at $2\theta = 34.43 \pm 0.02$ as the concentrations of In^{3+} ions were 0.8 and 1.2 at%. Even the radius of In^{3+} ions is larger than that of Zn^{2+} ions, but the concentration of In^{3+} ions used to dope into the In-doped ZnO nanowires was very low, and the 2θ value of the c-orientation (200) peak was almost unchanged. The results in Figures 1–6 prove that when the undoped ZnO and the Eu^{3+}-doped ZnO and In^{3+}-doped ZnO nanowires grew well, the synthesized ZnO-based nanowires had a good wurtzite hexagonal crystal structure. The undoped ZnO, the Eu^{3+}-doped ZnO, and the In^{3+}-doped ZnO had similar crystalline results; thus, we believe they will have different PLE and PL spectra.

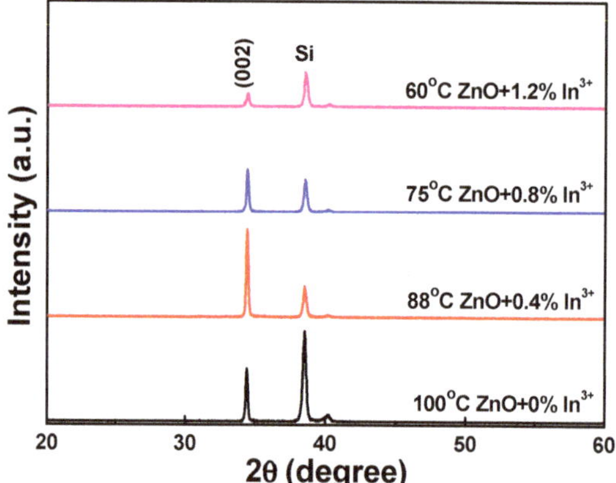

Figure 6. XRD patterns of the synthesized In^{3+}-doped ZnO nanowires as a function of the concentration of In^{3+} ions (or synthesis temperature).

The radii of 4-coordination and 6-coordination Zn^{2+} ions are 60 and 74 pm, and the radii of 4-coordination and 6-coordination In^{3+} ions are 62 and 80 pm; 4-coordination Eu^{3+} ions do not exist, and the radius of 6-coordination Eu^{3+} ions is 95 pm. The values of the calculated lattice constant, c, of the Eu^{3+}-doped and In^{3+}-doped ZnO nanowires were considerably smaller than the c value of the undoped ZnO nanowires. Because all the investigated ZnO-based nanowires had a hexagonal wurtzite structure, the Zn^{2+}, Eu^{3+}, and In^{3+} ions were in a 6-coordination structure. Comparison of the results in Figures 2, 5 and 6 shows that the radii of Eu^{3+} (0.95 nm) and In^{3+} (0.80 nm) are larger than that of Zn^{2+} (0.74 nm), and no variation in the lattice parameters of the undoped, Eu^{3+}-doped, and In^{3+}-doped ZnO nanowires was observable. However, comparison of the images in Figures 1a, 3 and 4 show that from the top view, the ZnO-based nanowires changed from an equilateral hexagon to a non-equilateral hexagon configuration as the concentrations of Eu^{3+} and In^{3+} ions increased. We believe this was the result of the differences between the radius of Zn^{2+} ions and the radius of Eu^{3+} ions.

Figure 7 shows the room-temperature PLE spectra of different ion-doped ZnO nanowires recorded in the wavelength range of 200 to 350 nm. The PLE spectra were measured at an emission wavelength of 395 nm, while the PL spectra were measured for maximum emission intensity. There are obvious differences between the PLE spectra of pure ZnO and different ion-doped ZnO nanowires. The emission intensity of the PLE spectra of the Eu^{3+}-doped ZnO nanowires increased with the concentration of Eu^{3+} ions and reached a maximum value in the 60 °C-grown ZnO-40-Eu, and that of In^{3+}-doped ZnO nanowires appeared in the 88 °C-grown ZnO-4-In. Comparison of the results in Figures 1–4 suggests that the photoluminescence properties of the ZnO-based nanowires were dependent on their crystalline properties. We, therefore, used the PLE and PL spectra of the 100 °C-grown ZnO, 60 °C-grown ZnO-40-Eu, and 88 °C-grown ZnO-4-In nanowires to compare their optical properties.

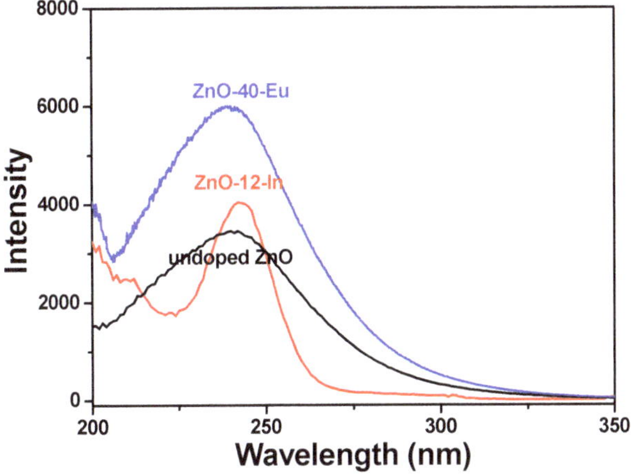

Figure 7. Photoluminescence excitation (PLE) spectra of 100 °C-grown ZnO, 60 °C-grown ZnO-40-Eu, and 88 °C-grown ZnO-4-In nanowires.

The PLE spectra in Figure 7 changed as different ions were added. As the wavelength in Figure 7 increased from 200 to 350 nm, the PLE emission intensities first decreased to a minimum, then increased to a maximum, then decreased again. Compared with that of pure ZnO nanowires, the wavelengths of 0 °C-grown ZnO-40-Eu, and 88 °C-grown ZnO-4-In nanowires reveal the minimum intensities were shifted to higher values and wavelengths of them reveal the maximum had no apparent changes. When different ions were added to synthesize the ZnO-based nanowires, they caused dissimilar energy level transitions in different energy bands, leading the phosphors to release light with different emission peaks (or wavelengths). This is why the PLE spectra of the 100 °C-grown ZnO, 60 °C-grown ZnO-40-Eu, and 88 °C-grown ZnO-4-In nanowires differed. The broadening of the emission peaks of the 100 °C-grown ZnO and 60 °C-grown ZnO-40-Eu nanowires can be interpreted by the formation of band tailing in the band gap, which often is induced by the formation of the defects in the semiconductor during the synthesis process or the introduction of an impurity into the semiconductor [15].

The room-temperature PL emission spectra of the 100 °C-grown ZnO, 60 °C-grown ZnO-40-Eu, and 88 °C-grown ZnO-4-In nanowires were excited using a wavelength of 242 nm and recorded in the range of 200 to 700 nm. These PL properties were obtained at room temperature, and the results are compared in Figure 8. In addition, there are obvious differences between the PL spectra of pure ZnO and different ion-doped ZnO nanowires. The wavelengths resulting in the maximum emission intensities were 395 ± 2, 396± 2, and 397 ± 2 nm for 100 °C-grown undoped ZnO, 60 °C-grown ZnO-40-Eu, and 88 °C-grown ZnO-4-In nanowires, respectively. These results suggest that the additions of Eu^{3+} or In^{3+} affect the maximum intensities of the emission spectra, but they almost cannot affect wavelengths resulting in the maximum emission intensities. The 100 °C-grown undoped ZnO and 60 °C-grown ZnO-40-Eu nanowires had two distinct, visible-light emission peaks, one centered at ~361 nm and the other at 395 or 396 nm, and the emission spectra ranged from 280 to 570 nm.

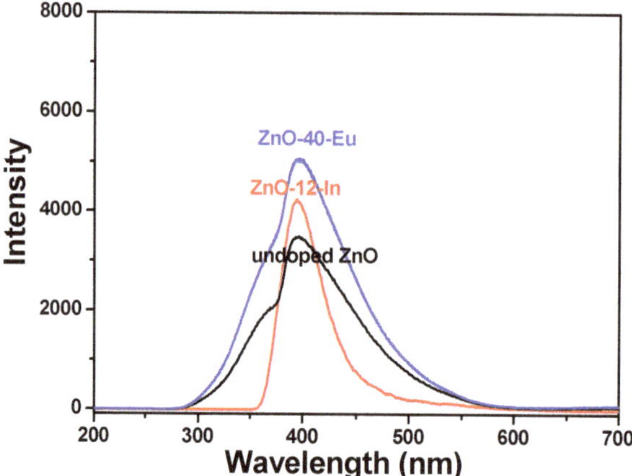

Figure 8. Photoluminescence emission (PL) spectra of 100 °C-grown ZnO, 60 °C-grown ZnO-40-Eu, and 88 °C-grown ZnO-4-In nanowires.

With the 88 °C-grown ZnO-4-In nanowires, only one peak was observed, and the emission spectra ranged from 360 to 520 nm. The maximum intensities of the PL emission spectra were 100 °C-grown ZnO < 88 °C-grown ZnO-4-In < 60 °C-grown ZnO-40-Eu. Our results suggest two important findings. First, when a hydrothermal method is used to grow ZnO-based nanowires, adding different concentrations of Eu^{3+} and In^{3+} ions can reduce the synthesis temperature, and this temperature decreases with the concentrations of Eu^{3+} and In^{3+} ions. Second, when different ions in different concentrations are added to grow ZnO-based nanowires, the optical properties of the synthesized ZnO-based nanowires can be controlled. The PLE and PL properties of these ZnO-based nanowires depend on their crystalline properties.

In the past, different researches or different methods to synthesize the undoped and ion-doped ZnO nanowires would have different emission results of PL spectra. Xu et al. measured a PL spectrum using a He-Cd laser of 260 nm as the excitation source at room temperature. They found that the undoped ZnO nanowires has a visible-light emission band with center at about 490 nm and the In-doped ZnO nanowires with an emission band centering at around 398 nm dominate in the PL spectrum [15]. Kim et al. used He-Cd (325 nm) laser as the excitation source, they found that the room-temperature PL spectroscopy of In-doped ZnO nanowires exhibits an unapparent UV emission centered at 378 nm and a broad emission centered at 561 nm [13]. Zeng et al. found that as a Xe lamp was used as the excitation light, the room-temperature PL properties of an undoped ZnO nanosheet featured a broad yellow band centered at 575 nm and they thought that the emission of yellow light is attributed to transitions of oxygen interstitials (Oi) [21]. They also found that as UV (464 nm) light was used as the excitation light, the spectra of Eu-doped ZnO nanowires emitted a pure red luminescence and consisted of a series of resolved emission peaks centered at 577, 589, 612, 619, and 654 nm, they can be assigned to the transitions of $^5D_0 \rightarrow {}^7F_0$ (577 nm), $^5D_0 \rightarrow {}^7F_1$ (589 nm), $^5D_0 \rightarrow {}^7F_2$ (612, 619 nm), and $^5D_0 \rightarrow {}^7F_3$ (654 nm), respectively [22].

Geburt et al. also found that the Eu-doped ZnO nanowires have the sharp and structured emission features between 1.5 and 2.1 eV, and these peaks can be clearly assigned to the $^5D_0 \rightarrow {}^7F_J$ (J = 0, 1, 2, 3, 4) transitions of Eu^{3+} ions [18]. Lupan et al. recorded the PL spectrum of Eu-doped ZnO nanowires under UV 266 nm excitation, no red emission due to the transitions of Eu^{3+} ions could be detected. The main emission peak centered at 382 nm was found and a weak green luminescence due to ZnO intrinsic defects could also be observed at about 530 nm [17]. Gomi et al. found earlier that interstitial Zn (Zn_i) causes the red emission [23] and Teke et al. found that vacancy Zn and oxygen (V_{Zn} and V_O) cause the green emission [24]. TekeXing et al. also found that V_{Zn} and V_O cause the green emission and excess oxygen cause the orange–red emission [25]. Chen et al. also found that the excess oxygen on the surfaces of ZnO nanowires causes the orange–red emission, and it is adjustable via the annealing process or the surfaces' modification of ZnO nanowires [26]. However, we had measured the emission spectrum of ZnO NPs excited by UV light with wavelength of 242 nm, and the main peak was at about 393 nm [27]. Usually, the PL emission of ZnO-based materials is attributed to different defects, such as oxygen vacancies (V_O), zinc vacancies (V_{Zn}), or is caused by the complex defects of involving interstitial zinc (Zn_i) and interstitial oxygen (O_i) [28]. The energy level of Zn_i is one type of shallow donor level below conduction band, and it locates 3.15 or 2.91 eV, whereas, the energies of V_{Zn}, O_i, and O_{Zn} are more acceptors level above the valence band. The 3.15 eV related to light wavelength can be calculated using 1240 (nm)/3.15 = 394 nm. In this study, the centered wavelengths of PL spectra for the grown undoped, Eu-doped, and In-doped ZnO nanowires are about 395 nm. These results suggest that Zn_i is the main reason to cause PL properties of the grown undoped, Eu-doped, and In-doped ZnO nanowires.

Thus, in our case, we used the hydrothermal method to synthesize the ZnO-based nanowires and used $Eu(NO_3)_2$-$6H_2O$ or $In(NO_3)_2$-$6H_2O$ as precursors of the dopants. We believe that the Eu^{3+} and In^{3+} ions will purely occupy the sites of Zn and act as a dopant to change the semiconducting characteristic and enhance the emission intensities of Eu-doped and In-doped ZnO nanowires, and the defects of Zn_i, V_{Zn}, and V_O will not happen in the synthesized ZnO-based nanowires. The emission peaks of 100 °C-grown undoped ZnO, 60 °C-grown ZnO-40-Eu, and 88 °C-grown ZnO-4-In nanowires centered at about 395 nm are characteristic of ZnO near band edge recombination. However, as the $Eu(NO_3)_2$-$6H_2O$ or $In(NO_3)_2$-$6H_2O$ are used as the dopants of ZnO-based nanowires and the hydrothermal method is used to synthesize the Eu-doped and In-doped ZnO nanowires, their emission intensities of PL spectra are really enhanced. When In is used as dopant in ZnO nanowires, an enhancement in near-band-edge emission peak (located at around 395 nm) occurs than for pure ZnO nanowires while visible-light emission is decreased significantly owing to change in the growth kinetics due to In supply which helps to reduce the number of defect states resulting in improved crystal quality [29]. XRD patterns in Figures 5 and 6 show that the 100 °C-grown undoped ZnO nanowires had high crystal quality and the crystal qualities of the 60 °C-grown ZnO-40-Eu and 88 °C-grown ZnO-4-In nanowires were higher than that of 100 °C-grown undoped ZnO nanowires. For that, the visible-light emission, which is caused by the number of defect states, cannot be found in Figure 8 and only near-band-edge emission peak is observed. We thought that it is the reason that the maximum PL intensities of the 60 °C-grown ZnO-40-Eu and 88 °C-grown ZnO-4-In nanowires are higher than that of the 100 °C-grown undoped ZnO nanowires. The synthesized 60 °C-grown ZnO-40-Eu and 88 °C-grown ZnO-4-In nanowires can be the promising active materials in UV detectors and opto-electronic devices.

4. Conclusions

In this study, we used a hydrothermal method to investigate a simple process for synthesizing the Eu^{3+}-doped and In^{3+}-doped ZnO nanowires at temperatures lower than 100 °C. We found that the requisite synthesis temperatures for undoped ZnO, ZnO-10-Eu, ZnO-20-Eu, ZnO-30-Eu, and ZnO-40-Eu nanowires were 100, 90, 80, 70, and 60 °C, and for ZnO-4-In, ZnO-8-In, and ZnO-12-In nanowires were 88, 75, and 60 °C, respectively. For the undoped, Eu^{3+}-doped, and In^{3+}-doped ZnO nanowires, the (200) peak was the main diffraction peak, and the 2θ values of the *c*-orientation (200) peak were almost unchanged as the concentration of Eu^{3+} ions increased from 0 to 4 at% and the concentration of In^{3+} ions increased from 0 to 1.2 at%. The (200) diffraction peak was observed when the concentrations of Eu^{3+} ions were 3 and 4 at%; its diffraction intensity increased, and the FWHM value of the (200) diffraction peak decreased with the concentration of Eu^{3+} ions. Two distinct emission peaks were discernible in the 100 °C-grown undoped ZnO and 60 °C-grown ZnO-40-Eu nanowires, one centered at ~300 nm and the other at 395 or 396 nm, and the emission spectra ranged from 280 to 570 nm. The visible-light emission the visible-light emission was not found in the 100 °C-grown undoped-ZnO, 60 °C-grown ZnO-40-Eu, and 88 °C-grown ZnO-4-In nanowires and only near-band-edge emission peak was observed because their crystal qualities were enhanced, and the number of defect states was reduced. The 88 °C-grown ZnO-4-In nanowires exhibited only one peak, and the emission spectra ranged from 360 to 520 nm. The PL emission spectra showed that the maximum intensities increase in the order 100 °C-grown ZnO < 88 °C-grown ZnO-4-In < 60 °C-grown ZnO-40-Eu.

Author Contributions: Y.-F.W. organized the paper and helped with writing it; W.-Y.C. and C.-F.Y. organized the paper and helped with editing the English; J.-R.S. and C.-C.C. carried out the experiments, analyzed the data, and measurements.

Acknowledgments: Financial support from the NSFC project titled "Fabrication of high efficiency ammonia gas sensors by using P3HT-zinc oxide nanowire hetero-junction for pilot detection liver cancer" is deeply appreciated.

Conflicts of Interest: The authors declare no conflict of interest.

References

1. Chung, J.; Lee, J.; Lim, S. Annealing effects of ZnO nanorods on dye-sensitized solar cell efficiency. *Physica B Condens. Matter* **2010**, *405*, 2593–2598. [CrossRef]

2. Saito, N.; Haneda, H.; Sekiguchi, T.; Ohashi, N.; Sakaguchi, I.; Koumoto, K. Low-temperature fabrication of light-emitting zinc oxide micropatterns using self-assembled monolayers. *Adv. Mater.* **2002**, *14*, 418–421. [CrossRef]

3. Kwak, G.; Seol, M.; Tak, Y.; Yong, K. Superhydrophobic ZnO nanowire surface: Chemical modification and effects of UV irradiation. *J. Phys. Chem. C* **2009**, *113*, 12085–12089. [CrossRef]

4. Wei, A.; Pan, L.H.; Huang, W. Recent progress in the ZnO nanostructure-based sensors. *Mater. Sci. Eng. B* **2011**, *176*, 1409–1421. [CrossRef]

5. Lupan, O.; Shishiyanu, S.; Chow, L.; Shishiyanu, T. Nanostructured zinc oxide gas sensors by successive ionic layer adsorption and reaction method and rapid photothermal processing. *Thin Solid Films* **2008**, *516*, 3338–3345. [CrossRef]

6. Niarchos, G.; Makarona, E.; Tsamis, C. Growth of ZnO nanorods on patterned templates for efficient, large-area energy scavengers. *Microsyst. Technol.* **2010**, *16*, 669–675. [CrossRef]

7. Kenanakis, G.; Vernardou, D.; Koudoumas, E.; Katsarakis, N. Growth of c-axis oriented ZnO nanowires from aqueous solution: The decisive role of a seed layer for controlling the wires' diameter. *J. Cryst. Growth* **2009**, *311*, 4799–4804. [CrossRef]

8. Mondal, B.; Maity, S.; Das, S.; Panda, D.; Saha, H.; Kundu, A. Fabrication and packaging of MEMS based platform for hydrogen sensor using ZnO–SnO$_2$ composites. *Microsyst. Technol.* **2016**, *22*, 2757–2764. [CrossRef]

9. Zhao, H.; Liu, X.J.; Chen, L.Z.; Chang, S.P.; Chang, M. Fabrication of ultra-high-sensitivity flexible strain sensor based on single ZnO nanowire. *Microsyst. Technol.* **2017**, *23*, 1703–1707. [CrossRef]

10. Thomas, D.; Sadasivuni, K.K.; Waseem, S.; Kumar, B.; Cabibihan, J.J. Synthesis, green emission and photosensitivity of Al-doped ZnO film. *Microsyst. Technol.* **2018**, *24*, 3069–3073. [CrossRef]

11. Bai, S.N.; Tsai, H.H.; Tseng, T.Y. Structural and optical properties of Al-doped ZnO nanowires synthesized by hydrothermal method. *Thin Solid Films* **2007**, *516*, 155–158. [CrossRef]

12. Chang, L.T.; Wang, C.Y.; Tang, J.S.; Nie, T.X. Electric-Field Control of Ferromagnetism in Mn-Doped ZnO Nanowires. *Nano Lett.* **2014**, *14*, 1823–1829. [CrossRef] [PubMed]

13. Kim, S.H.; Umar, A.; Hwang, S.W.; Al-Garni, H.; Abaker, M.; Al-Sayari, S.A.; Dar, G.N.; Al-Hajry, A. Growth of branched In-doped ZnO nanowires: Structural and Optical Properties. *AIP Conf. Proc.* **2011**, *1370*, 142–148.

14. Hsu, C.L.; Tsai, T.Y. Fabrication of Fully Transparent Indium-Doped ZnO Nanowire Field-Effect Transistors on ITO/Glass Substrates. *J. Electrochem. Soc.* **2011**, *158*, K20–K23. [CrossRef]

15. Xu, L.; Su, Y.; Chen, Y.Q.; Xiao, H.H.; Zhu, L.A.; Zhou, Q.T.; Li, S. Synthesis and Characterization of Indium-Doped ZnO Nanowires with Periodical Single-Twin Structures. *J. Phys. Chem. B* **2006**, *110*, 6637–6642. [CrossRef]

16. AlRifai, S.A.; Kulnitskiy, B.A. Microstructural and optical properties of europium-doped zinc oxide nanowires. *J. Phys. Chem. Sol.* **2013**, *74*, 1733–1738. [CrossRef]

17. Lupan, O.; Pauporte, T.; Viana, B.; Aschehoug, P.; Ahmadic, M.; Cuenya, B.R.; Rudzevich, Y.; Lin, Y.; Chow, L. Eu-doped ZnO nanowire arrays grown by electrodeposition. *Appl. Surf. Sci.* **2013**, *282*, 782–788. [CrossRef]

18. Geburt, S.; Lorke, M.; da Rosa, A.L.; Frauenheim, T.; Röder, R.; Voss, T.; Kaiser, U.; Heimbrodt, W.; Ronning, C. Intense Intrashell Luminescence of Eu-Doped Single ZnO Nanowires at Room Temperature by Implantation Created Eu–O_i Complexes. *Nano Lett.* **2014**, *14*, 4523–4528. [CrossRef] [PubMed]

19. Chen, Y.C.; Cheng, H.Y.; Yang, C.F.; Hsieh, Y.T. Investigation of the Optimal Parameters in Hydrothermal Method for the Synthesis of ZnO Nanorods. *J. Nanomater.* **2014**, *2014*, 430164. [CrossRef]

20. Guo, L.; Zhang, H.; Zhao, D.X.; Li, B.H.; Zhang, Z.H.; Jiang, M.M.; Shen, D.Z. High responsivity ZnO nanowires based UV detector fabricated by the dielectrophoresis method. *Sens. Actuators B Chem.* **2012**, *166*, 12–16. [CrossRef]

21. Zeng, X.Y.; Yuan, J.L.; Wang, Z.Y.; Zhang, L.D. Nanosheet-Based Microspheres of Eu^{3+}-doped ZnO with Efficient Energy Transfer from ZnO to Eu^{3+} at Room Temperature. *Adv. Mater.* **2007**, *19*, 4510–4514. [CrossRef]

22. Lin, C.Y.; Yang, S.H.; Lin, J.L.; Yang, C.F. Effects of the Concentration of Eu^{3+} Ions and Synthesizing Temperature on the Luminescence Properties of $Sr_{2-x}Eu_xZnMoO_6$ Phosphors. *Appl. Sci.* **2017**, *7*, 30. [CrossRef]

23. Gomi, M.; Oohira, N.; Ozaki, K.; Koyano, M. Photoluminescent and Structural Properties of Precipitated ZnO Fine Particles. *Jpn. J. Appl. Phys.* **2003**, *42*, 481–485. [CrossRef]

24. Janotti, A.; Van de Walle, C.G. Fundamentals of zinc oxide as a semiconductor. *Rep. Prog. Phys.* **2009**, *72*, 126501. [CrossRef]

25. Teke, A.; Ozgur, U.; Dogan, S.; Gu, X.; Morkocu, H.; Nemeth, B.; Nause, J.; Everitt, H.O. Excitonic fine structure and recombination dynamics in single-crystalline ZnO. *Phys. Rev. B* **2004**, *70*, 195207. [CrossRef]

26. Chen, T.; Xing, G.Z.; Zhang, Z.; Chen, H.Y.; Wu, T. Tailoring the photoluminescence of ZnO nanowires using Au nanoparticles. *Nanotechnology* **2008**, *19*, 435711. [CrossRef]

27. Yuan, Z.S.; Lin, C.Y.; Sun, S.H.; Liu, N.Y.; Yang, C.F.; Chen, C.C. Investigation of Undoped ZnO@SiO_2 Core-Shell Phosphors with Green Luminescence. *Microsyst. Technol.* **2018**. [CrossRef]

28. Mo, Z.J.; Hao, Z.H.; Wu, H.Z.; Yang, Q.; Zhuo, P.; Yang, H.; Xu, J.P.; Zhang, X.S.; Li, L. Synthetic and effect of annealing on the luminescent properties of ZnO nanowire. *J. Lumin.* **2016**, *175*, 232–236. [CrossRef]

29. Farid, S.; Mukherjee, S.; Sarkar, K.; Mazouchi, M.; Stroscio, M.A.; Dutta, M. Enhanced optical properties due to indium incorporation in zinc oxide nanowires. *Appl. Phys. Lett.* **2016**, *108*, 021106. [CrossRef]

 *electronics*

Article

Electronic Circuit with Controllable Negative Differential Resistance and its Applications

Vladimir Ulansky [1,2,*], **Ahmed Raza** [3] **and Hamza Oun** [2]

[1] Research and Development Department, Mathematical Modelling & Research Holding Limited, London W1W 7LT, UK

[2] Department of Electronics, National Aviation University, 03058 Kyiv, Ukraine; hamzahashoor@gmail.com

[3] Projects and Maintenance Section, The Private Department of the President of the United Arab Emirates, Abu Dhabi 000372, UAE; ahmed.awan786@gmail.com

* Correspondence: vulanskyi@mmrholding.org; Tel.: +44(0)2038236006 or +380632754982

Received: 5 March 2019; Accepted: 3 April 2019; Published: 8 April 2019

Abstract: Electronic devices and circuits with negative differential resistance (NDR) are widely used in oscillators, memory devices, frequency multipliers, mixers, etc. Such devices and circuits usually have an *N*-, *S*-, or Λ-type current-voltage characteristics. In the known NDR devices and circuits, it is practically impossible to increase the negative resistance without changing the type or the dimensions of transistors. Moreover, some of them have three terminals assuming two power supplies. In this paper, a new NDR circuit that comprises a combination of a field effect transistor (FET) and a simple bipolar junction transistor (BJT) current mirror (CM) with multiple outputs is proposed. A distinctive feature of the proposed circuit is the ability to change the magnitude of the NDR by increasing the number of outputs in the CM. Mathematical expressions are derived to calculate the threshold currents and voltages of the N-type current-voltage characteristics for various types of FET. The calculated current and voltage thresholds are compared with the simulation results. The possible applications of the proposed NDR circuit for designing single-frequency oscillators and voltage-controlled oscillators (VCO) are considered. The designed NDR VCO has a very low level of phase noise and has one of the best values of a standard figure of merit (FOM) among recently published VCOs. The effectiveness of the proposed oscillators is confirmed by the simulation results and the implemented prototype.

Keywords: negative differential resistance; current-voltage characteristics; multiple simple current mirror; threshold voltage; oscillator; voltage-controlled oscillator

---

1. Introduction

Nowadays, negative differential resistance devices and circuits are widely used in oscillators, memory, frequency dividers, and multiplier circuits [1–5]. The presence of negative resistance in an electrical circuit makes it possible not to dissipate electrical energy in the form of heat, but to generate electrical power, even if it has only two terminals and not three as in transistors. There are two types of negative resistance, namely, differential and static. Sometimes NDR is also called negative dynamic resistance. The NDR is the first derivative of the voltage relative to the current at the operating point. The current-voltage characteristics with NDR region can be created in two ways. The first way involves the use of special electronic devices, such as a Gunn diode, a tunnel diode, three-terminal graphene NDR devices [6], and others. In the second way, the current-voltage characteristics with NDR region are created artificially with the help of special electronic circuits. However, the known NDR electronic circuits have some disadvantages that limit their use. Let us consider the well-known electronic circuits with NDR. The studies [7,8] considered the NDR circuit based on complementary metal-oxide-semiconductor (CMOS) NDR inverters requiring two power supplies. Thus, to create the current-voltage characteristics with NDR region three terminals should be used. The study [9] considered an electronic oscillator based on a bipolar junction transistor (BJT)-metal-oxide-semiconductor field-effect transistor (MOSFET) structure with Λ-type current-voltage characteristics and two power supplies. The study [10] considered an NDR circuit composing three resistors and two BJT. The N-type current-voltage characteristics are achieved by selecting the appropriate resistor values. The study [11] considered a special connection of a BJT with a junction gate field-effect transistor (JFET) that has N-type current-voltage characteristics. However, this circuit is subject to thermal runaway, which makes it difficult to use the circuit in practice. The study [12] considered a novel sinusoidal NDR VCO for very high frequency band. The VCO circuit comprises a JFET in combination with a P–channel metal-oxide-semiconductor (MOS) improved Wilson current mirror (CM). The study [13] considered a new NDR circuit, which uses a FET and BJT transistors to create the S-type current-voltage characteristics. The study [14] considered a systematic method to design NDR circuits comprising two transistors and resistors only. The study [15] considered a comparison of five proposed NDR VCOs for microwave applications. The NDR circuits include a gallium-arsenide transistor and different BJT CMs. The study [16] considered a novel voltage-controlled NDR device, using complementary silicon-on-insulator four-gate transistors. The work experimentally demonstrated new circuits for the inductor-capacitor (LC) oscillator and Schmitt trigger based on the proposed NDR device. The study [17] considered a novel multiple NDR device with an ultra-high peak-to-valley current ratio by combining tunnel diode with a conventional MOSFET. The study [18] proposed complement double-peak NDR devices by combining tunnel diode with conventional CMOS and its compact five-state latch circuit by introducing standard ternary inverter. The study [19] considered four novel NDR circuits based on the combination of the standard n-channel MOS transistors and silicon-germanium heterojunction bipolar transistor (HBT). Depending on the design parameters, the proposed circuits can exhibit Λ- or N-type current-voltage characteristics. The study [20] considered a tri-valued memory circuit based on two cascoded MOS-BJT-NDR devices that can show the NDR current-voltage characteristic by adjusting the MOS transistor parameters. The study [21] considered a three-terminal voltage controlled Λ-type negative resistance MOSFET structure using the merged integrated circuit of a NELS (n-channel enhancement mode with load operated at saturation) inverter and an n-channel enhancement MOS driver.

It should be noted that in the reviewed NDR devices and circuits there is practically no possibility of increasing the negative resistance without changing the type of transistors or the dimensions of transistors. Moreover, some devices and circuits have three terminals assuming two power supplies. These circumstances significantly reduce the range of possible applications of NDR devices and circuits. For example, negative resistance may not be sufficient to start-up the oscillator circuit [22]. This paper proposes a new NDR circuit based on a FET in conjunction with multiple simple CMs, in which the magnitude of the negative resistance is easily controlled by changing the number of CM outputs. Description, mathematical and numerical modeling of the NDR circuit is given. The proposed NDR electronic circuit can be used in designing an oscillator, a VCO, an amplifier, etc. The most promising applications are related to generating ultrahigh-frequency signals with low phase noise.

2. Circuit Operation

Figure 1a shows the proposed NDR circuit, which consists of two bias resistors R_a and R_b, a FET (Q_0) and simple CM with $n - 1$ outputs (collectors of transistors $\overline{Q_2, Q_n}$). We further assume that transistors $\overline{Q_1, Q_n}$ are matched. Figure 1b shows the current-voltage characteristics of the NDR circuit where the total current I_0 is a function of the power supply voltage V_{xy}. The steepness of the current-voltage characteristic between points β and γ depends on the number of outputs of the CM. Curve 1 corresponds to the one output of the CM, i.e., only transistors Q_1 and Q_2 are used. In this particular case, the differential resistance between points β and γ is positive, and the circuit does not have an NDR region. When choosing the appropriate transistors, the circuit can have the NDR region even with one CM output.

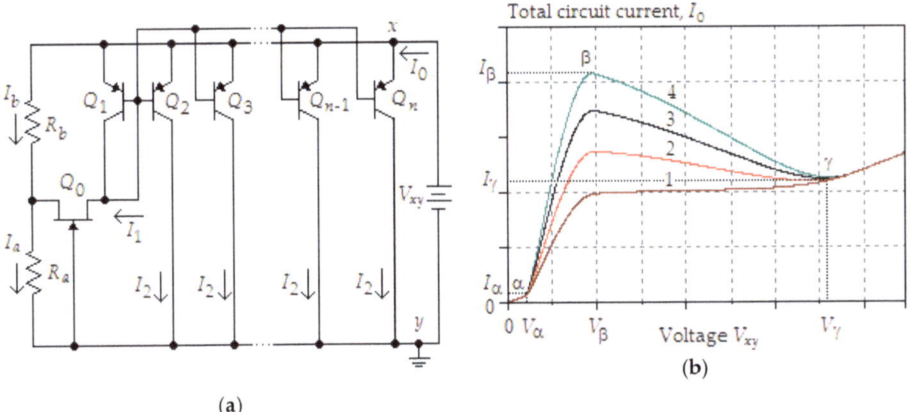

Figure 1. (**a**) Negative differential resistance circuit with multiple simple current mirror; (**b**) N-type current-voltage characteristics of the circuit.

As can be seen in Figure 1b, with an increase in the number of the CM outputs to two and further to four, the differential resistance becomes negative and the slope of the characteristic in the NDR region increases.

The current-voltage characteristics in Figure 1b include four regions, respectively, between points 0 and V_α, V_α and V_β, V_β and V_γ, and V_γ and ∞. The NDR region is located between points V_β and V_γ.

Let's look at the general principle of the circuit operation. When voltage V_{xy} varies from 0 to V_α, all transistors in the circuit are OFF. At the supply voltage V_α, all transistors are turning ON, and, up to the voltage V_β, the current I_1 rises, and the total current I_0 also increases. The current I_2 is a mirrored copy of I_1, and it behaves like I_1. When the supply voltage is V_β, the current I_1 reaches a maximum, which also corresponds to the maximum of the total current I_0. When the voltage V_{xy} changes from V_β to V_γ, the current I_1 decreases due to an increase in the negative voltage between the gate and the source of the transistor Q_0. At the same time, an increase in current I_b does not cover this decrease in current I_1. As a result, the total current I_0 decreases up to the voltage V_γ, at which the voltage between the gate and the source of the transistor Q_0 reaches a pinch-off voltage and Q_0 is turning OFF, and, consequently, the transistors $\overline{Q_1, Q_n}$ also turning OFF. When the supply voltage is higher than V_γ, the total current I_0 is entirely determined by the voltage V_{xy} and the resistances of R_a and R_b. Therefore, in the interval (V_γ, ∞) differential resistance of the current-voltage characteristics is positive.

Let us determine the coordinates of points α, β, and γ. In the first region of the current-voltage characteristics, between points 0 and V_α, all transistors are OFF, and the overall current depends on the resistor values R_a and R_b, and power supply voltage V_{xy}.

$$I_0 = \frac{V_{xy}}{R_a + R_b}. \tag{1}$$

The threshold voltage V_α is determined by applying Kirchhoff's voltage law (KVL) equation to the circuit of Figure 1a when transistor Q_0 is turning ON and $V_{DS0} = 0$, where V_{DS0} is the drain-source voltage of Q_0.

$$V_\alpha = V_{EB} + I_a R_a, \tag{2}$$

where V_{EB} is the emitter-base voltage of transistors $\left(\overline{Q_1, Q_n}\right)$ and I_a is the current through resistor R_a. Since at voltage V_α the current I_a is equal to I_0, then Equation (2) transforms to the following form:

$$V_\alpha = \frac{V_{EB}(R_a + R_b)}{R_b}. \tag{3}$$

Combining (1) and (3), we obtain:

$$I_\alpha = \frac{V_{EB}}{R_b}. \tag{4}$$

When voltage V_{xy} a little exceeds V_α transistors $\overline{Q_1, Q_n}$ are turning ON, and transistor Q_0 starts to operate in the triode region where:

$$V_{DS0} < V_{GS0} + V_{P0}, \tag{5}$$

where V_{GS0} and V_{P0} are, respectively, the gate-source and pinch-off voltage of Q_0.

By applying KVL from $+V_{xy}$ to ground, we get:

$$V_{xy} = V_{EB} + V_{DS0} + I_a R_a. \tag{6}$$

From (6) we find current I_a as follows:

$$I_a = \frac{V_{xy} - V_{EB} - V_{DS0}}{R_a} \tag{7}$$

As it follows from the circuit of Figure 1a:

$$I_b = I_a - I_1. \tag{8}$$

Substituting (7) to (8) gives:

$$I_b = \frac{V_{xy} - V_{EB} - V_{DS0}}{R_a} - I_1. \tag{9}$$

Applying KVL around the loop $(R_b,\ Q_0,\ Q_1)$, we obtain:

$$-I_b R_b + V_{DS0} + V_{EB} = 0. \tag{10}$$

Substituting (9) into (10) and performing some mathematical transformations, we get:

$$V_{DS0} = \frac{R_b V_{xy}}{R_a + R_b} - I_1(R_a \| R_b) - V_{EB}. \tag{11}$$

The gate-source voltage of transistor Q_0 is given by:

$$V_{GS0} = -I_a R_a. \tag{12}$$

Voltage V_{GS0} we obtain by combining (7), (11), and (12) and performing necessary transformations.

$$V_{GS0} = -\frac{R_a V_{xy}}{R_a + R_b} - I_1(R_a \| R_b). \tag{13}$$

Substituting V_{DS0} from (11) to (7), we determine that:

$$I_a = \frac{V_{xy}}{R_a + R_b} + \frac{I_1 R_b}{R_a + R_b}. \tag{14}$$

We find the current I_0 by applying Kirchhoff's current law (KCL) to the node y in the circuit of Figure 1a.

$$I_0 = I_a + (n-1)I_2. \tag{15}$$

By substitution of (14) into (15), we obtain general equation for the circuit total current.

$$I_0 = (n-1)I_2 + \frac{I_1 R_b}{R_a + R_b} + \frac{V_{xy}}{R_a + R_b}. \tag{16}$$

Since the current I_2 is close to the reference current I_1, we can assume that $I_2 \approx I_1$. In this case Equation (16) transforms into the following form:

$$I_0 \approx I_1\left(n - 1 + \frac{R_b}{R_a + R_b}\right) + \frac{V_{xy}}{R_a + R_b}. \tag{17}$$

The slope of the current-voltage characteristics in the region of negative resistance is determined by the first derivative of the current I_0 with respect to the voltage V_{xy}.

$$\frac{dI_0}{dV_{xy}} \approx \left(n - 1 + \frac{R_b}{R_a + R_b}\right)\frac{dI_1}{dV_{xy}} + \frac{1}{R_a + R_b}. \tag{18}$$

As can be seen from (18), the slope is indeed proportional to the number of outputs of the current-mirror n. Therefore, by changing n, it is possible to reduce or increase the slope of the current-voltage characteristics between the voltage thresholds V_β and V_γ.

Substituting (11) and (13) into (5), we find that when transistor Q_0 operates in the triode mode, the following relationship holds between the supply voltage V_{xy} and the voltages V_{EB} and V_{P0}:

$$V_{xy} < V_{EB} - V_{P0}. \tag{19}$$

When voltage V_{xy} increases more, FET Q_0 reaches the saturation region where:

$$V_{DS0} \geq V_{GS0} + V_{P0}. \tag{20}$$

The threshold voltage V_β is reached at the boundary between the triode and the saturation region of transistor Q_0, i.e., when:

$$V_{DS0} = V_{GS0} + V_{P0}. \tag{21}$$

Substituting (11) and (13) into (21), we obtain the threshold voltage V_β.

$$V_\beta = V_{EB} - V_{P0}. \tag{22}$$

We determine the threshold current I_β by substitution of (22) into (17).

$$I_\beta \approx I_1\left(n - 1 + \frac{R_b}{R_a + R_b}\right) + \frac{V_{EB} - V_{P0}}{R_a + R_b}. \tag{23}$$

As can be seen in Figure 1b, the slope of the current-voltage characteristics in the regions $(0,\ V_\alpha)$ and $(V_\gamma,\ \infty)$ is the same. It means that at the voltage threshold V_γ the FET Q_0 is OFF and $I_1 = 0$. Transistor Q_0 is turning OFF when $V_{GS0} = V_{P0}$. Substituting V_{P0} instead of V_{GS0} into (13) and solving the obtained equation with respect to V_{xy}, we get:

$$V_{xy} = V_\gamma = -\left(1 + \frac{R_b}{R_a}\right)V_{P0}. \tag{24}$$

The threshold current I_γ we find by substituting V_{xy} from (24) into (1).

$$I_\gamma = \frac{|V_{P0}|}{R_a}. \tag{25}$$

As can be seen from (11), (13), (14), (16), (17), and (23), to calculate the currents and voltages related to the thresholds of the circuit current-voltage characteristics, it is necessary to know the current I_1. Modeling the current I_1 depends on the type of FET Q_0. Transistor Q_0 can be an N-channel JFET, metal-semiconductor field-effect transistor (MESFET), high-electron-mobility transistor (HEMT), or pseudomorphic high-electron-mobility transistor (PHEMT). In the voltage region (V_β, V_γ) transistor Q_0 operates in the saturation mode. Therefore, we should model the current I_1 for the case when Q_0 operates in the saturation mode.

As is well known, the operation of a JFET in the saturation mode is quite good described by the Shockley equation [23].

Substituting V_{GS0} from (13) to the Shockley equation gives:

$$I_1 = I_{DSS}\left(1 + \frac{AV_{xy} + BI_1}{V_{P0}}\right)^2, \tag{26}$$

where I_{DSS} is the saturation drain-source current at zero gate–source voltage, A and B are determined as follows:

$$A = R_a/(R_a + R_b), B = R_a\|R_b.$$

Solving (26) with respect to current I_1, we obtain the following quadratic equation:

$$\frac{I_{DSS}B^2}{V_{P0}^2}I_1^2 + \left[2I_{DSS}\left(1 + \frac{AV_{xy}}{V_{P0}}\right)\frac{B}{V_{P0}} - 1\right]I_1 + I_{DSS}\left(1 + \frac{AV_{xy}}{V_{P0}}\right)^2 = 0. \tag{27}$$

Equation (27) has two positive roots. The acceptable root is the value of the current I_1 that is less than I_{DSS}.

Figure 2a shows the dependence of the drain current I_1 versus power supply voltage V_{xy} in the interval (V_β, V_γ) when BF245B is used as a JFET Q_0 and Positive-Negative-Positive (PNP) BJT transistors BFT92W are used in the CM. Assume that $R_a = 0.5$ kΩ, $R_b = 2$ kΩ, and $n = 5$, i.e., the CM has four outputs. From the simulation program with integrated circuit emphasis (SPICE) model of the selected JFET transistor follows that $V_{P0} = -2.31$ V and $I_{DSS} = 6$ mA.

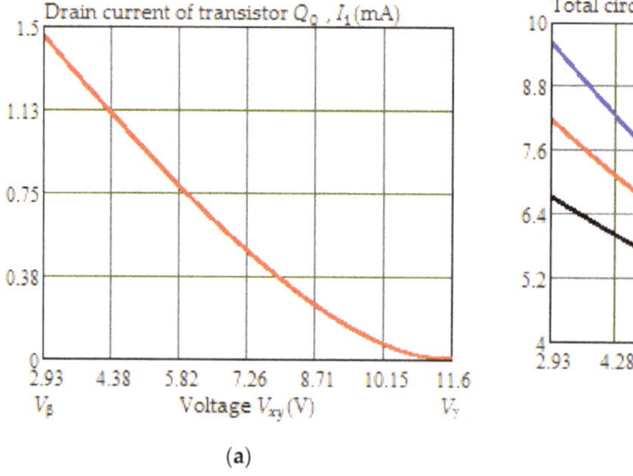

(a)

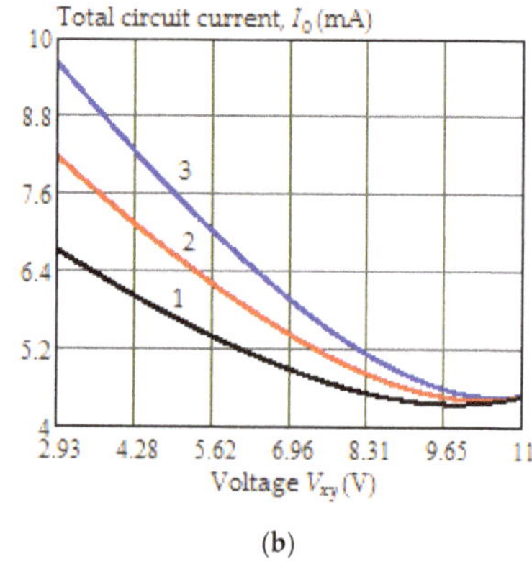

(b)

Figure 2. (a) Dependence of the drain current I_1 versus voltage V_{xy}; (b) Dependence of the total current I_0 versus voltage V_{xy} in the negative differential resistance region when $n = 4$ (curve 1), $n = 5$ (curve 2), and $n = 6$ (curve 3).

As can be seen in Figure 2a, in the NDR region the current I_1 changes from 1.47 mA to 0. Figure 2b shows the dependence of the total current I_0 versus voltage V_{xy} in the NDR region when $n = 4$ (curve 1), $n = 5$ (curve 2), and $n = 6$ (curve 3).

The curves in Figure 2b were calculated using (17) and (27). As can be seen in Figure 2b, the total current I_0 increases by the same value with each increase in the number of the CM outputs. At the same time, the threshold voltage V_γ shifts slightly to the right with increasing n.

Let us compare the calculated and simulated voltage and current thresholds for the circuit of Figure 1a for the same data as in Figure 2. From the interactive SPICE simulation of transistor BFT92W operation with the help of Multisim (ed. 14.1) follows that $V_{EB} = 0.52$ V (at threshold α) and $V_{EB} = 0.62$ V (at threshold β). The results of calculations and SPICE simulations are shown in Table 1.

Table 1. Calculated and simulated threshold voltages and currents with junction gate field-effect transistor (JFET).

Threshold	Calculated Value	Simulated Value	Error %
V_α (V)	0.65	0.62	−4.8%
I_α (mA)	0.26	0.28	7.1%
V_β (V)	2.93	2.89	−1.4%
I_β (mA)	8.22	8.30	1%
V_γ (V)	11.60	11.05	−5.0%
I_γ (mA)	4.62	4.50	−2.7%

As can be seen in Table 1, the absolute relative error for the calculated voltage thresholds of the NDR region does not exceed 5% and for current thresholds slightly exceed 2.5%. Such a high accuracy of calculating voltage and current thresholds testifies on the adequacy of the derived mathematical equations.

Let us now consider the case when Q_0 is a GaAs transistor, i.e., MESFET, HEMT, or PHEMT. We model the current of Q_0 by the Statz nonlinear model, which has a high accuracy in the approximation of the drain current [24].

Substituting V_{DS0} and V_{GS0} from (11) and (13) to the Statz model [24,25], we get the following nonlinear equations in respect to the drain current I_1:

$$\frac{\delta\left(AV_{xy}+BI_1+V_{P0}\right)^2}{1-\Delta\left(AV_{xy}+BI_1+V_{P0}\right)}\left[1+\lambda\left(CV_{xy}-BI_1-V_{EB}\right)\right]\left\{1-\left[1-\alpha\left(CV_{xy}-BI_1-V_{EB}\right)/3\right]^3\right\}-I_1=0,$$

$$\text{(28)}$$

$$\text{for } 0 < CV_{xy}-BI_1-V_{EB} < 3/\alpha,$$

$$\frac{\delta\left(AV_{xy}+BI_1+V_{P0}\right)^2}{1-\Delta\left(AV_{xy}+BI_1+V_{P0}\right)}\left[1+\lambda\left(CV_{xy}-BI_1-V_{EB}\right)\right]-I_1=0, \text{ for } CV_{xy}-BI_1-V_{EB}\geq 3/\alpha, \quad \text{(29)}$$

where α is the current saturation parameter, δ is the transistor transconductance, Δ is the doping profile parameter, λ is the channel length modulation coefficient, and C is determined as follows:

$$C = R_b/(R_a + R_b).$$

Let us again compare the calculated and simulated voltage and current thresholds for the circuit of Figure 1a when a low noise PHEMT ATF-33143 is used as transistor Q_0 and the same PNP transistors BFT92W are used in the CM. From the SPICE model of ATF-33143 [26] follows that $\alpha = 4$ [1/V], $\delta = 0.48$ $\left[\text{A/V}^2\right]$, $\Delta = 0.8$, $\lambda = 0.09$ [1/V], and $V_{P0} = -0.95$ [V]. The other circuit parameters have the same values as in the previous example.

To obtain the calculated values of the voltage and current thresholds, we solve Equations (28) and (29) and use Equations (3), (4), and (22)–(25). The results of calculations and SPICE simulations with the help of Multisim (ed. 14.1) are shown in Table 2.

Table 2. Calculated and simulated threshold voltages and currents with pseudomorphic high-electron-mobility transistor (PHEMT).

Threshold	Calculated Value	Simulated Value	Error %
V_α (V)	0.65	0.62	−4.8%
I_α (mA)	0.26	0.28	7.1%
V_β (V)	1.57	1.71	8.2%
I_β (mA)	7.18	6.92	−3.8%
V_γ (V)	4.75	4.73	−0.4%
I_γ (mA)	1.9	1.91	0.5%

As can be seen in Table 2, a good agreement exists between the theoretical and simulated values of the current and voltage thresholds, which proves the validity of the derived equations.

3. Circuit Applications

3.1. LC Oscillator

Oscillators are one of the main elements in modern communication, control, and navigation systems. Modern oscillators can be divided into two classes, namely oscillators with negative input impedance and oscillators with NDR. A distinctive feature of the first-class oscillators is the presence of a negative real part in the input impedance. Examples of such oscillators are numerous Colpitts, Clapp, Hartley oscillator circuits, and their modifications [27–30], as well as cross-coupled CMOS oscillators [31–34]. The second class of microwave oscillators supposes to use a tunnel diode or a Gunn diode [35,36], which have an NDR region in the N-type current-voltage characteristics. The location of the operating point in the NDR region leads to the creation of a negative resistance induced into the contour of the LC tank to compensate for its losses. The circuit of Figure 1a can also be used for designing an LC oscillator because it has an NDR region.

Figure 3 shows an LC oscillator on the base of the proposed NDR circuit. The oscillator tank circuit consists of an RF coil L and two series-connected capacitors C_1 and C_2. Small capacitor C_F is a feedback capacitor allowing to speed-up the oscillator start-up. Large capacitor C_0 reduces the noise level significantly at the nodes y, z, d, and s of the oscillator. This allows increasing the slope of the noise skirt around the fundamental harmonic, which in-turn reduces the oscillator phase noise.

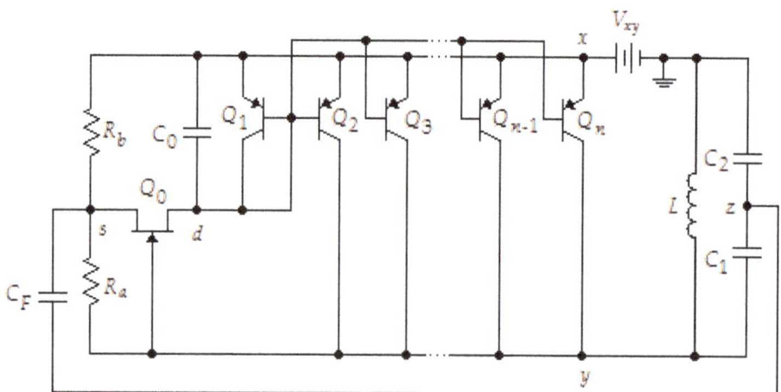

Figure 3. LC oscillator with NDR.

3.1.1. Simulation Results

Let us perform a SPICE simulation of the proposed oscillator circuit with the help of Multisim (ed. 14.1). Assume that $n = 3$, $R_a = 0.15$ kΩ, $R_b = 1$ kΩ, transistor Q_0 is a PHEMT ATF-33143, and all transistors in the CM are BFT92W. Figure 4 shows the simulated current-voltage characteristics. As can be seen in Figure 4, the NDR region has the following voltage and current thresholds: $V_\beta = 1.58$ V, $I_\beta = 15$ mA, $V_\gamma = 7.26$ V, and $I_\gamma = 6.33$ mA. We set the operating point at $\overline{V}_{xy} = 3.75$ V and $\overline{I}_0 = 12.5$ mA.

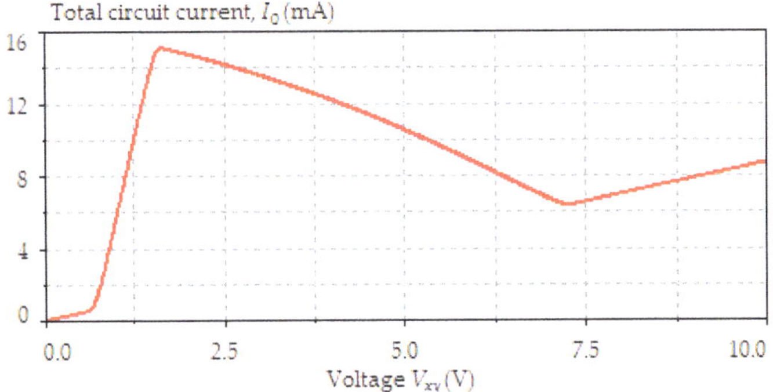

Figure 4. Oscillator current-voltage characteristics.

The selected circuit components have the following values: $L = 5$ nH, $C_0 = 10$ μF, and $C_1 = C_2 = 5$ pF. Figure 5 shows the oscillator starting voltage waveforms at the output node y for different values of the feedback capacitor C_F. The frequency of oscillations is 1.096 GHz. We can observe from Figure 5, that the oscillations reach the steady-state amplitude of 2.2 V at $t = 130$ ns and 2.4 V at $t = 55$ ns when $C_F = 5$ pF and $C_F = 10$ pF, respectively. Thus, an increase in the capacitance C_F leads to a significant reduction in the self-excitation time of the oscillator and an increase in the amplitude of the steady-state oscillations.

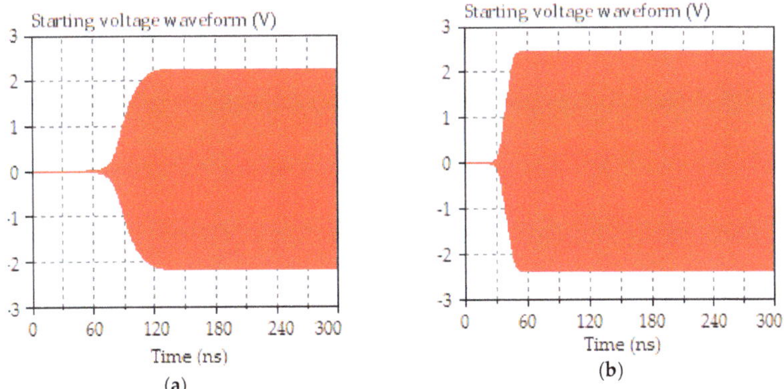

Figure 5. Oscillator start-up behavior at $C_F = 5$ pF (**a**) and $C_F = 10$ pF (**b**).

Figure 6 shows the amplitude spectrum of the oscillated voltage for $C_0 = 1$ nF and $C_F = 5$ pF (blue line) and $C_0 = 10$ μF and $C_F = 10$ pF (red line). As can be seen from comparison of two spectrums in Figure 6, the noise skirt of the fundamental harmonic at the level of -80 dBm is significantly narrower for larger values of capacitances C_0 and C_F. Moreover, the total harmonic distortion (THD) is 3.3% for

$C_0 = 10\ \mu\text{F}$ and $C_F = 10$ pF and 3.7% for $C_0 = 1$ nF and $C_F = 5$ pF, i.e., larger capacitances C_0 and C_F provide a smaller level of THD.

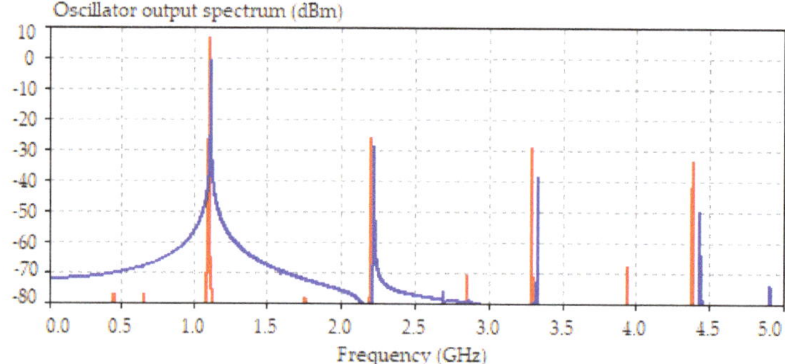

Figure 6. Oscillator output spectrum when $C_0 = 1$ nF and $C_F = 5$ pF (blue line) and $C_0 = 10\ \mu$F and $C_F = 10$ pF (red line).

The decrease in the noise level of the oscillator output voltage in Figure 6 is explained by the fact that with an increase in the capacitances C_0 and C_F, the simulated spectral density of noise decreases significantly at the nodes y, z, d, and s of the oscillator circuit as shown in Figure 7. As can be seen in Figure 7c, the most substantial decrease in noise spectral density occurs at the drain of transistor Q_0, i.e., just where there is a large capacitance C_0. This observation confirms a similar conclusion concerning the effect of capacitance on noise in CMOS LC oscillators [37].

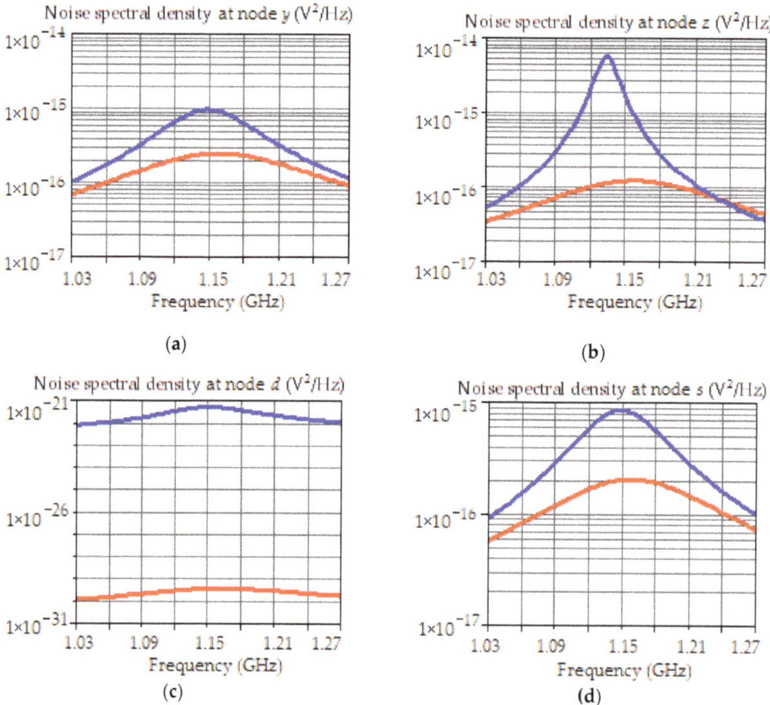

Figure 7. Noise spectral density at nodes y (a), z (b), d (c), and s (d) when $C_0 = 1$ nF and $C_F = 5$ pF (blue line) and $C_0 = 10\ \mu$F and $C_F = 10$ pF (red line).

3.1.2. Oscillator Prototype Implementation

The oscillator circuit of Figure 3 was implemented using a JFET BF245B (NXP Semiconductors, Eindhoven, Netherlands) as transistor Q_0 and five transistors BFT92W (NXP Semiconductors, Eindhoven, Netherlands) in the CM, i.e., $n = 5$. We selected the following component values: $R_a = 0.5\ \text{k}\Omega$, $R_b = 2\ \text{k}\Omega$, $C_1 = C_2 = 82$ pF, $C_F = 2.2$ pF, $C_0 = 0$, and $L = 330$ nH.

Figure 8 shows the printed circuit board (PCB) assembly of the implemented NDR oscillator.

Figure 8. Photograph of the NDR oscillator printed circuit board assembly.

Figure 9 shows the measured current-voltage characteristics of the implemented NDR oscillator. The measured values of the threshold voltages in the NDR region are $V_\beta^* = 3$ V and $V_\gamma^* = 12$ V. The calculation of the theoretical voltage thresholds by Equations (22) and (24) gives $V_\beta = 2.93$ V and $V_\gamma = 11.55$ V. As can be seen, there is a perfect agreement between the measured and theoretical results. The dc operating point has the following coordinates: $\overline{V}_{xy} = 5.2$ V and $\overline{I}_0 = 7.6$ mA.

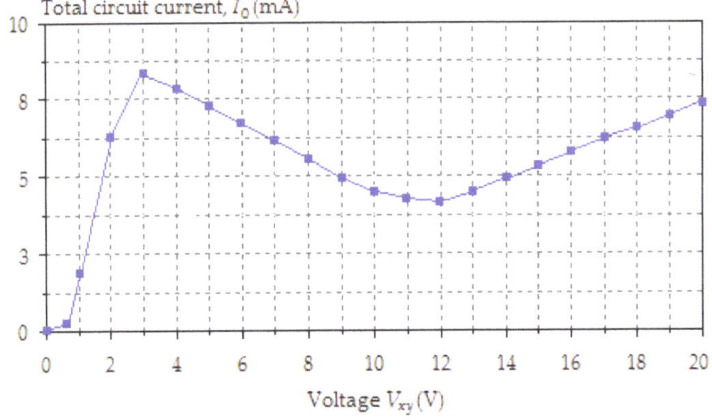

Figure 9. Measured current-voltage characteristics of the implemented NDR oscillator.

Figure 10 shows the photographs of the output voltage (a) and output power spectrum (b) of the implemented oscillator. We used the HMO1002 oscilloscope (Rohde & Swartz, Munich, Germany) and the HMS3000 spectrum analyzer (Rohde & Swartz, Munich, Germany) to measure the oscillator's output voltage in the time and frequency domain. To connect the oscillator output to oscilloscope and spectrum analyzer, we used, respectively, RF probes HZ154 (Rohde & Swartz, Munich, Germany) and P-20A (Auburn Technology Corporation, Wichita, Kansas, USA) with 20 dB attenuation. We can see from Figure 10 that the frequency and the peak-to-peak amplitude of oscillations are 16.1 MHz and

4.12 V, respectively. We can also observe in Figure 10b that the noise-floor power level is more than 75 dB below the fundamental harmonic power.

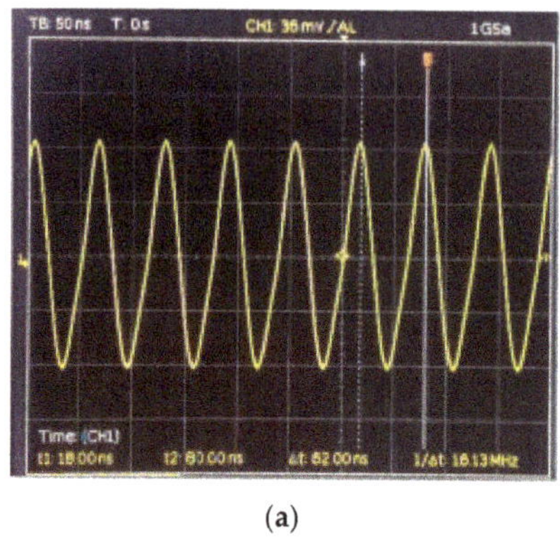
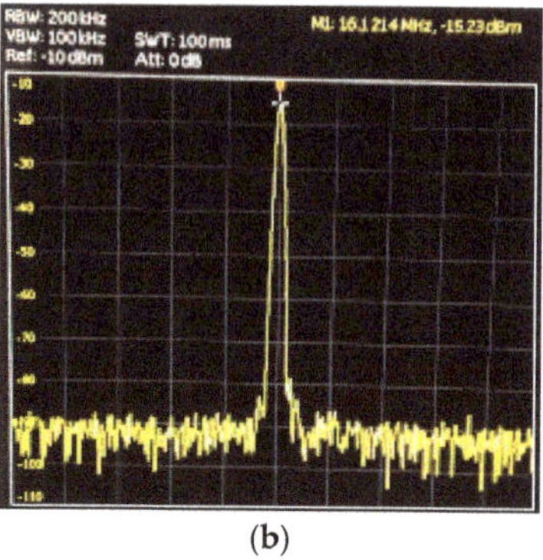

(a)   (b)

Figure 10. (**a**) The oscillogram of the oscillator output voltage; (**b**) The measured spectrum of the oscillator output power with span of 20 MHz.

3.2. LC Voltage Controlled Oscillator

Voltage-controlled oscillators are fundamental building units in modern phase-locked loop synthesizers used in communication and navigation transceivers [38–40]. The NDR circuit of Figure 1a can also be used to design an LC VCO. Figure 11 shows the proposed LC VCO with a controllable slope of the NDR region. The varactor diodes VC_1 and VC_2 replace the capacitors C_1 and C_2 in the circuit of Figure 3. The control voltage V_c is applied to cathodes of varactor diodes VC_1 and VC_2 providing a frequency tuning of the VCO. Resistor R_c isolates the variable power supply from the VCO tank circuit.

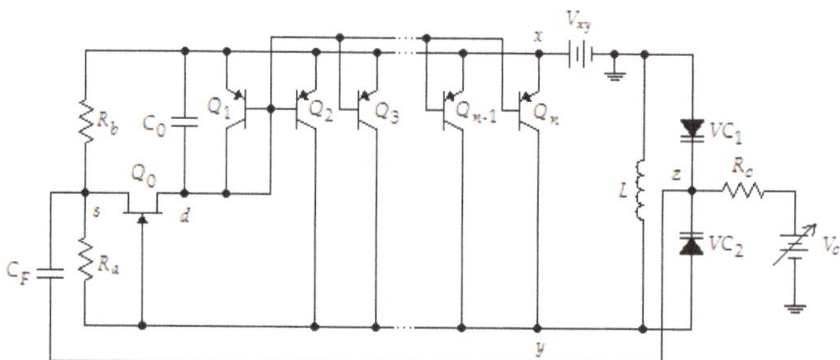

Figure 11. LC voltage-controlled oscillator with NDR.

The SPICE simulation of the VCO circuit with the help of Multisim (ed. 14.1) was conducted using varactor diodes ZC820 (Zetex) and the same transistors, inductor L, and resistors R_a and R_b as in Section 3.1.1. We set the VCO circuit elements C_0, C_F, and R_c to 10 μF, 10 pF and 10 kΩ, respectively. The control voltage V_c was varied from 1 to 25 V. Figures 12 and 13 show the VCO starting voltage waveform (a) and steady-state voltage waveform (b) when $V_c = 1$ V and $V_c = 25$ V, respectively. The oscillation frequency varied from 775 MHz (at $V_c = 1$ V) to 1.375 GHz (at $V_c = 25$ V). The THD is 4.6% at $V_c = 1$ V and 3.8% at $V_c = 25$ V.

From a comparison of voltage starting waveforms in Figures 12a and 13a, we can observe that voltage oscillations reach the steady-state mode at 160 ns and 80 ns, respectively. In the steady state operation mode, the voltage amplitude is around 2 V over the entire control voltage range.

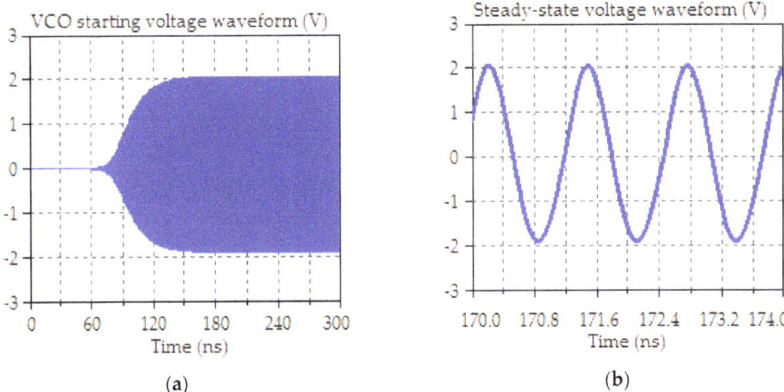

Figure 12. (**a**) Voltage-controlled oscillators (VCO) starting voltage waveform when $V_c = 1$ V; (**b**) VCO steady-state voltage when $V_c = 1$ V.

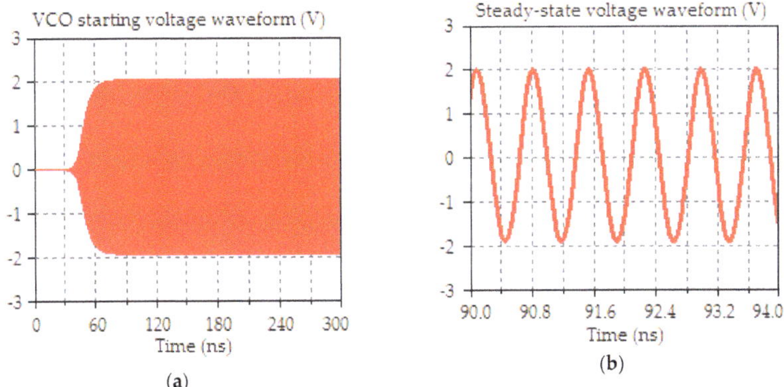

Figure 13. (**a**) VCO starting voltage waveform when $V_c = 25$ V; (**b**) VCO steady-state voltage when $V_c = 25$ V.

Figure 14 shows the tuning characteristic of the VCO. As can be seen in Figure 14, the simulated VCO covers a wide frequency range. The ratio of the maximum VCO frequency to minimum exceeds 1.75.

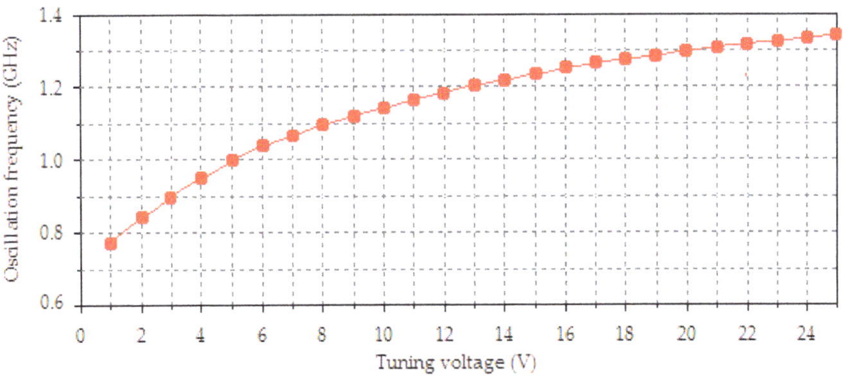

Figure 14. VCO tuning characteristic.

LC Voltage-Controlled Oscillator Performance

Let us compare the overall performance of the proposed NDR VCO with state of the art VCOs. The conventional FOM is used to evaluate the overall performance of the designed VCO, which includes phase noise at a particular frequency offset from the carrier $PN(\Delta f)$, power dissipation P_d, and the ratio of the carrier frequency f_c to the frequency offset Δf for comparing VCOs operating at different frequencies [41,42].

$$FOM = PN(\Delta f) - 20\log\left(\frac{f_c}{\Delta f}\right) + 10\log\left(\frac{P_d}{1\text{ mW}}\right). \tag{30}$$

Table 3 presents the part numbers of the VCO elements. In the simulation, we used the VCO circuit of Figure 11 when $n = 2$. The simulated values of the NDR threshold voltages are as follows: $V_\beta = 2.84$ V and $V_\gamma = 20.7$ V. The calculated thresholds are $V_\beta = 2.84$ V and $V_\gamma = 20.7$ V. The selected DC operating point has the following coordinates: $\overline{V}_{xy} = 6.5$ V and $\overline{I}_0 = 8.5$ mA. Thus, the VCO power dissipation is 55.25 mW. The control voltage V_c applied to the cathodes of the SMV1104-34 varactors varied from 2 V to 6 V. The frequency tuning range is from 1.225 GHz to 1.620 GHz. Figure 15 shows the dependence of the VCO phase noise versus Δf when $V_c = 2$ V. As can be seen in Figure 15, the use of large C_0 reduces phase noise for more than 20 dB in all range of offset frequencies. The low level of the phase noise is also due to the use of a high-Q coil L [43]. In the tuning VCO range, the inductor quality factor is varied from 50 to 60.

Table 3. Part numbers used in the designed voltage-controlled oscillator.

Circuit Elements	Part Numbers
Transistor Q_0	NE722S01
Transistors $\overline{Q_1, Q_n}$	MRF5211LT1
Inductor L	0201DS-3N3XJEU
Capacitor C_0	C1608X5R1E105K
Capacitor C_F	C0603C0G1E030C
Varactors	SMV1104-34
Resistor R_a	ERJ1GEJ471
Resistor R_b	ERJ2GEJ392

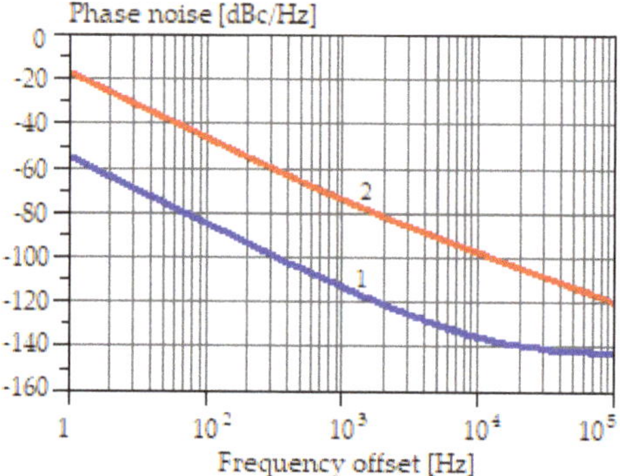

Figure 15. Phase noise versus offset frequency when $C_0 = 1\ \mu$F (curve 1) and $C_0 = 1$ nF (curve 2).

Table 4 shows a comparison of the designed VCO with the VCOs in recently published studies in terms of the FOM (30). As can be seen in Table 4, the designed NDR VCO has one of the best FOM.

It should be noted that all previously published VCOs in Table 4, except [58,59], fabricated in CMOS or BiCMOS technologies.

Table 4. Performance of designed VCO and some recently published VCOs.

VCO	Frequency GHz	Frequency Offset MHz	Phase Noise dBc/Hz	Power Dissipation mW	FOM dBc/Hz
[44]	1.61	0.1	−121	2.7	−202
[45]	2.5	1	−119.7	0.515	−190.3
[46]	11.58	1	−112.62	6	−198.6
[47]	8	1	−134.3	6.6	−204
[48]	2.7	0.1	−121.3	3.9	−204
[49]	3.6	1	−124	2.05	−192
[50]	15.57	1	−116.6	6	−192.7
[51]	2.4	1	−120	0.267	−193.3
[52]	2.4	1	−135.6	6.17	−195.3
[53]	12.67	1	−120.6	17.7	−190
[54]	1.94	1	−153	20	−205.7
[55]	2.38	3	−132.7	1	−190.7
[56]	2.4	1	−124	2.86	−187.25
[57]	7	1	−132	198	−185.9
[58]	7.9	1	−135	1456	−181.3
This work	1.225	0.1	−141.1	55.25	−205.4

Oscillators manufactured using MESFET, HEMT, and PHEMT have significantly lower FOM values due to substantially higher power consumption [58,59]. However, as follows from Table 4, the proposed NDR VCO can even compete with the best CMOS oscillators due to the low level of phase noise and despite the significantly higher power consumption.

4. Discussion and Conclusions

There is a large number of electronic circuits [1–22], in which the current-voltage characteristics have an NDR region. The basis of these circuits is formed by various combinations of BJT and FET. Conventionally, the circuits with NDR can be classified into the following groups: circuits with MOS transistors [7,8,16–18,21], circuits with BJT transistors [10,14], circuits with BJT and JFET [11–15], circuits with JFET and MOS transistors [12,22], circuits with BJT and MOS transistors [9,20], and circuits with BJT and HBT [19]. It should be noted that in the known NDR devices and circuits it is practically impossible to change the angle of inclination of the current-voltage characteristics in the area of negative resistance. Therefore, it is not possible to increase the NDR of the device or circuit without changing the type or the size of the transistors.

This paper proposes a new NDR circuit based on the connection of a FET and a BJT simple current mirror with multiple outputs. A JFET, MESFET, HEMT, or PHEMT can be used as a FET. A distinctive feature of this circuit is the ability to control the NDR without changing the types or the size of transistors. This feature is based on the property of a simple current mirror to increase the current gain due to the parallel connection of transistors at the output of the mirror [59]. In the proposed circuit, the current-voltage characteristics are of the N-type with three threshold voltages. General mathematical equations for calculating the threshold voltages and currents have been derived. Since the threshold current related to the beginning of the NDR region depends on the FET drain current, a mathematical modeling of this current has been performed for a JFET and a gallium-arsenide FET, such as MESFET,

HEMT, or PHEMT. A comparison of the calculated voltage and current thresholds with the SPICE simulations showed perfect convergence as for the case of using a JFET as well as for PHEMT, which was modeled by the Statz nonlinear model. The latter indicates the adequacy of mathematical expressions derived for the calculation of current and voltage thresholds.

The proposed NDR circuit can be used to design various oscillators. By connecting a parallel LC tank to the output of the proposed NDR circuit, one can get a sinusoidal oscillator, which can operate in different frequency bands. When using an ultra-high frequency JFET, the maximum frequency is limited to several hundred MHz. When using a gallium-arsenide transistor such as MESFET, HEMT, or PHEMT, the maximum oscillation frequency lies in the region of several GHz. Self-excitation of the oscillator by the proposed NDR circuit depends on the magnitude of the negative resistance introduced into the parallel LC tank circuit. If the value of the introduced negative resistance is sufficient to compensate for losses in the tank circuit, then the amplitude of oscillations increases and reaches a steady-state value. However, if the magnitude of the introduced negative resistance is insufficient, the oscillator does not self-excite. In any other NDR oscillator, in this case, it is necessary to change the transistors or their sizes for increasing negative resistance. However, in the proposed NDR circuit, it is enough to add one or more transistors in the current mirror as shown in Figure 1a and in this case, according to formula (18), the NDR will increase, and hence the absolute value of the negative resistance introduced into the tank circuit will also increase. Then the oscillator will oscillate. A SPICE simulation of the LC oscillator with a PHEMT and a BJT current-mirror at the frequency of 1.096 GHz showed that the generated signal has a low level of distortion, as well as a low noise level when using additional capacitances in the positive feedback circuit and between the drain of the PHEMT and ground. The implemented LC oscillator prototype operating in the high-frequency band has confirmed theoretical results. The proposed NDR circuit can also be used to design a VCO. Depending on the transistors used, such VCO can operate in different frequency bands, ranging from high-frequencies and up to microwaves. Thus, the simulated VCO circuit with PHEMT covers the frequency range from 775 MHz to 1.375 GHz, i.e., the frequency overlap ratio is higher than 1.75. Moreover, the amplitude of oscillation is about 2 V and practically does not change in the whole range of tunable frequencies. A comparison of the performance characteristics of the designed VCO with VCOs in previously published studies has shown that it is about 20 dB more efficient than the HEMT VCOs and is not inferior to the best CMOS VCOs.

The proposed NDR circuit can also be used in laboratory works in the electronics departments of universities to study the properties of negative resistance, to model various oscillators and to analyze the conditions for self-excitation of oscillators.

Our future work will include an analysis of the use of various bipolar and MOS current mirrors instead of the simple current mirror in the proposed NDR circuit.

Author Contributions: This article presents the collective work of all authors. The first two authors (V.U. and A.R.) jointly participated in the conceptualization of the problem, development of mathematical models, simulation of the developed circuits, numerical calculations, and writing the article. The third author (H.O.) developed a printed circuit board of the oscillator prototype, soldered electronic components, and made the necessary measurements.

Funding: This research received no external funding.

Conflicts of Interest: The authors declare no conflict of interest.

Abbreviations

The following abbreviations exist in the manuscript:

BJT	Bipolar junction transistor
CM	Current mirror
CMOS	Complementary metal-oxide-semiconductor
FET	Field-effect transistor
FOM	Figure of merit
HBT	Heterojunction bipolar transistor
HEMT	High-electron-mobility transistor
JFET	Junction gate field-effect transistor
KCL	Kirchhoff's current law
KVL	Kirchhoff's voltage law
MESFET	Metal-semiconductor field-effect transistor
MOS	Metal-oxide-semiconductor
MOSFET	Metal-oxide-semiconductor field-effect transistor
NDR	Negative differential resistance
NENS	N-channel enhancement mode with load operated at saturation
PCB	Printed circuit board
PHEMT	Pseudomorphic high-electron-mobility transistor
PNP	Positive-negative-positive
RF	Radio frequency
SPICE	Simulation program with integrated circuit emphasis
THD	Total harmonic distortion
VCO	Voltage controlled oscillator

References

1. Tseng, P.; Chen, C.H.; Hsu, S.A.; Hsueh, W.J. Large negative differential resistance in graphene nanoribbon superlattices. *Phys. Lett. A* **2018**, *382*, 1427–1431. [CrossRef]
2. Hwu, R.J.; Djuandi, A.; Lee, S.C. Negative differential resistance (NDR) frequency conversion with gain. *IEEE Trans. Microw. Theory Tech.* **1993**, *41*, 890–893. [CrossRef]
3. Liang, D.S.; Gan, K.J. New D-Type Flip-Flop Design using negative differential resistance circuits. In Proceedings of the 4th IEEE International Symposium on Electronic Design, Test and Applications, Hong Kong, China, 23–25 January 2008; pp. 1–8.
4. Chen, S.L.; Griffin, P.B.; Plummer, J.D. Negative differential resistance circuit design and memory applications. *IEEE Trans. Electron Devices* **2009**, *56*, 634–640. [CrossRef]
5. Wang, S.; Pan, A.; Grezes, C.; Amiri, P.K.; Wang, K.L.; Chui, C.O.; Gupta, P. Leveraging nMOS negative differential resistance for low power, high-reliability magnetic memory. *IEEE Trans. Electron Devices* **2017**, *64*, 4084–4090. [CrossRef]
6. Wu, Y.; Farmer, D.B.; Zhu, W.; Han, S.J.; Dimitrakopoulos, C.D.; Bol, A.A. Three-terminal graphene negative differential resistance devices. *ACS Nano* **2012**, *6*, 2610–2616. [CrossRef]
7. Gan, K.J.; Hsiao, C.C.; Tsai, C.S.; Chen, Y.H.; Wane, S.Y.; Kuo, S.H. A novel voltage-controlled oscillator design by MOS-NDR devices and circuits. In Proceedings of the Fifth International Workshop on System-on-Chip for Real-Time Applications, Banff, AB, Canada, 20–24 July 2005; pp. 372–375.
8. Tsai, C.S.; Hsiao, C.C.; Gan, K.J.; Wu, J.M.; Hsieh, M.Y.; Liao, C.C. An Oscillator Design Based on MOS-NDR Inverter. In Proceedings of the International Conference on Systems and Signals, Kaohsiung, Taiwan, 28–29 April 2005; pp. 1–5.
9. Semenov, A. Mathematical model of the microelectronic oscillator based on the BJT-MOSFET structure with negative differential resistance. In Proceedings of the 2017 IEEE 37th International Conference on Electronics and Nanotechnology (ELNANO), Kiev, Ukraine, 18–20 April 2017; pp. 146–151.
10. Gan, K.J.; Chun, K.Y.; Yeh, W.K.; Chen, Y.H.; Wang, W.S. Design of dynamic frequency divider using negative differential resistance circuit. *Int. J. Recent Innov. Trends Comput. Commun.* **2015**, *3*, 5224–5228.
11. Stanley, I.W.; Ager, D.J. Two-terminal negative dynamic resistance. *Electronic Lett.* **1970**, *6*, 1–2. [CrossRef]
12. Ulansky, V.V.; Ben Suleiman, S.F. Negative differential resistance based voltage-controlled oscillator for VHF band. In Proceedings of the 2013 IEEE International Scientific Conference on Electronics and Nanotechnology, Kiev, Ukraine, 16–19 April 2013; pp. 80–84.

13. Kumar, U. Simulation of a novel bipolar-FET type-S negative resistance circuit. *Act. Passiv. Electron. Compon.* **2003**, *26*, 129–132. [CrossRef]

14. Chua, L.O.; Yu, J.; Yu, Y. Bipolar-JFET-MOSFET negative resistance devices. *IEEE Trans. Circuits Syst.* **1985**, *32*, 46–61. [CrossRef]

15. Ulansky, V.V.; Ben Suleiman, S.F.; Elsherif, H.M.; Abusaid, M.F. Optimization of NDR VCOs for microwave applications. In Proceedings of the 2016 IEEE 36th International Conference on Electronics and Nanotechnology (ELNANO), Kyiv, Ukraine, 19–21 April 2016; pp. 353–357.

16. Akarvardar, K.; Chen, S.; Vandersand, J.; Blalock, B.; Schrimpf, R.; Prothro, B. Four-gate transistor voltage-controlled negative differential resistance device and related circuit applications. In Proceedings of the 2006 IEEE International SOI Conference, Niagara Falls, NY, USA, 2–5 October 2006; pp. 1–6.

17. Shin, S.; Kim, K.R. Multiple negative differential resistance devices with ultra-high peak-to-valley current ratio for practical multi-valued logic and memory applications. *Jpn. J. Appl. Phys.* **2015**, *54*, 1–7. [CrossRef]

18. Shin, S.; Kim, K.R. Novel five-state latch using double-peak negative differential resistance and standard ternary inverter. *Jpn. J. Appl. Phys.* **2016**, *55*, 1–6. [CrossRef]

19. Gan, K.J.; Tsai, C.S.; Liang, D.S. Design and characterization of the negative differential resistance circuits using the CMOS and BiCMOS process. *Analog Integr. Circuits Signal Process.* **2010**, *62*, 63–68. [CrossRef]

20. Gan, K.J.; Tsai, C.S.; Liang, D.S.; Wen, C.M.; Chen, Y.H. Tri-valued memory circuit using MOS-BJT-NDR circuits fabricated by standard SiGe process. *Jpn. J. Appl. Phys.* **2006**, *45*, 1–4. [CrossRef]

21. Wu, C.Y.; Lai, K.N. Integrated Λ-type differential negative resistance MOSFET device. *IEEE J. Solid-State Circuits* **1979**, *14*, 1094–1101.

22. Ulansky, V. Low phase-noise HEMT microwave voltage-controlled oscillator. In Proceedings of the IEEE Microwaves, Radar and Remote Sensing Symposium (MRRS), Kiev, Ukraine, 25–27 August 2011; pp. 55–58.

23. Jagger, R.C.; Blalock, T.N. *Microelectronic Circuit Design*, 2nd ed.; McGraw-Hill: New York, NY, USA, 2004; pp. 251–254.

24. Statz, H.; Newman, P.; Smith, W.; Pucel, R.A.; Haus, H.A. GaAs FET device and circuit simulation in Spice. *IEEE Trans. Electron Devices* **1987**, *34*, 160–169. [CrossRef]

25. Converting GaAs FET Models for Different Nonlinear Simulators. Available online: https://docplayer.net/21705870-California-eastern-laboratories-an1023-converting-gaas-fet-models-for-different-nonlinear-simulators.html (accessed on 2 April 2003).

26. ATF-33143. Low Noise Pseudomorphic HEMT in a Surface Mount Plastic Package. Data Sheet. Available online: https://cdn.datasheetspdf.com/pdf-down/A/T/F/ATF-33143-AVAGO.pdf (accessed on 6 January 2006).

27. Conan Zhan, J.H.; Maurice, K.; Duster, J.; Kornegay, K.V. Analysis and design of negative impedance LC oscillators using bipolar transistors. *IEEE Trans. Circuits Syst. I Fundam. Theory Appl.* **2003**, *50*, 1461–1464. [CrossRef]

28. Ulansky, V.V.; Fituri, M.S.; Machalin, I.A. Mathematical modeling of voltage-controlled oscillators with the Colpitts and Clapp topology. *Electron. Control Syst.* **2009**, *19*, 82–90.

29. Chung, C.; Chao, S. Robust Colpitts and Hartley oscillator design. In Proceedings of the 2014 IEEE International Frequency Control Symposium (FCS), Taipei, Taiwan, 19–24 May 2014; pp. 1–5.

30. Ulansky, V.V.; Elsherif, H.M. A new method of designing UHF FET Colpitts oscillator. In Proceedings of the 2014 IEEE 34th International Scientific Conference on Electronics and Nanotechnology (ELNANO), Kyiv, Ukraine, 15–18 April 2014; pp. 388–392.

31. Daliri, M.; Maymandi-Nejad, M. Analytical model for CMOS cross-coupled LC-tank oscillator. *IET Circuits Devices Syst.* **2014**, *8*, 1–5. [CrossRef]

32. Hajimiri, A.; Lee, T.H. Design issues in CMOS differential LC oscillators. *IEEE J. Solid-State Circuits* **1999**, *34*, 717–724. [CrossRef]

33. Hou, J.A.; Wang, Y.H. A 5 GHz differential Colpitts CMOS VCO using the bottom PMOS cross-coupled current source. *IEEE Microw. Wirel. Compon. Lett.* **2009**, *19*, 401–403.

34. Grebennikov, A. *RF and Microwave Transistor Oscillator Design*; John Wiley & Sons Ltd.: Chichester, UK, 2007; 458p.

35. Poole, C.; Darwazeh, I. *Microwave Active Circuit Analysis and Design*; Elsevier: Amsterdam, The Netherlands, 2016; 664p.

36. Sze, S.M. *Active Microwave Diodes in Modern Semiconductor Device Physics*; Wiley: New York, NY, USA, 1997; pp. 343–407.

37. Hegazi, E.; Sjoland, H.; Abidi, A.A. A filtering technique to lower LC oscillator phase noise. *IEEE J. Solid-State Circuits* **2001**, *36*, 1921–1930. [CrossRef]

38. Bianchi, G. *Phase-Locked Loop Synthesizer Simulation*; McGraw-Hill: New York, NY, USA, 2005; pp. 64–78.

39. Luong, H.C.; Leung, G.C.T. *Low-Voltage CMOS RF Frequency Synthesizers*; Cambridge University Press: Cambridge, UK, 2004; pp. 28–44.

40. Leenaerts, D.; van der Tang, J. *Circuit Design for RF Transceivers*; Springer: Berlin, Germany, 2001; 323p.

41. Kinget, P. Integrated GHz voltage-controlled oscillators. In *Analog Circuit Design*; Sansen, W., Huijsing, J., van de Plassche, R., Eds.; Springer: Berlin, Germany, 1999; pp. 353–381.

42. Tiebout, M. *Low Power VCO Design in CMOS*; Springer: Berlin, Germany, 2006; 128p.

43. Coilcraft. RF Inductor Comparison Tool. Q vs Frequency. Available online: https://www.coilcraft.com/apps/compare/compare_rf.cfm (accessed on 11 October 2018).

44. Cai, H.L.; Yang, Y.; Qi, N. A 2.7-mW 1.36–1.86-GHz LC-VCO with a FOM of 202 dBc/Hz enabled by a 26%-size-reduced nano-particle-magnetic-enhanced inductor. *IEEE Trans. Microw. Theory Tech.* **2014**, *62*, 1221–1228. [CrossRef]

45. Rout, P.K.; Nanda, U.K.; Acharya, D.P.; Panda, G. Design of LC VCO for optimal figure of merit performance using CMODE. In Proceedings of the 1st International Conference on Recent Advances in Information Technology (RAIT), Dhanbad, India, 15–17 March 2012; pp. 1–6.

46. Kim, S.J.; Seo, D.I.; Kim, J.S. Compact CMOS LiT VCO achieving 198.6 dBc/Hz FoM. *Electron. Lett.* **2018**, *54*, 175–177. [CrossRef]

47. Zailer, E.; Belostotski, L.; Plume, R. 8-GHz, 6.6-mW LC-VCO with small die area and FOM of 204 dBc/Hz at 1-MHz offset. *IEEE Microw. Wirel. Compon. Lett.* **2016**, *26*, 936–938. [CrossRef]

48. Chung, T.W.; Huang, T.C.; Chung, S. A 2.7GHz 3.9mW mesh-BJT LC-VCO with −204dBc/Hz FOM in 65nm CMOS. In Proceedings of the IEEE Custom Integrated Circuits Conference, San Jose, CA, USA, 9–12 September 2012; pp. 1–6.

49. Narayanan, A.T.; Kimura, K.; Deng, W. A pulse-driven LC-VCO with a figure-of-merit of −192dBc/Hz. In Proceedings of the 40th European Solid-State Circuits Conference (ESSCIRC), Venice Lido, Italy, 22–26 September 2014; pp. 1–5.

50. Lin, Y.C.; Yeh, M.L.; Chang, C.C. A high figure-of-merit low phase noise 15-GHz CMOS VCO. *J. Mar. Sci. Technol.* **2013**, *21*, 82–86.

51. Ghorbel, I.; Haddad, F.; Rahajandraibe, W. Optimization of voltage-controlled oscillator VCO using current-reuse technique. In Proceedings of the 26th International Conference on Microelectronics (ICM), Doha, Qatar, 14–17 December 2014; pp. 1–5.

52. Bhat, M.V.; Jain, S.; Srivatsa, M.P. Design of low phase noise voltage-controlled oscillator for phase locked loop. In Proceedings of the International Conference on Microelectronic Devices, Circuits and Systems (ICMDCS), Vellore, India, 10–12 August 2017; pp. 1–5.

53. Mirajkar, P.; Chand, J.; Aniruddhan, S.; Theertham, S. Low phase noise Ku-band VCO with optimal switched-capacitor bank design. *IEEE Trans. Very Larg. Scale Integr. (VLSI) Syst.* **2018**, *26*, 589–593. [CrossRef]

54. Zuo, C.; der Spiegel, J.V.; Piazza, G. Dual-mode resonator and switchless reconfigurable oscillator based on piezoelectric AlN MEMS technology. *IEEE Trans. Electron Devices* **2011**, *58*, 3599–3603. [CrossRef]

55. Rottava, R.E.; Camara Santes Junior, C.; Rangel de Sousa, F.; Nunes de Lima, R. Ultra-low-power 2.4 GHz Colpitts oscillator based on double feedback technique. In Proceedings of the IEEE International Symposium on Circuits and Systems, Beijing, China, 19–23 May 2013; pp. 1785–1788.

56. Sachan, D.; Kumar, H.; Goswami, M.; Misra, P.K. A 2.4 GHz low power low phase-noise enhanced FOM VCO for RF applications using 180 nm CMOS technology. *Wirel. Pers. Commun.* **2018**, *101*, 391–403. [CrossRef]

57. Thi Do, T.N.; Szhau Lai, S.; Horberg, M. A MMIC GaN HEMT voltage-controlled-oscillator with high tuning linearity and low phase noise. In Proceedings of the IEEE Compound Semiconductor Integrated Circuit Symposium (CSICS), New Orleans, LA, USA, 11–14 October 2015; pp. 1–4.

58. Liu, H.; Zhu, X.; Boon, C.C. Design of ultra-low phase noise and high power integrated oscillator in 0.25 μm GaN-on-SiC HEMT technology. *IEEE Microw. Wirel. Compon. Lett.* **2014**, *24*, 120–122. [CrossRef]

59. Gray, P.R.; Hurst, P.J.; Lewis, S.H.; Meyer, R.G. *Analysis and Design of Analog Integrated Circuits*; John Wiley & Sons, Inc.: New York, NY, USA, 2001; 875p.

Article

Low Cost Test Pattern Generation in Scan-Based BIST Schemes

Guohe Zhang [1], Ye Yuan [1], Feng Liang [1], Sufen Wei [1,2] and Cheng-Fu Yang [3,*

[1] School of Microelectronics, Xi'an Jiaotong University, Xi'an 710049, China; zhangguohe@xjtu.edu.cn (G.Z.); yy492138848@stu.xjtu.edu.cn (Y.Y.); fengliang@xjtu.edu.cn (F.L.); weisufen@jmu.edu.cn (S.W.)

[2] School of Information and Engineering, Jimei University, Fujian 361021, China

[3] Department of Chemical and Materials Engineering, National University of Kaohsiung, No. 700, Kaohsiung University Rd., Nan-Tzu District, Kaohsiung 811, Taiwan

* Correspondence: cfyang@nuk.edu.tw

Received: 26 January 2019; Accepted: 8 March 2019; Published: 12 March 2019

Abstract: This paper proposes a low-cost test pattern generator for scan-based built-in self-test (BIST) schemes. Our method generates broadcast-based multiple single input change (BMSIC) vectors to fill more scan chains. The proposed algorithm, BMSIC-TPG, is based on our previous work multiple single-input change (MSIC)-TPG. The broadcast circuit expends MSIC vectors, so that the hardware overhead of the test pattern generation circuit is reduced. Simulation results with ISCAS'89 benchmarks and a comparison with the MSIC-TPG circuit show that the proposed BMSIC-TPG reduces the circuit hardware overhead about 50% with ensuring of low power consumption and high fault coverage.

Keywords: test pattern generation; built-in self-test; broadcast circuit; low cost

---

1. Introduction

Nowadays VLSI testing is always used to ensure the correctness and reliability of the finished chip [1], but we encountered some problems during VLSI testing. In the process of chip testing, the test power consumption is two to four times greater compared with the normal power consumption [2,3]. This excessive power consumption will limit the stability of the circuit and it will also increase the cost of packaging [4]. In consideration of the economics of design for testability, we need to balance the cost and interest [5]. Therefore, this paper aims to find a low-cost test pattern generation method based on our previous work multiple single-input change (MSIC)-TPG [6].

The built-in self-test (BIST) method can effectively reduce the difficulty and complexity of VLSI testing. The BIST technology can be roughly divided into two categories: logic BIST (LBIST) and memory BIST (MBIST) [7,8]. The test pattern generation method proposed in this paper is based on the LBIST method. The traditional LBIST technology is based on the pseudo-random test patterns generated by the linear feedback shift register (LFSR) [9]. This will lead to a large test power consumption during the test. To solve this problem, a method of MSIC test pattern generation combining a pseudo-random sequence with a low-transition sequence has been proposed in paper [6]. It can consider both high fault coverage and low power consumption [10,11]. However, this method increases the area of circuits. To overcome the limitations, a novel low-cost BIST architecture using test pattern broadcast circuits called broadcast-based multiple single input charge (BMSIC)-TPG has been developed. This method reduces the area overhead, and it scores well in power consumption and fault coverage.

The rest of this paper is organized as follows. In Section 2, the proposed BMSIC-TPG scheme is presented. The mathematical features of the new BMSIC sequences is described in Section 3. In Section 4, the performance of the BMSIC sequences are analyzed. Conclusions are given in Section 5.

2. BMSIC-TPG Structure

2.1. BMSIC-TPG

The BMSIC-TPG structure includes a clock control module, an original scan chain generation module, and a broadcast module [12–15], as shown in Figure 1. The square A is the clock control module, which is used to generate a slow clock (CLK1) and a fast clock (CLK2). CLK1 is used to drive LFSR to update seed vectors [16], and CLK2 is used to drive reconstructed Johnson counter to update Johnson vectors and generate the vector J [17]. The square B is the original scan chain generation module, which is composed of the LFSR, the reconfigurable Johnson counter, and the exclusive OR (XOR) network [18]. The LFSR is used to generate the seed vector. The reconfigurable Johnson counter is used to generate the Johnson vector, and the XOR network generates the original scan chain vector by bitwise XOR operation of the seed vector S and the vector J. The square C is the broadcast module, which will broadcast m original scan chains to 4m broadcast scan chains [14]. The specific circuit is shown in Figure 3.

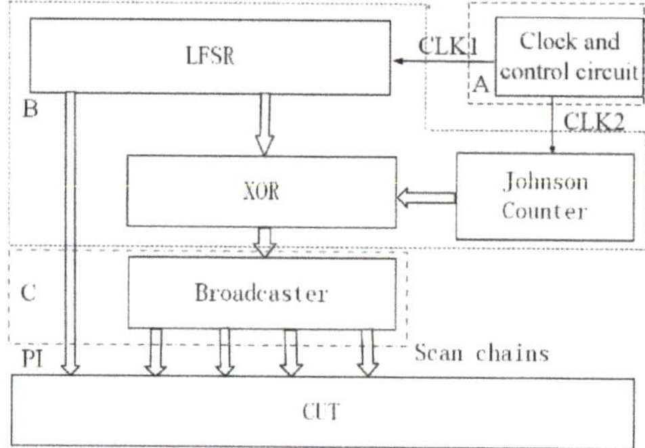

Figure 1. Broadcast-based multiple single input charge (BMSIC)-TPG structure.

Suppose that there are m scan chains before broadcast and M scanning chains after broadcast, and each scanning chain has l scanning cells in a full scan design. The bits of seed vector S generated by LFSR is m. The bits of the vector J generated by reconfigurable Johnson counter is L.

Through many mathematical analyses and experiments, we know the solution to optimize the configuration of BMSIC-TPG is $L = l$, which is called the test convention constraint. Due to test pattern generation algorithm, the test pattern generator is constrained to have $L \geq m$, which is called the test generation constraint, and must be satisfied compared with the test convention constraint. The test convention constraint is an optimal to generate the test pattern under the premise of satisfying the test generation constraint. Obviously, the two constraints are satisfied if $L \geq m$. The configuration is the optimal configuration if $L = l$. If $l < m$, it just satisfies the test generation constraint $L \geq m$ (in this article $L = m$). According to these constraints and the broadcast of test patterns we studied, we can find $M = 4m$, the bits of the seed vector is m, the bits of the Johnson vector is $L(L = l)$, this is the premise of the follow-up contents. The above analysis result is also easy to understand. Under the premise that $L \geq m$, the filling of the scan chain is realized by cyclic shifting of the updated current Johnson vector. The period of the cyclic shift is L. If $L < l$, the cyclic shift of the Johnson vector needs several shifting cycles, and the filling value will correspondingly appear repetitive parts. So the possibility of transition between adjacent bits of vectors generated after encoding will increase. If $L = l$, the cyclic shift of the Johnson vector is exactly a shift period, and the filled value is just all the bits of the Johnson vector. So this configuration can ensure low possibility of transition. If $L > l$, the cyclic shift of the Johnson vector is less than one shift period, and the filling value is part of the bits of Johnson vector. Thus, the possibility of transition will decrease. But this configuration will increase the area of the circuits. In this paper, the bits of seed vector and Johnson vector are always equal which is called the same scan configuration.

2.2. LFSR Structure and Johnson Counter Structure

The LFSR is composed of multiple shift registers and XOR gates connected in a certain way. The m-bit linear feedback shift register can generate $(2^m - 1)$ different states at most [19]. If the m-bit linear feedback shift register generates $(2^m - 1)$ different states and begins to repeat periodically, the $(2^m - 1)$ different states of the sequence is called the maximum length sequence, which is also known as the M sequence.

Because the number of transitions between adjacent bits of the test vector is positively correlated with the power consumption of the test [20], the Johnson counter can generate Johnson vectors that has low transition properties between adjacent vectors and adjacent bits of the same vector. The Johnson sequence is a single input change sequence (SIC). The vector generated by the next clock in the sequence is a one-bit change from the previous clock generation vector. Johnson sequences consist of a series of "0" and a series of "1". So we choose a Johnson counter to reduce power consumption. But a simple Johnson counter can not complete the data shift loading process. We reconstruct the Johnson counter according to the test pattern generation method.

The L-bit reconfigurable Johnson counter is shown in Figure 2. When the mode is set to one, the counter implements the counting function. Under this mode, the initialization of all flip-flops will be completed after running L clocks if the Rst signal is set to zero. If the Rst signal is set to one, the counter implements the normal counting function. When the mode is set to zero, the counter implements a shift function and feeds the last bit of the counted vector back to the first bit. The adjacent bit of each Johnson vector jumps to zero or one, so the sequence generated by the reconfigurable Johnson counter still holds the single-hop characteristic.

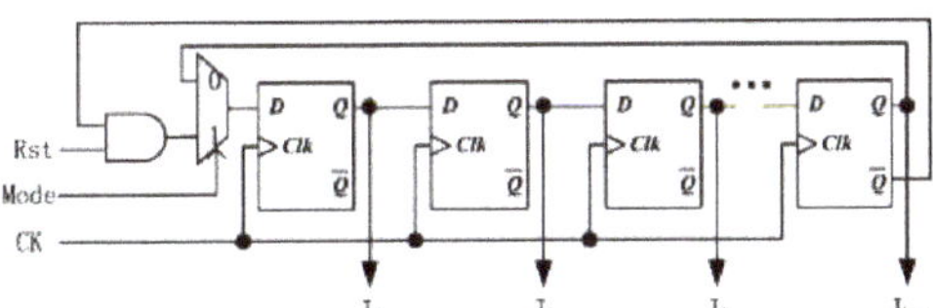

Figure 2. L-bit reconfigurable Johnson counter structure.

2.3. XOR Network

The XOR network generates the original scan chain vector by bitwise XOR operation of the seed vector S which is generated by the LFSR and the vector J which is generated by the reconfigurable Johnson counter. The LFSR generates an m-bit seed vector $S = [S_0, S_1, S_2, \ldots, S_{m-1}]$. The reconfigurable Johnson counter generates a L-bit vector $J = [J_0, J_1, J_2, \ldots, J_{L-1}]$. The result of the bit-wise XOR operation is $X = [X_1, X_{L+1}, X_{2L+1}, \ldots, X_{(m-1)L+1}]$.

2.4. Broadcaster

A broadcaster [12–14] distributes test patterns from a MSIC-TPG [6] module to fill multiple scan chains in a minimally constrained manner. The specific structure is shown in Figure 3. The broadcast circuit extends the original scan chains from two to eight. S_1 and S_2 are original scan chains. B_0 and B_1 are broadcast vectors which are generated by the two-bit LFSR. The post-broadcast seed vectors $S_{11}, S_{12}, S_{13}, S_{14}$ and $S_{21}, S_{22}, S_{23}, S_{24}$ to be applied to the scan chains are generated by bit-XOR the original scan chains S_1 and S_2 and the broadcast vectors B_0 and B_1. Suppose the number of original scan chains is m, so the number of seed vectors after broadcast is M and $M = 4m$.

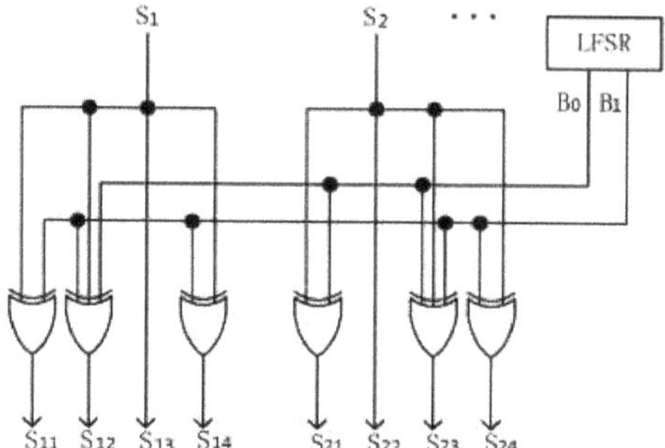

Figure 3. Broadcast circuit.

2.5. The Process of BMSIC-TPG

What follows are the operation mode of the BMSIC-TPG.

1. Clock control module generates CLK1 and CLK2. CLK1 drives the LFSR to update the seed vector, and CLK2 drives the reconfigurable Johnson counter to update the J vector and enables the scan to move in.
2. Original scan chain generation module is made up of the LFSR, reconfigurable Johnson counter, and XOR network. The LFSR generates the S vector. The reconfigurable Johnson counter generates the J vector. The XOR network operates the bit-XOR between the S vector and the J vector to generate the original scan chain data.
3. Broadcast module is used to extend the original scan chain.

3. BMSIC-TPG Mathematical Features

3.1. Periodicity

Since the seed vectors, broadcast vectors, and original scan chain vectors before the broadcast are all periodic and the XOR operation is a linear operation, the BMSIC test pattern is also assumed to have periodic characteristics [6]. Suppose the seed vector is $S = [S_0, S_1, S_2, \ldots, S_{m-1}]$. The vector J is $J = [J_0, J_1, J_2, \ldots, J_{L-1}]$, and the broadcast vector B is $B = [B_0, B_1]$. Then at time t, S, J, and B can be expressed as:

$$\begin{aligned}
\mathbf{S}(t, x) &= S_0(t)x^0 + S_1(t)x^1 + \ldots + S_m(t)x^m \\
\mathbf{J}(t, x) &= J_0(t)x^0 + J_1(t)x^1 + \ldots + J_{L-1}(t)x^{L-1} \\
\mathbf{B}(t, x) &= B_0(t)x^0 + B_1(t)x^1.
\end{aligned} \tag{1}$$

According to the generation algorithm, the original input test vector consists of parts or all bits of the seed vector or multiplexing of seed vectors, which can be expressed as:

$$\begin{aligned}
\mathbf{V}_{in}(t, x) &= S_0(t)x^0 + S_1(t)x^1 + \ldots + S_{m-1}(t)x^{m-1} + \\
&\quad S_0(t)x^m + S_1(t)x^{m+1} + \ldots + S_{m-1}(t)x^{2m-1} + \ldots + S_h(t)x^{N-1}.
\end{aligned} \tag{2}$$

In Equation (2), $1 \leq h \leq m$ and h is an integer. The specific value depends on the number of the original inputs N and the number of seed vectors m. At the same time, the k-th original scan chain vector can be expressed as:

$$\mathbf{C}_k(t, x) = [\sum_{i=0}^{L-1} S_{k-1}(t)x^i \oplus J_k(t, x)]x^{N+(k-1)L}. \tag{3}$$

The vector in Equation (3) represents the J vector applied to the k-th scan chain.

Assume that the two pre-broadcast original scan chain vectors S_1, S_2 of the broadcaster shown in Figure 3 are denoted as $C_q(t, x)$, $C_{q+1}(t, x)$. The test vector of the i-th scan chain after broadcasting is $V_i(t, x)$, then the eight scan chain vectors $S_{11}, S_{12}, S_{13}, S_{14}, S_{21}, S_{22}, S_{23}, S_{24}$ after broadcasting can be expressed as:

$$\mathbf{S}_{11} = V_{4q-3}(t, x) = C_q(t, x) \oplus \sum_{j=1}^{L} B_1(t)x^j$$

$$\mathbf{S}_{12} = V_{4q-2}(t, x) = C_q(t, x) \oplus \sum_{j=1}^{L} [B_0(t) \oplus B_1(t)]x^j$$

$$\mathbf{S}_{13} = V_{4q-1}(t, x) = C_q(t, x)$$

$$\mathbf{S}_{14} = V_{4q-3}(t, x) = C_q(t, x) \oplus \sum_{j=1}^{L} B_0(t)x^j$$

$$\mathbf{S}_{21} = V_{4q+1}(t, x) = C_{q+1}(t, x) \oplus \sum_{j=1}^{L} B_0(t)x^j \tag{4}$$

$$\mathbf{S}_{22} = V_{4q+2}(t, x) = C_{q+1}(t, x)$$

$$\mathbf{S}_{23} = V_{4q+3}(t, x) = C_{q+1}(t, x) \oplus \sum_{j=1}^{L} [B_0(t) \oplus B_1(t)]x^j$$

$$\mathbf{S}_{24} = V_{4q+4}(t, x) = C_{q+1}(t, x) \oplus \sum_{j=1}^{L} B_1(t)x^j.$$

Considering the above, the ω complete scan chain vector loaded into the circuit under test can be expressed as:

$$\mathbf{P}(\omega) = P(t_\omega, x) = V_{in}(t_\omega, x) + \sum_{i=1}^{M} V_i(t_\omega, x). \tag{5}$$

Bit-XOR the ωth test pattern with the dth test pattern can be expressed as:

$$\mathbf{P}(\omega) \oplus \mathbf{P}(d) = V_{in}(t_\omega, x) \oplus V_{in}(t_d, x) + \sum_{i=1}^{M} [V_i(t_\omega, x) \oplus V_i(t_d, x)]. \tag{6}$$

Only if $S(t_\omega, x) = S(t_d, x)$, $B(t_\omega, x) = B(t_d, x)$, and $\sum_{l=1}^{m} C_l(t_\omega, x) = \sum_{l=1}^{m} C_l(t_d, x)$ are established at the same time, then $\mathbf{P}(\omega) \oplus \mathbf{P}(d) = 0$ is established. It is known that the period of seed vector S is $T_S = 2^m - 1$. The period of the broadcast vector B is $T_B = 2^2 - 1 = 3$. It is known from the literature [6] that the period of original scan chain vector S before broadcast is $T_{MSIC} = (2^m - 1)2L$. So the BMSIC test pattern is also periodic, and the period is the least common multiple of the period of seed vector, the broadcast vector and the original scan chain vector. It can be expressed as:

$$\mathbf{T}_{BMSIC} = \begin{cases} (2^m - 1)2L & (T_{MSIC}\%3 = 0) \\ (2^m - 1)6L & (T_{MSIC}\%3 \neq 0). \end{cases} \tag{7}$$

From Equation (7), the period of the BMSIC test pattern is related to the number of bits of the seed vector and the J vector. Under the same configuration, the T_{BMSIC} is larger than the T_{MSIC} [6]. The greater the period of the test pattern is, the better the pseudo-randomness of the test pattern sample is, and the higher the fault coverage is. The number of bits in the seed vector S and J vectors directly affects the hardware overhead. The exponential relationship and multiple relationship of the Equation (7) make it possible to obtain a test pattern with large period and good pseudo-randomness with fewer vector bits, so BMSIC test patterns can reduce hardware overhead on the premise of achieving satisfactory fault coverage.

3.2. Transition

The number of transitions between adjacent bits of the test pattern is positively correlated with the power consumption of the test [20], so it can be used to quantitatively analyze the transition properties of the BMSIC test pattern. We take some test patterns under different scanning configurations as the samples and count the transition numbers for the BMSIC test generation method. We have obtained some statistical laws after our analysis. The results are shown in Table 1, and the "transition period" indicates how many test patterns the transition characteristics will repeat. "Pattern transitions" indicates the total transitions of a single test pattern.

Table 1. Transition of the broadcast-based multiple single input charge (BMSIC)-TPG.

Seed Vector Bits	Johnson Vector Bits	Transition Period	Pattern Numbers	Pattern Transitions
8	8	8	1	0
			7	56
8	17	17	1	0
			8	60
			8	64
8	20	20	1	0
			8	60
			11	64
10	10	10	1	0
			9	90
10	36	36	1	0
			10	76
			25	80
10	44	44	1	0
			10	76
			33	80

We can make a conclusion from Table 1: (1) the transitions of per test pattern repeats with L for the transition period. (2) If $L = m$, the transition of one pattern is zero, and the transition of $L - 1$ patterns is $8(m - 1)$. (3) If $L > m$, the transition of one pattern is zero. The transition of m patterns is $8(m - 1) + 4$, and the transition of $(L - m - 1)$ patterns is $8m$. The above conclusion is derived because the BMSIC test pattern needs to satisfy both the test generation constraints and the test convention constraints. The average transition between the adjacent slices of the BMSIC test pattern is calculated as Equation (8), which is almost equivalent to the average transition of the MSIC [4] test pattern (shown in Equation (9)).

$$C_{BMSIC\_ave} = \begin{cases} \frac{8(m-1)}{L} & (L = m) \\ \frac{8m(L-1)-4m}{L(L-1)} & (L > m). \end{cases} \tag{8}$$

$$C_{MSIC\_ave} = \begin{cases} \frac{2(M-1)}{L} & (L = m) \\ \frac{2M(L-1)-M}{L(L-1)} & (L > m). \end{cases} \tag{9}$$

3.3. Randomness

The random sequence can detect most of the faults in CUT. Therefore, this paper discusses the randomness of the "0", "1" distribution of BMSIC test patterns that have been generated to evaluate its capability to detect faults. The generated test pattern is evaluated from the scanning moving direction and the test pattern direction respectively according to the scanning test scheme and the scanning design technique.

As shown in Figure 4a, the randomness of scan moving direction is to calculate the probability of "0" or "1" of a given scan chain under each test pattern, which reflects the randomness of the same test pattern between its scan units on this scan chain. However, the randomness of scan moving direction does not reflect the randomness of the same scanning unit in the designated scan chain being filled with "0" or "1" under different test patterns, thus introducing the randomness of test pattern direction as shown in Figure 4b. The randomness of test pattern direction is to calculate the probability of "0" or "1" being filled in different test patterns for each scan unit in the specified scan chain, which reflects the difference between test patterns. We take 10,000 BMSIC test patterns and each test pattern has 32 scan chains as the samples and choose one chain to study. Then compare with LFSR and MSIC test patterns in the same configuration to reflect the performance of randomness. Other chains also has the similar result. The probability distribution of logic "0" in the scan moving direction is shown in Figure 5. It can be concluded BMSIC has a large fluctuation in randomness and has periodicity. The probability distribution of logic "0" in the test pattern direction is shown in Figure 6. It can be concluded the randomness of the test sequence arranged from good to bad is MSIC, BMSIC and LFSR, but the distributions are basically between 0.495 and 0.505, all have good randomness. So we consider the BMSIC test patterns have good fault detection capability.

Scanning Moving Direction →

	D0	D1	D2	D3	PO
Test Pattern 1	0	1	0	1	0.50
Test Pattern 2	1	0	1	1	0.33
Test Pattern 3	1	1	0	0	0.50
Test Pattern 4	0	1	1	0	0.50

(a)

Test Pattern Direction

	D0	D1	D2	D3
Test Pattern 1	0	1	0	1
Test Pattern 2	1	0	1	1
Test Pattern 3	1	1	0	0
Test Pattern 4	0	1	1	0
PO	0.50	0.33	0.50	0.50

(b)

Figure 4. (a) Scan moving direction randomness analysis diagram. (b) Test pattern direction randomness analysis diagram.

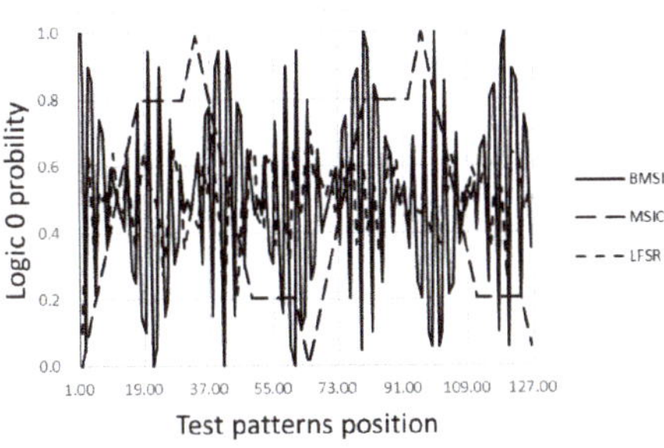

Figure 5. Scan moving direction Logic "0" probability distribution.

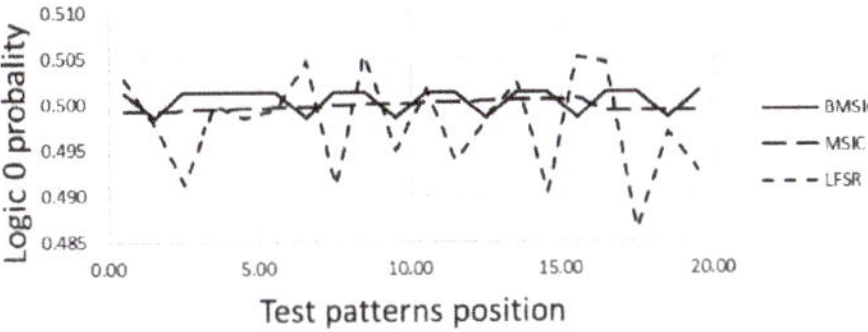

Figure 6. Test pattern direction Logic "0" Probability distribution.

The above analysis shows that BMSIC test patterns are good in randomness. Although it is based on a statistical approximation, it is necessary to evaluate the fault detection capability of test pattern. Therefore, it is speculated that BMSIC test patterns can achieve satisfactory fault coverage.

4. BMSIC-TPG Performance Analysis

BMSIC test patterns had low power consumption and low area overhead and it can achieve satisfactory fault coverage from theoretical analysis. This section verifies its fault coverage, power consumption, and area cost performance through the specific simulation experiments and performance estimates. The circuit under test (CUT) in the experiment are five circuits in ISCAS'89 series, using Nangate 45 nm process library. The synthesis of CUT were carried out with DFT_Compiler of Synopsys. Test generation and test application were carried out with Perl. The fault simulation was carried out with TetraMAX. The power consumption simulation was carried out with the Synopsys Design Analyzer and Prime Power. Because the BMSIC-TPG proposed in this paper was designed to overcome the drawback of the previous method [6], we compare the performance of BMSIC with our previous method [6].

4.1. Fault Simulation

The fault simulation results of the BMSIC test generation method are shown in Table 2. DFF represents the number of scanning units in the circuit under test. Chain represents the number of scan chains. Depth represents the number of scanning units per scan chain. TL represents the number of test patterns, and SFC and TFC represent stuck fault coverage and transition fault coverage respectively.

We used DFT_Compiler of synopsys to synthesize the CUT, Perl to implement test patterns generation algorithm to achieve test generation, test application, and TetraMAX to complete the fault simulation, the results are shown in Table 2. Comparing with the literature [6], BMSIC test program can achieve higher fault coverage under the same configuration. Comparing with the literature [21], we needed less test patterns to achieve high fault coverage. At the same time, we found that the same CUT under different test generation configuration resulted in different fault coverage, indicating that the fault coverage is related to test generation configuration.

Table 2. Comparison of fault coverage of the three test generation methods.

CUT	DFF	Chain	Depth	TL	LFSR SFC	LFSR TFC	MSIC [6] SFC	MSIC [6] TFC	BMSIC SFC	BMSIC TFC
		32	20	10,000	91.44	80.75	90.51	74.3	91.42	74.02
S13207	638	36	18	10,000	92.01	80.33	86.63	71.11	92.6	73.57
		40	16	10,000	95.24	84.05	89.22	70.82	93.52	77.71
		32	17	10,000	93.9	85.55	92.01	78.36	89.45	75.12
S15850	534	36	15	10,000	93.82	85.55	91.23	76.79	91.79	77.01
		40	14	10,000	94.69	86.59	90.11	73.19	93.59	80.60
		40	44	10,000	99.55	97.04	97.34	86.28	99.97	95.99
S35932	1728	48	36	10,000	99.60	96.66	99.94	93.99	99.98	92.04
		56	31	10,000	99.56	96.34	97.77	88.7	99.98	94.53
		40	41	10,000	93.48	83.67	83.69	59.11	84.65	60.4
S38417	1636	48	35	10,000	93.68	83.18	85.33	61.68	85.22	61.58
		56	30	10,000	93.66	82.87	84.34	61.09	83.46	52.74
		40	36	10,000	95.99	90.51	93.39	76.6	97.32	81.31
S38584	1426	48	30	10,000	96.01	90.99	95.36	82.4	98.00	84.67
		56	26	10,000	97.17	92.15	95.23	82.11	98.16	90.04

4.2. Power Consumption Simulation

The power simulation results of the BMSIC test generation method are shown in Table 3. The test frequency was 100 MHz, and the power supply voltage was 1.1 V. Table 3 shows the total power consumption and peak power consumption caused by the three test generation methods: LFSR, MSIC, and BMSIC. From Table 3, the BMSIC test pattern generation circuit has obvious advantages in terms of the total power consumption and the peak power consumption compared with the LFSR generation method. The MSIC generation method was better in power consumption compared with BMSIC generation method. But the difference is not particularly obvious.

Table 3. Comparison of power simulation results of the three test generation methods.

CUT	DFF	Chain	Depth	Total (μW)			Peak (μW)		
				LFSR	MSIC [6]	BMSIC	LFSR	MSIC [6]	BMSIC
S13207	638	32	20	116.61	105.89	107.78	6891.83	5535.75	5582.29
		36	18	116.39	104.75	107.72	6818.38	5737.82	5617.09
		40	16	116.59	104.51	108.20	7337.72	5536.47	5617.74
S15850	534	32	17	109.26	96.4	99.31	6633.73	5137.78	5429.68
		36	15	108.59	95.13	99.41	6518.92	5026.07	5396.98
		40	14	109.09	94.8	99.13	6656.53	5442.83	5345.9
S35932	1728	40	44	320.86	276.83	272.26	20,835.3	17,317.1	19,303.6
		48	36	322.54	275.91	277.25	21,014.8	14,695.9	19,864.5
		56	31	322.22	274.54	282.07	21,380.5	25,411.3	20,922.4
S38417	1636	40	41	347.57	280.81	286.98	20,349.8	17,630.4	17,503.9
		48	35	347.32	282.52	286.92	20,578.5	17,081.9	16,952.8
		56	30	346.68	280.41	289.17	19,979	17,425.5	17,150.1
S38584	1426	40	36	335.26	286.89	292.9	21,125.5	15,583.2	18,163.6
		48	30	335.46	287.13	293.95	20,058.9	15,768.9	17,835.2
		56	26	335.43	285.67	291.39	20,011.9	16,110.7	15,692.3

4.3. Area Overhead Evaluation

The hardware area cost of the three test patterns are shown in Table 4. The unit is the area of a two-input XOR gate. The "reduction" indicates the percentage reduction in area of BMSIC compared with MSIC. From the analysis of Table 4, BMSIC method had a great advantage in an equivalent scan configuration, and it can reduce the area overhead by about 50% in the best case.

Table 4. Area Overhead Comparison of the three test generation methods.

CUT	Chain	Depth	LFSR			MSIC [6]			BMSIC			Reduction (%)
			S	L	Area	S	L	Area	S	L	Area	
S13207	32	20	32	0	94	32	32	218	8	20	131	39.9
	36	18	36	0	103	36	36	243	9	18	131	46.09
	40	16	40	0	116	40	40	271	10	16	133	50.92
S15850	32	17	32	0	94	32	32	218	8	17	122	44.04
	36	15	36	0	103	36	36	243	9	15	122	49.79
	40	14	40	0	116	40	40	271	10	14	127	53.14
S35932	40	44	40	0	116	40	44	283	10	44	212	25.09
	48	36	48	0	139	48	48	325	12	36	207	36.31
	56	31	56	0	161	56	56	378	14	31	209	44.71
S38417	40	41	40	0	116	40	41	274	10	41	204	25.55
	48	35	48	0	139	48	48	325	12	35	205	36.92
	56	30	56	0	161	56	56	378	14	30	206	45.5
S38584	40	36	40	0	116	40	40	271	10	36	190	29.89
	48	30	48	0	139	48	48	325	12	30	190	41.54
	56	26	56	0	161	56	56	378	14	26	195	48.41

5. Conclusions

This paper proposes a low-cost test pattern generation method BMSIC-TPG based on our previous work MSIC-TPG, which can take into account both low power consumption and satisfactory fault coverage [6]. The hardware overhead of the proposed MSIC-TPG is reduced by inserting a broadcaster between the MSIC-TPG module and the CUT. The inserted broadcaster is responsible for distributing test patterns from a MSIC-TPG module to fill a larger number of scan chains. By the introduction of the broadcaster, one original scan chain can be split into several shorter scan chains in a balanced way. Analysis results show that BMSIC sequences have the favorable features of uniform distribution and low input transition density. Compared with MSIC-TPG, experimental results show that in most cases, hardware overhead is reduced by 50% and fault coverage is higher. This is achieved with a little increase in test power and no increase in test length to hit a target fault coverage. For the larger CUT, the performance of the proposed BMSIC-TPG in area overhead is better.

Author Contributions: Conceptualization, G.Z. and F.L.; formal analysis, Y.Y.; funding acquisition, F.L.; investigation, S.W.; methodology, S.W.; project administration, C.-F.Y.; resources, C.-F.Y.; validation, G.Z.; writing—original draft, Y.Y.; writing—review and editing, F.L. and Y.Y.

Funding: This research was partly supported by the Core Electronic Devices, High-end General Chips and Basic Software Products Projects of China (2017ZX01030204) and the National Natural Science Foundation of China under Grant 61474093.

Conflicts of Interest: The authors declare no conflict of interest.

References

1. Takahashi, Y.; Maeda, A. Multi Domain Test: Novel test strategy to reduce the Cost of Test. In Proceedings of the 29th VLSI Test Symposium, Dana Point, CA, USA, 1–5 May 2011; pp. 303–308.
2. Dutta, A.; Kundu, S.; Chattopadhyay, S. Thermal Aware Don't Care Filling to Reduce Peak Temperature and Thermal Variance during Testing. In Proceedings of the 2013 22nd Asian Test Symposium, Jiaosi Township, Taiwan, 8–21 November 2013; pp. 25–30.
3. Puczko, M. Low power test pattern generator for BIST. In Proceedings of the 2015 Selected Problems of Electrical Engineering and Electronics (WZEE), Kielce, Poland, 7–19 September 2015; pp. 1–6.
4. Bushnell, M.; Agrawal, V. *Essentials of Electronic Testing for Digital, Memory and Mixed-Signal VLSI Circuits*; Springer Publishing Company, Incorporated: New York City, NY, USA, 2013.
5. Lei, S.C.; Shao, Z.B.; Liang, F. *VLSI Testing*; Publishing House of Electronic Industry: Beijing, China, 2008.
6. Liang, F.; Zhang, L.; Lei, S.; Zhang, G.; Gao, K.; Liang, B. Test Patterns of Multiple SIC Vectors: Theory and Application in BIST Schemes. *IEEE Trans. Very Large Scale Integr. (VLSI) Syst.* **2013**, *21*, 614–623. [CrossRef]
7. McLaurin, T. Periodic Online LBIST Considerations for a Multicore Processor. In Proceedings of the 2018 IEEE International Test Conference in Asia (ITC-Asia), Harbin, China, 15–17 August 2018; pp. 37–42.
8. Shirur, Y.J.M.; Bhimashankar, B.C.; Chakravarthi, V.S. Performance analysis of low power microcode based asynchronous P-MBIST. In Proceedings of the 2015 International Conference on Advances in Computing, Communications and Informatics (ICACCI), Kochi, India, 10–13 August 2015; pp. 555–560.
9. Datta, D.; Datta, B.; Dutta, H.S. Design and implementation of multibit LFSR on FPGA to generate pseudorandom sequence number. In Proceedings of the 2017 Devices for Integrated Circuit (DevIC), Kalyani, India, 23–24 March 2017; pp. 346–349.
10. Seo, S.; Lee, Y.; Lee, J.; Kang, S. A scan shifting method based on clock gating of multiple groups for low power scan testing. In Proceedings of the Sixteenth International Symposium on Quality Electronic Design, Santa Clara, CA, USA, 2–4 March 2015; pp. 162–166.
11. Xiang, D.; Wen, X.; Wang, L. Low-Power Scan-Based Built-In Self-Test Based on Weighted Pseudorandom Test Pattern Generation and Reseeding. *IEEE Trans. Very Large Scale Integr. (VLSI) Syst.* **2017**, *25*, 942–953. [CrossRef]
12. Lee, K.J.; Chen, J.J.; Huang, C.H. Broadcasting test patterns to multiple circuits. *IEEE Trans. Comput.-Aided Des. Integr. Circuits Syst.* **1999**, *18*, 1793–1802.

13. Jeníček, J.; Novák, O.; Chloupek, M. Advanced scan chain configuration method for broadcast decompressor architecture. In Proceedings of the 2011 9th East-West Design Test Symposium (EWDTS), Sevastopol, Ukraine, 9–12 September 2011; pp. 140–143.

14. Gu, X.; Sheu, B.; Wang, Z.; Wang, L.; Wu, S.; Wen, X.; Jiang, Z. VirtualScan: Test Compression Technology Using Combinational Logic and One-Pass ATPG. *IEEE Des. Test Comput.* **2008**, *25*, 122–130.

15. Chen, J.; Lee, K. Test Stimulus Compression Based on Broadcast Scan With One Single Input. *IEEE Trans. Comput.-Aided Des. Integr. Circuits Syst.* **2017**, *36*, 184–197. [CrossRef]

16. Pomeranz, I. Computing Seeds for LFSR-Based Test Generation From Nontest Cubes. *IEEE Trans. Very Large Scale Integr. (VLSI) Syst.* **2016**, *24*, 2392–2396. [CrossRef]

17. Singh, S.; Kaur, S.; Kaur, R.; Kaler, R.S. Photonic processing of all-optical Johnson counter using semiconductor optical amplifiers. *IET Optoelectron.* **2017**, *11*, 8–14. [CrossRef]

18. Stoev, I.I.; Borodzhieva, A.N.; Mutkov, V.A. FPGA Implementation of Johnson Counters Applied in the Educational Process. In Proceedings of the 2018 IEEE 24th International Symposium for Design and Technology in Electronic Packaging (SIITME), Iasi, Romania, 25–28 October 2018; pp. 99–103.

19. Touba, N.A. Survey of Test Vector Compression Techniques. *IEEE Des. Test Comput.* **2006**, *23*, 294–303. [CrossRef]

20. Girard, P.; Landrault, C.; Pravossoudovitch, S.; Severac, D. Reducing power consumption during test application by test vector ordering. In Proceedings of the 1998 IEEE International Symposium on Circuits and Systems (Cat. No.98CH36187), ISCAS '98, Monterey, CA, USA, 31 May–3 June 1998; Volume 2, pp. 296–299.

21. Nourani, M.; Tehranipoor, M.; Ahmed, N. Low-Transition Test Pattern Generation for BIST-Based Applications. *IEEE Trans. Comput.* **2008**, *57*, 303–315. [CrossRef]

Review

Overview: Types of Lower Limb Exoskeletons

Daniel S Pamungkas [1], **Wahyu Caesarendra** [2,3,*], **Hendawan Soebakti** [4], **Riska Analia** [1] and **Susanto Susanto** [1]

1   Mechatronics Study Program, Politeknik Negeri Batam 29432, Indonesia; daniel@polibatam.ac.id (D.S.P.); riskaanalia@polibatam.ac.id (R.A.); susanto@polibatam.ac.id (S.S.)
2   Faculty of Integrated Technologies, Universiti Brunei Darussalam, Jalan Tungku Link BE1410, Brunei
3   Department of Mechanical Engineering, Diponegoro University, Jl. Prof.H.Soedarto S.H, Semarang 50275, Indonesia
4   Robotics Study Program, Politeknik Negeri Batam 29432, Indonesia; hendawan@polibatam.ac.id
*   Correspondence: wahyu.caesarendra@ubd.edu.bn; Tel.: +62-673-7345-623

Received: 30 September 2019; Accepted: 28 October 2019; Published: 4 November 2019

Abstract: Researchers have given attention to lower limb exoskeletons in recent years. Lower limb exoskeletons have been designed, prototype tested through experiments, and even produced. In general, lower limb exoskeletons have two different objectives: (1) rehabilitation and (2) assisting human work activities. Referring to these objectives, researchers have iteratively improved lower limb exoskeleton designs, especially in the location of actuators. Some of these devices use actuators, particularly on hips, ankles or knees of the users. Additionally, other devices employ a combination of actuators on multiple joints. In order to provide information about which actuator location is more suitable; a review study on the design of actuator locations is presented in this paper. The location of actuators is an important factor because it is related to the analysis of the design and the control system. This factor affects the entire lower limb exoskeleton's performance and functionality. In addition, the disadvantages of several types of lower limb exoskeletons in terms of actuator locations and the challenges of the lower limb exoskeleton in the future are also presented in this paper.

Keywords: actuator; lower limb exoskeleton; wearable robot

---

1. Introduction

Nowadays, people work even more strenuously and require stronger and longer lasting muscle movements. However, human muscles have a fatigue limit when doing regular and repetitive activities. To help overcome this fatigue limit, some researchers have suggested that humans use an external wearable device, i.e., an exoskeleton. An exoskeleton, also known as a wearable robotic, is a system that can be worn to help human beings to support and protect parts of their bodies [1]. Such a device has been used for many applications, including enhancing workers when doing their jobs or as medical tools for rehabilitation. In many industries, exoskeletons have been used to increase worker strength for walking on long journeys [2] or lift heavy items [3]. In the medical field, exoskeletons have been used to assist patients who have lost their ability to walk due to spinal cord injuries, stroke, and other trauma [4]. Coenen et al. [5] reported that rehabilitation exoskeletons can improve the quality of exercises during rehabilitation and can accelerate recovery process.

The application of the exoskeleton to the human body can be divided into three locations: (1) throughout the human body [6], (2) at the upper part of human body, such as the torso and arms [7], and (3) at the lower part of the human body, i.e., from the waist down [8]. Various parts of the human body simultaneously play certain functions in supporting movement during walking. However, the lower limbs of the human body have more important roles than the other parts. This is because the lower limbs generate more torque than other parts while walking. This paper reviews a number of existing published papers related to lower limb exoskeletons. However, this paper limits its discussion to the classification of joint motions and types of actuators of the exoskeleton.

The joints in the lower limb of the human body are the hips, knees, and ankles. Each joint has different abilities to move or degrees of freedom (DoF), as shown in more detail in Table 1. The types of lower limb exoskeletons based on joint motions are differentiated into several types based on how the actuators drive the exoskeleton. The actuators can drive just the hips, the knees, or the ankles. In a small number of studies, exoskeletons have multiple actuators to drive a combination of joints. These combinations of actuators are hips and knees, knees and ankles, and all three joints (hips, knees, and ankles).

Table 1. Degrees of freedom (DoF) of each joint in the lower limb.

No	Joints	DoF	Movement
1	Hips	3	Flexion–extension
			Abduction–adduction
			Internal–external rotation
2	Knees	2	Flexion–extension
			rotation
3	Ankles	3	plantar flexion–dorsiflexion
			Abduction–adduction
			Eversion–inversion

The source of motion of the actuator can be distinguished by whether it is an active actuator or a passive one. An active exoskeleton is one that uses a power source to activate the actuators. Moreover, an actuator for active motion can be electric, pneumatic [9], or hydraulic [3]. On the other hand, a passive exoskeleton is a device that has no power source. This type of exoskeleton exploits kinematic forces, e.g., by using springs and dampers [10].

This paper is organized as follows. Section 2 presents a brief overview of the biomechanics of the walking process. Section 3 presents a number of exoskeletons grouped according to the locations of the actuators. Section 4 discusses the present state and the future of the lower limb exoskeleton. Finally, Section 5 provides concluding remarks.

2. The Walking Process

Prior to designing the kinematics of a lower limb exoskeleton, it is necessary to study the cycle of the human walking gait. The human walking gait can be modeled as represented in Figure 1 (adapted from [11]). The human walking gait cycle starts from the right heel contact and ends with the same situation; the steps in the gait cycle are described in Table 2. This cycle determines the movement of the exoskeleton. Moreover, the relationship between the human walking gait cycle with the angle of the hips and the torques needs to be considered. In [12], the authors performed tests on 15 subjects to examine the relationship between angles and torques within the human walking gait cycle. The subjects consisted of 11 males and four females, with heights between 166 and 184 cm. The results of the experiments are presented in Figure 2. The experiment was only conducted for walking on a level surface. The results of the relationships would have differed in the case of other activities, such as running. In addition, angles and torques generated by the human gait also depend on the condition of the surface, e.g., walking on a ramp [13]. The movement of an exoskeleton can be determined by using the relationship between the angles and the cycle gait. Ultimately, torque can be calculated based on the movements of an exoskeleton.

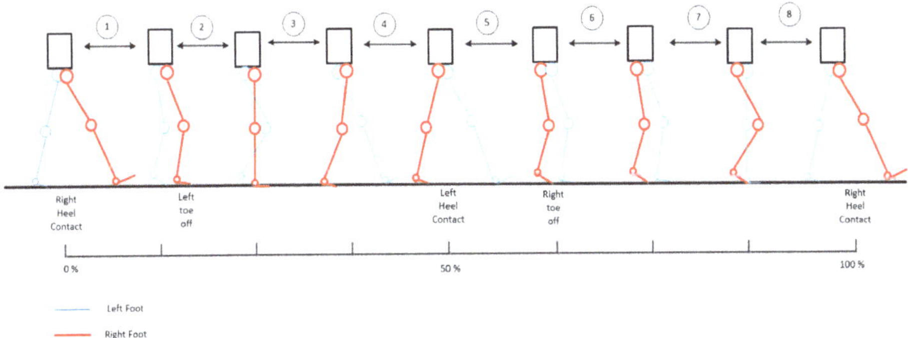

Figure 1. Human walking gait cycle (adapted from [11]).

Table 2. Gait cycle position.

No	Right Leg	Left Leg
1	Heel strike	Pre swing
2	Loading response	Toe off
3	Mid Stance	Mid Swing
4	Terminal Stance	Terminal Swing
5	Pre swing	Heel strike
6	Toe off	Loading response
7	Mid Swing	Mid Stance
8	Terminal Swing	Terminal Stance

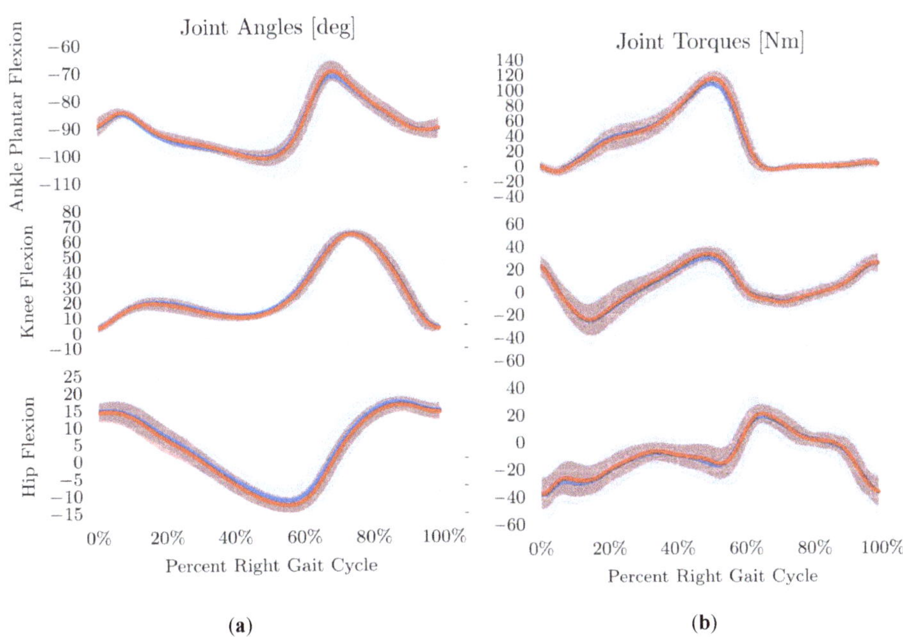

Figure 2. (**a**) Angles of joints and (**b**) torques of the joint versus the gait cycle (adapted from [12]).

Dominant movement takes place in the sagittal plane (the plane that divides the human body into right and left parts) during the gait. Lower human limbs have three main joints, namely ankles, knees and hips. Therefore, researchers have focused on the movement of exoskeletons in these joints.

For research purposes, there have been several simplifications of the DoF of lower human limb. Some studies have simplified the lower limbs into seven DoF, i.e., three on hip, one on the knee, and three on the ankle; five DoF, i.e., two on the hip, one on the knee, and two on the ankle; and even only three DoF for the main DoF only, as presented in Figure 3. The main DoF are the flexion (positive direction) and extension (negative direction) movements of the hip, knee and ankle joints. For each hip and ankle, one or two DoF can be added. The other DoF in the hip and ankle are abduction (movement of pulling away from the center of the body) vs. adduction (movement of pushing toward the center of the body) and eversion (movement away from midline of the body) vs. inversion (movement toward the midline of the body). The simplified DoF of the exoskeleton cause a reduction in the ability of the device to be utilized by the user. However, researchers have mainly focused on the basic human movements that are used for daily activities. The various types of locations/joint movements of the actuators of exoskeletons are discussed in Section 3.

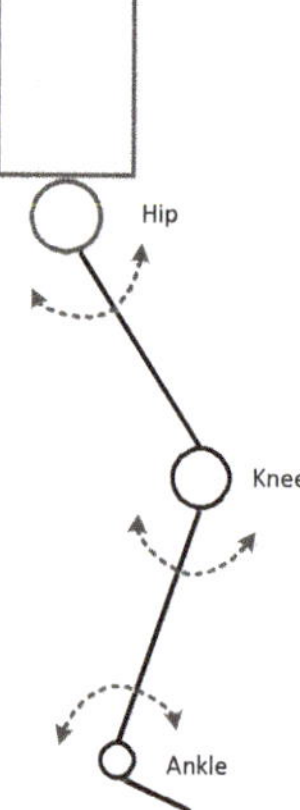

Figure 3. The dominant movements of the human walking gait.

3. Joints of the Exoskeleton

This section provides a discussion about the various types of exoskeletons. Exoskeletons are classified by the location of the power sources of the lower exoskeleton. The actuators of the lower exoskeleton are usually placed on the joints of the human. The purpose of this part of the exoskeleton is to move the joint of the user. The actuator can be placed only one joint on each leg, such as the hip, knee or ankle; a combination between two joints (hip and knee, or knee and ankle); or a combination between three joints (hip, knee and ankle).

3.1. Hip Exoskeleton

Hips connect the upper limbs and the lower limbs. Human hips enable their owner to perform the motions of flexion/extension, abduction/adduction, and medial/lateral rotation (three DoF motions). These motions are required for a human to walk or run. Most researchers have placed actuators on the hips of the users for their exoskeletons. Moreover, Lenzi et al. [14] concluded that the hip exoskeleton enables a reduction of hip and ankle muscle activities.

Honda developed an exoskeleton called the Honda Walking Assist [15]. This device has one Direct Current (DC) motor on each hip. The force from the motor is passed to the thigh of the user through the straps, resulting in a light and neat exoskeleton. In another design similar to Honda

Walking Assist [15], Giovacchini et al. [16] developed a hip orthotic whose actuator is located near the user's hips. This exoskeleton helps the user to move their hip in the flexion and extension directions (Figure 4a). Moreover, this wearable robot is equipped with a passive actuator that enables the user to move in the abduction–adduction direction, resulting in user comfort.

The HiBSO (hip ball screw orthosis) exploits a ball screw in each leg for transmitting the force from a DC motor [17]. At the end of the ball screw is a strap that passes the actuation movement to the thigh (see Figure 4b). This structure has other movements beside the flexion/extension action. The HiBSO enables the user to move in the abduction/adduction direction in the hips, while allowing for the rotation of the thigh. Another wearable robot, Powered Hip Exoskeleton or PH-EXOS also added abduction/adduction motions and internal and external rotation [18]. These motions enrich its primary movements, namely flexion/extension, with the abduction and adduction motions being passive actions. The motors are placed on the waist of the user and are connected to the pulley through a Bowden cable, as seen in Figure 4c.

Asbeck et al. [19] constructed an exoskeleton called the Exosuit. They constructed the webbing straps with a geared motor that is carried on the user's back. These straps are linked to the thigh of the user, as seen in Figure 4d. These straps perform by contracting and expanding on the leg during the heel strike until the terminal stance.

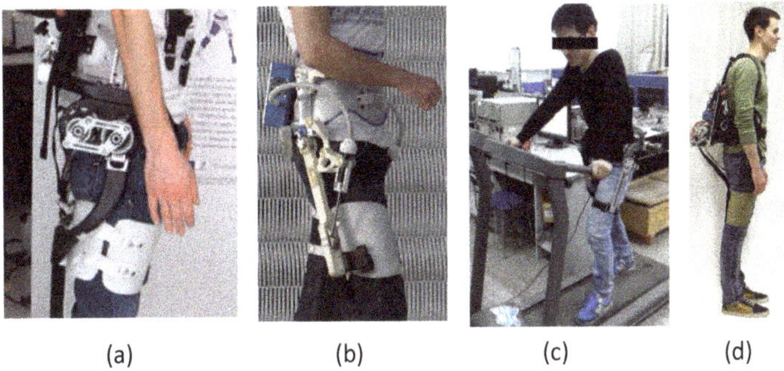

(a)   (b)   (c)   (d)

Figure 4. (**a**) exoskeleton developed by Giovacchini et al. [16]; (**b**) HiBSO (hip ball screw orthosis) [17]; (**c**) PH-EXOS [18]; (**d**) Exosuit [19].

3.2. Knee Exoskeleton

The human knee is an important object of study by researchers, because this part of the human body generates significant torque for walking [12], running [20], and movement from squatting to standing and vice versa [21,22]. Moreover, knees also restrain impact during those activities. Additionally, the position of the knee is between the hip and the ankle. Compared to the hip and ankle, the knee has a more straightforward movement—flexion/extension as well as rotation movements. However, for the sake of simplification, most studies of exoskeletons have modeled only one DoF for the knee exoskeleton, dedicated entirely to moving the knee in the flexion/extension actions.

A soft inflatable cushion is used as the actuator in an exoskeleton [23]. The inflatable part is placed behind the knee of the user. This component allows for the reduction of the weight of the exoskeleton. To inflate and deflate the component, a pneumatic system is used. This exoskeleton is inflated during the swing phase of the walking gait and deflated during the other phases of the walking gait cycle. Figure 5a shows this exoskeleton. Two DC motors are used to actuate two Bowden cables [24]. One of the cables is connected to a strap behind the lower thigh, while another cable is connected to a strap in front of the top of the thigh, as presented in Figure 5b.

Another activity that is often used while working is squatting. Human knees have an essential role in the squatting motion. The squatting action requires a high torque from the knees [21]. Moreover, several activities are required in the squatting movement. However, most wearable robots are not designed for this action. A passive one-DoF knee exoskeleton was developed by Ranaweera et al. [10] to help humans lift loads from the squatted position. This device employed two helical elastic springs on each knee. This component was connected to the pulley disk, which was placed behind the knees of the user. Figure 5c shows the prototype of this exoskeleton. Huang et al. [25] designed an exoskeleton to prevent injury because they observed that the squatting motion makes one susceptible to personal injury. This device can help the user to squat and walk without carrying any load. They utilized a motor and transmitted the torque to the gear and pulley by a flexion cable, as presented in Figure 5d. Meanwhile, an exoskeleton was designed for walking and kneeling. Wang et al. [26] developed a knee exoskeleton actuated with a motor and transmitted to a double pulley on the user's knee. This configuration helps the user to walk and to assume a kneeling posture. The design of this exoskeleton is presented in Figure 5e.

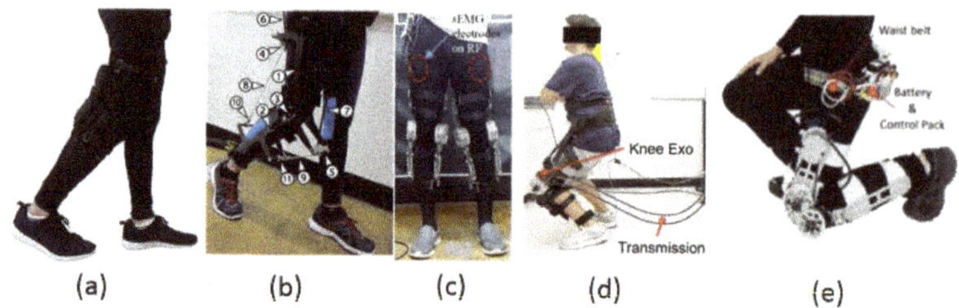

Figure 5. Knee exoskeletons using (**a**) an inflatable actuator [23], (**b**) a Bowden cable put in the front and back of the leg [24], (**c**) springs [10], (**d**) a DC motor and pulley [25], (**e**) double pulleys [26].

3.3. Ankle Exoskeleton

According to Figure 2, the ankle has the most significant torque during the walking gait compared to other joints. This has caused most studies to place actuators in the ankle. This joint has four bones in three planar motions (three DoF). However, plantar or dorsiflexion movement is the primary movement during the gait cycle. Some researchers have simplified the motions on their exoskeletons to only one DoF for this main ankle movement.

Mooney and Herr [27] developed a one-DoF exoskeleton, as seen in Figure 6a. This equipment was intended to help the user to walk while carrying a load. This exoskeleton uses a brushless DC motor (BLDC) placed in the shank of the user, while the motor controller and batteries are attached to a vest. The device activates a fiberglass strut to pull the ankle of the user; the struts are connected to the boots. Another design called a soft exosuit was proposed by Asbeck et al. [28]. This equipment's purpose is to assist the user in walking while carrying a load. Their exoskeleton is actuated by an electric motor; the motor, batteries and controller are placed in a backpack. This motor is connected to a Bowden cable. The purpose of the cable is to pull the heel of the user (see Figure 6b). The cables behave similarly to the human calf muscle. Another exoskeleton was designed by Bai et al. [29]. This device is used for the therapy of subjects suffering from ankle injuries. Figure 6c shows this exoskeleton. An electric motor is mounted in front of the shinbone, while the motor control and batteries are carried on the subject's waist. The belt is used to transmit the movement of the motor to the gear. This system functions to activate the subject's ankle. An ankle exoskeleton powered pneumatically was devised by Shorter et al. [9]. The movement of the subject's ankles is actuated by a rotary pneumatic actuator. This device is attached at the ankle of the user. This exoskeleton utilizes two valves, with the air source placed on the subject's waist.

Some studies have attempted to add additional DoF of the ankle exoskeleton, such those of Carberry et al. [30], Agrawal et al. [31], and Park et al. [32]. A two-DoF ankle exoskeleton was proposed by Carberry et al. [30]. They enhanced their exoskeleton with eversion/inversion movements. This device is intended for post-stroke rehabilitation. The mechanism of the exoskeleton is pneumatically actuated, with the air source placed separately from the exoskeleton, as shown in Figure 6d. Agrawal et al. [31]

developed another two-DoF exoskeleton which has a motion similar to Carberry's. They combined an active joint with a passive joint. The flexion/extension ankle motions are controlled by a DC servomotor, while the inversion/eversion motions are controlled use a spring and damper mechanism. Three pneumatic synthetic muscles are used to simulate the human leg muscles [32], as seen in Figure 6f. This system enables the user to move their ankle with two DoF using these devices with movements similar to the natural ankle.

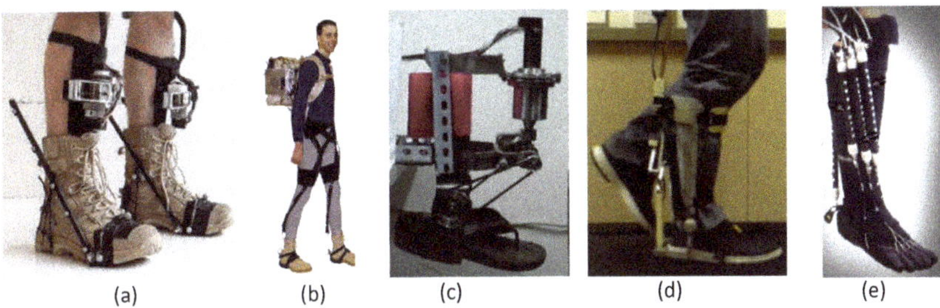

(a)   (b)   (c)   (d)   (e)

Figure 6. One-DoF ankle exoskeleton using: (**a**) DC motor and struts [27]; (**b**) motor and Bowden cable [28]; (**c**) DC motor and belt [29]; (**d**) two-DoF with active actuators [30]; and (**e**) two-DoF active pneumatic actuators [32].

3.4. Multiple Joints Exoskeleton

To actuate a joint, more than one muscle that passes through multiple joints may be required. Some studies have utilized more than one actuator to actuate joints in their exoskeletons. The actuators actuate a combinations of joints, namely hip and knee; knee and ankle; and hip, knee and ankle. These multiple joint exoskeletons need to have a more advanced control system than single joint exoskeletons, because the movement of the joints has to be controlled in such a way that results in a harmonious gait. These exoskeletons are discussed in this section.

Some researchers have proposed a knee–ankle, two-DoF exoskeleton for rehabilitation. The National University of Singapore (NUS) developed a device to rehabilitate stroke patients. They used a series elastic actuator (SEA) to actuate the knee and ankle joints [33]. The SEA is a connection between a motor with a serial spring. The actuators are placed between the joint and the thigh or shank of the user, with each actuator using a crank and a connecting rod to move the leg of the user. Figure 7a shows this exoskeleton. The WAKE-up (wearable ankle knee exoskeleton) also utilized an SEA [34]. The actuators are chosen to prevent direct contact between the user and the actuator. To transmit power, a timing belt is used. This device is a modular exoskeleton, which can be worn for single joints or for multiple joints at once.

In 2006, a prototype hip–knee exoskeleton was developed by University of Twente. This exoskeleton is called the LOPES (lower extremity-powered exoskeleton) and is intended as a rehabilitation device [35]. This device has actuators on the hip and knee. This design enables the user to move the hip on the flexion/extension and abduction/adduction directions, as well as the flexion/extension direction on the knee. The actuator is actuated by a motor and transmitted to the SEA using a Bowden cable. The LOPES is shown in Figure 7b. For rehabilitation purposes, the clearance of the foot must be sufficient so that this exoskeleton does not actuate the ankle during the gait.

Another type of combined exoskeleton is the hip, knee and ankle exoskeleton, in which all joints of the user's lower limb are actuated. In 2004, the University of California developed an exoskeleton called the Berkeley Lower Extremity Exoskeleton (BLEEX) [3]. This exoskeleton is actuated using linear hydraulics and has the ability to provide additional power for carrying heavy loads. The two DoF of the hip are actuated using active actuators, while the rotation of the hip is passively actuated by using springs and elastomers. One motion each is actively actuated in the knee joint and the ankle joint. The BLEEX is shown in Figure 7c. The exoskeleton constructed by the University of Salford is an example of an exoskeleton that uses pneumatic muscle actuators (PMA) [36]. This equipment is intended for paraplegic patient rehabilitation. They equipped their exoskeleton with PMA to actuate the hip with three DoF—only one each for the knee and ankle.

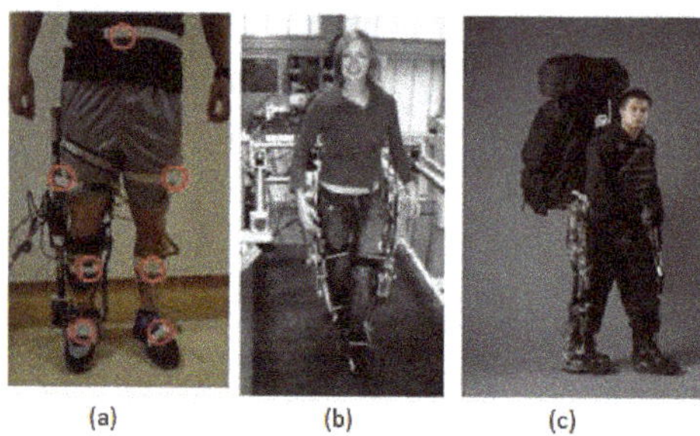

Figure 7. Multiple joints exoskeletons: (**a**) the National University of Singapore (NUS) exoskeleton [33]; (**b**) the lower extremity-powered exoskeleton (LOPES) (all lower limb joints) [35]; and (**c**) the Berkeley Lower Extremity Exoskeleton BLEEX [3].

4. Discussion

To achieve the aims of the exoskeletons, these devices are equipped with actuators to amplify human gait and power by actuating on joints. The selection of the joints to actuate depends on the objective of the research. Some studies have been intended for the development of exoskeletons whose joint systems are focused on rehabilitation, e.g., ankle exoskeletons developed for the rehabilitation of ankle injuries [29]. To study how an exoskeleton may help squatting activity, the knee joint has been focused on by some researchers [25]. In another example, to help the user carry a heavy load and walk long distances, some researchers considered an exoskeleton in which all the joints in the device were actuated with powerful actuators [3].

The simplest exoskeletons only use a single DoF for actuating the lower human limb joint. This type of manipulator is lightweight and solid. Moreover, this design has advantages, e.g., it is easy to wear, can be worn by different users, and is comfortable. However, such exoskeletons are not ergonomic. In addition, their use can also lead to injury for users [37]. Furthermore, these devices have a limited ability and are not versatile.

On the other hand, exoskeletons with high DoF resembling human DoF will enable users to walk normally and naturally because the design imitates human anatomy. These designs allow users to easily control their devices; moreover, these users have a high maneuverability. However, high-DoF exoskeletons are complex, costly, and cumbersome. In addition, their construction is challenging to wear.

Some exoskeletons combine active and passive actuators. Active actuators are used for actuating the main DoF, e.g., the flexion/extension motion, while additional movements, which are not too significant to the human gait, use passive actuators. This combination is intended to reduce the weight of the device and also to reduce the cost of production. Even though this design lowers power and the ability to maneuver, it is relatively suitable for rehabilitation, a goal which does not need a high level of maneuverability or power.

Other methods to reduce weight include selecting light and reliable materials, such as carbon fiber composites, and using small active actuators. The power source of the active actuator (e.g., batteries) has significant weight costs, although the weight of components has lessened in recent decades. Non-portable exoskeletons, such as exoskeletons for rehabilitation, can have a power source for active actuators outside the exoskeleton. This can minimize the actual weight of the device.

Active exoskeletons can be pneumatically, hydraulically, or electrically powered. The active pneumatic actuator uses a pressurized air source. This type of actuator has several advantages, namely, a light weight and flexibility similar to the human muscle. A pneumatic ankle exoskeleton weighs below 1 kg [32]. However, pneumatic actuators have limited power and are difficult to control. Hydraulic actuators have high power and stability. A hydraulic actuator is able to generate power up to 1.3 kW to actuate a joint. However, this type of actuator is cumbersome and costly. Electric actuators have the benefit of being controllable. This property can be utilized to allow for precise movements of the exoskeleton. Electric motors are often used for this reason. Power transmission from an electric actuator can be done through various ways, such as using Bowden cables, struts, ball screws or belts. Moreover, several types of electric motors can be used, such as a BLDC. However, the drive from an electric motor sometimes needs a gear system to amplify its power.

In the future, exoskeleton research should develop applicable devices that are lightweight, inexpensive, compact, convenient, durable, and efficient. Thus, research should focus on the design of exoskeletons, the choice of actuators and frames, and the harmonization of exoskeleton movement with the natural human gait. Table 3 presents exoskeletons with their types of actuators.

Table 3. Actuators in joints of lower limb exoskeletons.

No	Name/Institution	Joint	DoF	Actuator	Type of Actuator	Weight (kg)
1	Honda Walking Assist [15]	Hip	1	Active flexion–extension	DC motor	2.7
2	Exoskeleton/The BioRobotics Institute, Scuola Superiore Sant'Anna [16]	Hip	2	Active flexion–extension Passive abduction–adduction	DC motor (SEA) carbon fiber linkage	8.5
3	HIPSO [17]	Hip	2	Active flexion–extension Passive abduction–adduction	DC Motor ball-bearing	9.5
4	PH-EXOS [18]	Hip	3	Active flexion–extension Passive abduction–adduction and internal–external rotation	AC motor mechanical structure	n/a
5	Hip exosuit/Harvard University [19]	Hip	1	Active flexion–extension	DC motor	n/a
6	Soft inflatable exosuit [23]	Knee	1	Active flexion–extension	Pneumatics	0.16
7	Knee Exo/Carnegie Mellon University [24]	Knee	1	Active flexion–extension	DC motor	0.76
8	University of Moratuwa, Katubedd [10]	Knee	1	Passive flexion–extension	Spring	n/a
9	Soft hybrid EXO/The City University of New York [25]	Knee	1	Active flexion–extension	DC motor	n/a
10	The City University of New York [26]	Knee	1	Active flexion–extension	DC motor	3.2
11	MIT ankle exoskeleton [27]	Ankle	1	Active flexion–dorsiflexion	DC motor	4
12	Knee soft exosuit/Harvard University [28]	Ankle	1	Active flexion–dorsiflexion	DC motor	12.15
13	Beijing Institute of Technology [29]	Ankle	1	Active flexion–dorsiflexion	DC motor	n/a
14	University of Illinois [9]	Ankle	1	Active flexion–dorsiflexion	Pneumatics	n/a
15	University of Bristol [30]	Ankle	2	Active flexion–dorsiflexion Active Inversion–eversion	Pneumatics Pneumatics	n/a

Table 3. *Cont.*

16	University of Delaware [31]	Ankle	2	Active flexion–dorsiflexion	DC motor	n/a
				Passive Inversion–eversion	Spring and damper	
17	Harvard University [32]	Ankle	2	Active flexion–dorsiflexion	Pneumatics	0.95
				Active Inversion–eversion	Pneumatics	
18	National University of Singapore [33]	Knee–ankle	2	Active knee flexion–extension	DC motor	n/a
				Active ankle flexion–dorsiflexion	DC motor	
19	WAKE-up [34]	Knee–ankle	2	Active knee flexion–extension	DC motor	2.5
				Active ankle flexion–dorsiflexion	DC motor	
20	LOPES [35]	Hip–ankle	3	Active Hip Flexion–extension	DC motor	n/a
				Active hip abduction–adduction	DC motor	
				Active knee flexion–extension	DC motor	
21	BLEEX [3]	Hip–Knee–ankle	5	Active hip Flexion–extension	Hydraulic	14
				Active hip Abduction–adduction	Hydraulic	
				Passive hip rotation	Spring	
				Active knee Flexion–extension	Hydraulic	
				Active ankle flexion–dorsiflexion	Hydraulic	
22	University of Salford [36]	Hip–Knee–ankle	5	Active hip Flexion–extension	Pneumatic	12
				Active hip Abduction–adduction	Pneumatic	
				Passive hip rotation	Pneumatic	
				Active knee Flexion–extension	Pneumatic	
				Active ankle flexion–dorsiflexion	Pneumatic	

5. Conclusions

This paper presents a review of current studies focused on prototype lower limb exoskeletons. Some of them have had specific joints actuated, while others have used combinations of actuators to actuate multiple joints. The selection of the placement of the actuators to assist the joints is dependent on the aim of the research. Lower limb exoskeletons with DoF similar to human biomechanics enable users to move more naturally. However, the utilization of multiple DoF is complicated and cumbersome. To reduce the weight of devices, some exoskeletons use passive actuators instead of active ones. Some studies have utilized passive actuators for joints or movements that are not significantly involved in the human gait; however, passive actuators are more challenging to control. In the future, the development of research on exoskeletons is expected to significantly increase. The potential for use of these device is broadening, and in the future, the exoskeleton will have a vital role in daily human activities.

Author Contributions: Conceptualization, D.S.P., R.A., and S.S.; methodology, H.S.; validation, W.C. and D.S.P.; formal analysis, W.C. and H.S.; investigation, W.C., and D.S.P.; resources, W.C., S.S. and R.A.; data curation, W.C.; writing—original draft preparation, W.C. and D.S.P.; writing—review and editing, W.C., D.S.P. and R.S.; visualization, W.C. and D.S.P.; supervision, D.S.P.; project administration, W.C.; funding acquisition, D.S.P., and W.C.

Funding: This research received no external funding.

Acknowledgments: The research for this paper was financially supported by the DRPM KemenristekDikti Indonesia.

Conflicts of Interest: The authors declare that there is no conflict of interest regarding the publication of this paper. The funders had no role in the design of the study; in the collection, analyses, or interpretation of data; in the writing of the manuscript, or in the decision to publish the results.

References

1. Dollar, A.M.; Herr, H. Lower extremity exoskeletons and active orthoses: Challenges and state-of-the-art. *IEEE Trans. Robot.* **2008**, *24*, 144–158. [CrossRef]
2. Malcolm, P.; Derave, W.; Galle, S.; de Clercq, D. A Simple Exoskeleton That Assists Plantarflexion Can Reduce the Metabolic Cost of Human Walking. *PLoS ONE* **2013**, *8*, e56137. [CrossRef] [PubMed]
3. Kazerooni, H.; Steger, R.; Huang, L. Hybrid control of the Berkeley Lower Extremity Exoskeleton (BLEEX). *Int. J. Robot. Res.* **2006**, *25*, 561–573. [CrossRef]
4. Banchadit, W.; Temram, A.; Sukwan, T.; Owatchaiyapong, P.; Suthakorn, J. Design and implementation of a new motorized-mechanical exoskeleton based on CGA Patternized Control. In Proceedings of the 2012 IEEE International Conference on Robotics and Biomimetics, ROBIO 2012—Conference Digest, Guangzhou, China, 11–14 December 2012; pp. 1668–1673.
5. Coenen, P.; van Werven, G.; van Nunen, M.P.M.; van Dieën, J.H.; Gerrits, K.H.L.; Janssen, T.W.J. Robot-assisted walking vs overground walking in stroke patients: an evaluation of muscle activity. *J. Rehabil. Med.* **2012**, *44*, 331–337. [CrossRef] [PubMed]
6. Guizzo, E.; Goldstein, H. The rise of the body bots. *IEEE Spectr.* **2005**, *42*, 50–56. [CrossRef]
7. Hessinger, M.; Pingsmann, M.; Perry, J.C.; Werthschutzky, R.; Kupnik, M. Hybrid position/force control of an upper-limb exoskeleton for assisted drilling. In Proceedings of the IEEE International Conference on Intelligent Robots and Systems, Vancouver, AB, Canada, 24–28 September 2017; Volume 2017, pp. 1824–1829.
8. Christensen, S.; Bai, S.; Rafique, S.; Isaksson, M.; O'Sullivan, L.; Power, V.; Virk, G.S. AXO-SUIT—A modular full-body exoskeleton for physical assistance. In *Mechanisms and Machine Science*; Springer: Dordrecht, The Netherlands, 2019; Volume 66, pp. 443–450.
9. Shorter, K.A.; Kogler, G.F.; Loth, E.; Durfee, W.K.; Hsiao-Wecksler, E.T. A portable powered ankle-foot orthosis for rehabilitation. *J. Rehabil. Res. Dev.* **2011**, *48*, 459–472. [CrossRef]
10. Ranaweera, R.K.P.S.; Gopura, R.A.R.C.; Jayawardena, T.S.S.; Mann, G.K.I. Development of A Passively Powered Knee Exoskeleton for Squat Lifting. *J. Robot. Netw. Artif. Life* **2018**, *5*, 45. [CrossRef]
11. Gait | Joint Structure and Function: A Comprehensive Analysis, 5e | F.A. Davis PT Collection | McGraw-Hill Medical. Available online: https://fadavispt.mhmedical.com/content.aspx?bookid=1862§ionid=136086727 (accessed on 16 September 2019).
12. Moore, J.K.; Hnat, S.K.; van den Bogert, A.J. An elaborate data set on human gait and the effect of mechanical perturbations. *Peer J.* **2015**, *3*, e918. [CrossRef]
13. Voloshina, A.S.; Ferris, D.P. Biomechanics and energetics of running on uneven terrain. *J. Exp. Biol.* **2015**, *218*, 711–719. [CrossRef]
14. Lenzi, T.; Carrozza, M.C.; Agrawal, S.K. Powered hip exoskeletons can reduce the user's hip and ankle muscle activations during walking. *IEEE Trans. Neural Syst. Rehabil. Eng.* **2013**, *21*, 938–948. [CrossRef]
15. Walking Assist Device with Stride Management System | Research paper site of Honda R&D Co., Ltd. Available online: https://www.hondarandd.jp/point.php?pid=122&lang=en (accessed on 5 September 2019).
16. Giovacchini, F.; Vannetti, F.; Fantozzi, M.; Cempini, M.; Cortese, M.; Parri, A.; Vitiello, N. A light-weight active orthosis for hip movement assistance. *Robot. Auton. Syst.* **2015**, *73*, 123–134. [CrossRef]
17. Baud, R.; Ortlieb, A.; Olivier, J.; Bouri, M.; Bleuler, H. HIBSO hip exoskeleton: Toward a wearable and autonomous design. *Mech. Mach. Sci.* **2018**, *48*, 185–195.
18. Wu, Q.; Wang, X.; Du, F.; Zhang, X. Design and control of a powered hip exoskeleton for walking assistance. *Int. J. Adv. Robot. Syst.* **2015**, *12*, 18. [CrossRef]
19. Asbeck, A.T.; Schmidt, K.; Walsh, C.J. Soft exosuit for hip assistance. *Robot. Auton. Syst.* **2015**, *73*, 102–110. [CrossRef]
20. Grimmer, M.; Eslamy, M.; Seyfarth, A. Energetic and peak power advantages of series elastic actuators in an actuated prosthetic leg for walking and running. *Actuators* **2014**, *3*, 1–19. [CrossRef]

21. Hirata, R.; Duarte, M. Effect of relative knee position on internal mechanical loading while squatting. *Braz. J. Phys. Ther.* **2007**, *11*, 107–111.

22. Slater, L.V.; Hart, J.M. The influence of knee alignment on lower extremity kinetics during squats. *J. Electromyogr. Kinesiol.* **2016**, *31*, 96–103. [CrossRef]

23. Sridar, S.; Nguyen, P.H.; Zhu, M.; Lam, Q.P.; Polygerinos, P. Development of a soft-inflatable exosuit for knee rehabilitation. In Proceedings of the IEEE International Conference on Intelligent Robots and Systems, Vancouver, AB, Canada, 24–28 September 2017; pp. 3722–3727.

24. Witte, K.A.; Fatschel, A.M.; Collins, S.H. Design of a lightweight, tethered, torque-controlled knee exoskeleton. In Proceedings of the IEEE International Conference on Rehabilitation Robotics, London, UK, 17–20 July 2017; pp. 1646–1653.

25. Yu, S.; Huang, T.H.; Wang, D.; Lynn, B.; Sayd, D.; Silivanov, V.; Su, H. Design and Control of a Quasi-Direct Drive Soft Hybrid Knee Exoskeleton for Injury Prevention during Squatting. *arXiv* **2019**, arXiv:1902.07106.

26. Wang, J.; Li, X.; Huang, T.H.; Yu, S.; Li, Y.; Chen, T.; Su, H. Comfort-Centered Design of a Lightweight and Backdrivable Knee Exoskeleton. *IEEE Robot. Autom. Lett.* **2018**, *3*, 4265–4272. [CrossRef]

27. Mooney, L.M.; Herr, H.M. Biomechanical walking mechanisms underlying the metabolic reduction caused by an autonomous exoskeleton. *J. Neuroeng. Rehabil.* **2016**, *13*, 4. [CrossRef]

28. Asbeck, A.T.; de Rossi, S.M.M.; Holt, K.G.; Walsh, C. A Biologically Inspired Soft Exosuit for Walking Assistance. *Int. J. Robot. Res.* **2015**, *34*, 744–762. [CrossRef]

29. Bai, Y.; Gao, X.; Zhao, J.; Jin, F.; Dai, F.; Lv, Y. A portable ankle-foot rehabilitation orthosis powered by electric motor. *Open Mech. Eng. J.* **2015**, *9*, 982–991. [CrossRef]

30. Carberry, J.; Hinchly, G.; Buckerfield, J.; Tayler, E.; Burton, T.; Madgwick, S.; Vaidyanathan, R. Parametric design of an active ankle foot orthosis with passive compliance. In Proceedings of the IEEE Symposium on Computer-Based Medical Systems, Bristol, UK, 27–30 June 2011.

31. Agrawal, A.; Banala, S.K.; Agrawal, S.K.; Binder-Macleod, S.A. Design of a two degree-of-freedom ankle-foot orthosis for robotic rehabilitation. In Proceedings of the 2005 IEEE 9th International Conference on Rehabilitation Robotics, Chicago, IL, USA, 28 June–1 July 2005; Volume 2005, pp. 41–44.

32. Park, Y.L.; Chen, B.R.; Young, D.; Stirling, L.; Wood, R.J.; Goldfield, E.; Nagpal, R. Bio-inspired active soft orthotic device for ankle foot pathologies. In Proceedings of the 2011 IEEE/RSJ International Conference on Intelligent Robots and Systems, Francisco, CA, USA, 25–30 September 2011; pp. 4488–4495.

33. Chen, G.; Qi, P.; Guo, Z.; Yu, H. Mechanical design and evaluation of a compact portable knee-ankle-foot robot for gait rehabilitation. *Mech. Mach. Theory* **2016**, *103*, 51–64. [CrossRef]

34. Rossi, S.; Patane, F.; del Sette, F.; Cappa, P. WAKE-up: A wearable ankle knee exoskeleton. In Proceedings of the 5th IEEE RAS/EMBS International Conference on Biomedical Robotics and Biomechatronics, São Paulo, Brazil, 12–15 August 2014; pp. 504–507.

35. Veneman, J.F.; Kruidhof, R.; Hekman, E.E.G.; Ekkelenkamp, R.; van Asseldonk, E.H.F.; van der Kooij, H. Design and evaluation of the LOPES exoskeleton robot for interactive gait rehabilitation. *IEEE Trans. Neural Syst. Rehabil. Eng.* **2007**, *15*, 379–386. [CrossRef]

36. Costa, N.; Caldwell, D.G. Control of a biomimetic 'soft-actuated' 10DoF lower body exoskeleton. In Proceedings of the First IEEE/RAS-EMBS International Conference on Biomedical Robotics and Biomechatronics, Pisa, Italy, 20–22 February 2006; Volume 2006, pp. 495–501.

37. Wang, D.; Lee, K.M.; Guo, J.; Yang, C.J. Adaptive knee joint exoskeleton based on biological geometries. *IEEE/ASME Trans. Mechatron.* **2014**, *19*, 1268–1278. [CrossRef]